T0315055

Genomics

Genomics

Essential Methods

Edited by

Mike Starkey

Animal Health Trust

Ramnath Elaswarapu

LGC Ltd.

WILEY-BLACKWELL

A John Wiley & Sons, Ltd., Publication

Library of Congress Cataloguing-in-Publication Data
Genomics : essential methods/edited by Mike Starkey and Ramnath Elaswarapu.
    p. cm.
  ISBN 978-0-470-71157-6 (cloth)
  1. Genomics. I. Starkey, Mike. II. Elaswarapu, Ramnath.
  QH447.G4653486 2011
  572.8′6 – dc22

                                                    2010028093

A catalogue record for this book is available from the British Library.

This book is published in the following electronic formats: ePDF [9780470711620]; Wiley Online Library [9780470711675]

Set in 10/12 TimesRoman by Laserwords Private Limited, Chennai, India

First Impression   2011

Contents

7 Gene Expression in Mammalian Cells   155

Félix Recillas-Targa, Georgina Guerrero, Martín Escamilla-del-Arenal and Héctor Rincón-Arano

8 Using Yeast Two-Hybrid Methods to Investigate Large Numbers of Binary Protein Interactions   173

Panagoula Charalabous, Jonathan Woodsmith and Christopher M. Sanderson

9 Prediction of Protein Function   191

Hon Nian Chua

10 Elucidating Gene Function through Use of Genetically Engineered Mice

Mary P. Heyer, Cátia Feliciano, João Peca and Guoping Feng

11 Delivery Systems for Gene Transfer

Charlotte Lawson and Louise Collins

12 Gene Therapy Strategies: Constructing an AAV Trojan Horse

M. Ian Phillips, Edilamar M. de Oliveira, Leping Shen, Yao Liang Tang and Keping Qian

List of Contributors

Mary Barbara
Ontario Cancer Institute,
Princess Margaret Hospital,
University Health Network,
101 College Street,
TMDT, 8-356,
Toronto,
ON M5G 1L7,
Canada

Ted Brown
Samuel Lunenfeld Research Institute,
Mount Sinai Hospital,
Joseph and Wolf Lebovic Centre,
60 Murray Street,
6th Floor,
Toronto,
ON M5T 3L9,
Canada

Stephen A. Bustin
Institute of Cell and Molecular Science,
Barts and The London,
Queen Mary's School of Medicine
and Dentistry,
University of London,
Whitechapel,
London E1 1BB,
UK

Panagoula Charalabous
Department of Physiology,
School of Biomedical Sciences,
University of Liverpool,
Crown Street,
Liverpool L69 3BX, UK

Hon Nian Chua
Data Mining Department,
Institute for Infocomm Research,
1 Fusionopolis Way,
#21-01 Connexis (South Tower),
Singapore 138632,
Republic of Singapore

Jordy Coffa
Department of Pathology,
VU University Medical Center,
De Boelelaan, 1117,
1081 HV,
Amsterdam,
The Netherlands

Louise Collins
Department of Clinical Sciences,
Kings's College London School of Medicine,
James Black Centre,
125 Coldharbour Lane,
London,
SE5 9NU, UK

Edilamar M. de Oliveira
Laboratory of Biochemistry,
School of Physical Education and Sport,
Sao Paulo University,
Mello Moraes, 65,
Cidade Universitária,
Sao Paulo,
05508-9000,
Brazil

Toni Di Berardino
Mount Sinai Hospital Centre for Fertility and
Reproductive Health,
250 Dundas Street,
West 7th Floor,
Toronto,
ON M5T 2Z5,
Canada

Martín Escamilla-del-Arenal
Instituto de Fisiología Celular,
Departamento de Genética Molecular,
Universidad Nacional Autónoma de México,
Apartado Postal 70-242,
México D.F. 04510,
Mexico

Cátia Feliciano
Department of Neurobiology,
Duke University Medical Center,
401F Bryan Research Building,
Research Drive,
Durham, NC 27710,
USA

Guoping Feng
Department of Neurobiology,
Duke University Medical Center,
401F Bryan Research Building,
Research Drive,
Durham, NC 27710,
USA

David B. Friedman
Vanderbilt University School of Medicine,
215 Light Hall,
Nashville,
TN 37232,
USA

Ellen Greenblatt
Mount Sinai Hospital Centre for Fertility and
Reproductive Health,
250 Dundas Street,
West 7th Floor,
Toronto,
ON M5T 2Z5,
Canada

Georgina Guerrero
Instituto de Fisiología Celular,
Departamento de Genética Molecular,
Universidad Nacional Autónoma de México,
Apartado Postal 70-242,
México D.F. 04510,
Mexico

Mario Hermsen
Dept. Otorrinolaringología,
Instituto Universitario de Oncología del
Principado de Asturias,
Edificio H Covadonga 1ª Planta Centro,
Lab 2,
Hospital Universitario Central de Asturias,
Celestino Villamil s/n,
33006 Oviedo,
Spain

Mary P. Heyer
Department of Neurobiology,
Duke University Medical Center,
401F Bryan Research Building,
Research Drive,
Durham, NC 27710,
USA

Norman N. Iscove
Ontario Cancer Institute,
Princess Margaret Hospital,
University Health Network,
101 College Street,
TMDT, 8-356,
Toronto,
ON M5G 1L7,
Canada

Daniel C. Koboldt
Department of Genetics,
Washington University School of Medicine,
4444 Forest Park Avenue,
Box 8501, St. Louis,
MO 63108,
USA

Charlotte Lawson
Veterinary Basic Sciences,
Royal Veterinary College,
Royal College Street,
London NW1 0TU,
UK

Esther H. Lips
Department of Pathology,
Leiden University Medical Center,
Leiden,
PO Box 9600,
2300RC,
The Netherlands

Gerrit Meijer
Department of Pathology,
VU University Medical Center,
De Boelelaan, 1117,
1081 HV,
Amsterdam,
The Netherlands

Anneke Middeldorp
Department of Pathology,
Leiden University Medical Center,
Leiden,
PO Box 9600, 2300RC,
The Netherlands

Raymond D. Miller
Department of Genetics,
Washington University School of Medicine,
4566 Scott Ave.,
St. Louis,
MO 63110,
USA

Carolyn Modi
University Health Network Microarray Centre,
101 College Street,
TMDT, 9-301,
Toronto,
ON M5G 1L7,
Canada

Hans Morreau
Department of Pathology,
Leiden University Medical Center,
Leiden,
PO Box 9600,
2300RC,
The Netherlands

Tania Nolan
Sigma-Aldrich House,
Homefield Business Park,
Homefield Road,
Haverhill,
Suffolk CB9 8QP,
UK

Jan Oosting
Department of Pathology,
Leiden University Medical Center,
Leiden,
PO Box 9600,
2300RC,
The Netherlands

João Peca
Department of Neurobiology,
Duke University Medical Center,
401F Bryan Research Building,
Research Drive,
Durham, NC 27710,
USA

M. Ian Phillips
Keck Graduate Institute,
Claremont University Colleges,
535 Watson Drive,
Claremont,
CA 91711,
USA

Keping Qian
Keck Graduate Institute,
Claremont University Colleges,
535 Watson Drive,
Claremont,
CA 91711,
USA

Félix Recillas-Targa
Instituto de Fisiología Celular,
Departamento de Genética Molecular,
Universidad Nacional Autónoma de México,
Apartado Postal 70-242,
México D.F. 04510,
Mexico

Héctor Rincón-Arano
Instituto de Fisiología Celular,
Departamento de Genética Molecular,
Universidad Nacional Autónoma de México,
Apartado Postal 70-242,
México D.F. 04510,
Mexico

Nancy L. Saccone
Department of Genetics,
Division of Human Genetics,
Washington University School of Medicine,
St. Louis,
MO 63110,
USA

Christopher M. Sanderson
Department of Physiology,
School of Biomedical Sciences,
University of Liverpool,
Crown Street,
Liverpool L69 3BX,
UK

Leping Shen
Keck Graduate Institute,
Claremont University Colleges,
535 Watson Drive,
Claremont,
CA 91711, USA

Natalie Stickle
University Health Network Microarray Centre,
101 College Street,
TMDT, 9-301,
Toronto,
ON M5G 1L7,
Canada

Yao Liang Tang
Keck Graduate Institute,
Claremont University Colleges,
535 Watson Drive,
Claremont,
CA 91711, USA

Ronald van Eijk
Department of Pathology,
Leiden University Medical Center,
Leiden, PO Box 9600,
2300RC,
The Netherlands

Marjo van Puijenbroek
Department of Pathology,
Leiden University Medical Center,
Leiden, PO Box 9600,
2300RC,
The Netherlands

Tom van Wezel
Department of Pathology,
Leiden University Medical Center,
Leiden,
PO Box 9600,
2300RC,
The Netherlands

Carl Virtanen
University Health Network Microarray Centre,
101 College Street,
TMDT, 9-301,
Toronto,
ON M5G 1L7,
Canada

Neil Winegarden
University Health Network Microarray Centre,
101 College Street,
TMDT, 9-301,
Toronto,
ON M5G 1L7,
Canada

Jonathan Woodsmith
Department of Physiology,
School of Biomedical Sciences,
University of Liverpool,
Crown Street,
Liverpool L69 3BX,
UK

Bauke Ylstra
Microarray Facility,
VU University Medical Center,
De Boelelaan, 1117,
1081 HV,
Amsterdam,
The Netherlands

Preface

The scope of the field of genomics has expanded rapidly with the completion of the human genome sequencing project. Current understanding of biological systems has changed dramatically due to the combination of new technologies and the amount of data available, allowing for experimentation on a scale previously unimaginable.

The post-genomic era has opened up a plethora of opportunities for academic and commercial exploitation of these novel technologies. As increasing numbers of investigators seek to harness the fruits of genomics knowledge, it is essential that well-tested protocols are made available to researchers. With this in mind, it is apposite to launch this collection of protocols written by experts who are routinely employing these techniques in their laboratories.

This book represents more than a collation of step-by-step laboratory techniques, as it outlines the concepts underlying the techniques, and where possible provides alternative methods. This aspect adds considerable value to this book, and differentiates it from other 'protocol books'. All contributing authors have taken great care in explaining the principles of the techniques described, prior to presenting a detailed protocol.

Understandably, this volume does not purport to be a comprehensive collection of protocols in genomics per se; instead, the focus has been placed on key techniques in genomics and its derivative disciplines. The range of techniques presented is broad, and includes the detection of genetic variation, mRNA and genomic DNA copy number profiling, analysis of proteins by experimental and *in silico* methods, and the application of genomic strategies for therapeutic intervention.

Chapters 1–3 present procedures for genome analysis, including alternative strategies for the detection of chromosomal copy number alterations. The anomaly that it is often difficult to collect fresh tissues for research, and yet huge tissue banks exist in histopathology departments, is acknowledged by the description of procedures able to use archival samples (Chapters 1 and 3). The identification of single nucleotide polymorphisms (Chapter 2) has been instrumental to strategies for high-resolution, genome-wide association analysis (Chapter 4) in studies of disease susceptibility and pharmacogenetics.

Techniques for the analysis of gene expression are presented in Chapters 5–7. The trend in transcriptome analysis is towards the profiling of more strictly defined cell populations, necessitating the use of RNA amplification techniques that are discussed in Chapter 5. Real-time quantitative reverse transcription–PCR techniques (Chapter 6) are widely used for the detection and quantification of RNA as a consequence of their sensitivity and specificity. At the other end of the spectrum, there are many applications that require the capability to express transgenes *in vivo*, and Chapter 7 describes approaches for facilitating this via novel gene transfer methods.

The study of protein–protein interactions assists the understanding of biological function and elucidation of biochemical pathways, and Chapter 8 details use of the yeast two-hybrid system to generate high-confidence binary protein interaction data. The determination of gene function is central to functional genomics, and Chapters 9 and 10 explain alternative strategies towards attaining this objective.

The use of gene therapy for the modification of defective genes associated with patho-genesis is increasingly considered to be a viable approach for the treatment of disease. The penultimate two chapters address the issues involved and the strategies deployed.

Proteomic analysis is often a natural adjunct to transcriptional investigations, and the final chapter affords an introduction to protein profiling for the genomics specialist. This chapter is presented in a different format to the preceding chapters in that it offers a strategic guide, including a description of how to deal with some of the major issues and problems arising with protein profiling technologies.

We sincerely hope that these protocols, together with the summaries of their rationale, will help both experienced and new entrants in this field to carry out their experiments successfully.

Finally, we would like to thank all the contributing authors, David Hames, Clare Boomer and Jonathan Ray, the staff at Wiley-Blackwell, and most importantly our families for their support and endurance during the editing period.

Mike Starkey
Ramnath Elaswarapu

1
High-Resolution Analysis of Genomic Copy Number Changes

Mario Hermsen[1], Jordy Coffa[2], Bauke Ylstra[3], Gerrit Meijer[2], Hans Morreau[4], Ronald van Eijk[4], Jan Oosting[4] and Tom van Wezel[4]

[1]*Department Otorrinolaringología, Instituto Universitario de Oncología del Principado de Asturias, Oviedo, Spain*
[2]*Department of Pathology, VU University Medical Center, Amsterdam, The Netherlands*
[3]*Microarray Facility, VU University Medical Center, Amsterdam, The Netherlands*
[4]*Department of Pathology, Leiden University Medical Center, Leiden, The Netherlands*

1.1 Introduction

The analysis of DNA copy number changes throughout the whole genome started with the introduction of comparative genomic hybridization (CGH), first described in 1992 by Kallioniemi *et al.* [1]. This elegant technique was based on the competitive hybridization of two pools of fluorescent-labeled probes, one made up of whole-genome DNA of a test and another of a control sample, to a metaphase preparation of normal chromosomes. Along each chromosome, the fluorescent intensity of the test DNA was quantified and compared with the control intensity, resulting in 'relative copy number karyotypes.'

It appeared to be very difficult to reproduce the method in laboratories not specialized in chromosome techniques. Only after the publication of an article reviewing all steps in great detail did CGH become more widely applied [2], especially in cancer genetics. The possibility of using DNA obtained from formalin-fixed and paraffin-embedded (FFPE) samples opened up the way for retrospective studies of tumors with clinical follow-up data, enabling the identification of genetic changes related to tumor progression, invasion and metastasis [3].

Genomics: Essential Methods   Edited by Mike Starkey and Ramnath Elaswarapu
© 2011 John Wiley & Sons, Ltd.

The resolution of what is now called classical CGH is limited to a chromosomal band, approximately 5–10 Mb. This was overcome by the introduction of array comparative genomic hybridization (aCGH) in 1997 [4, 5]. The method is essentially the same, but now an array of genomic DNA clones or oligonucleotides serves as hybridization target, rather than metaphase chromosomes. The resolution of aCGH is now defined by the choice and/or the number of DNA clones and later oligonucleotides, and another advantage is that it does not require karyotyping. At the present moment, aCGH using oligonucleotide or single nucleotide polymorphism (SNP) arrays is most widely applied [6].

Multiple ligation-dependent probe amplification (MLPA), developed and first published in 2002 by Schouten *et al.* [7], is an alternative DNA copy number analysis technique, especially when specific genes or chromosomal regions are already known to be of interest. MLPA requires only 20 ng of DNA, enough to allow the simultaneous quantification of up to 50 different targets, which may be as small as 50 nucleotides. Another advantage of MLPA lies in its reproducibility and specificity, allowing application in a routine diagnostic setting while remaining time- and cost-efficient.

One increasingly important application is genomic profiling of FFPE samples. Across the world, large collections of FFPE samples with clinical follow-up exist. However, the DNA from FFPE samples shows varying levels of degradation depending largely on the length and the method of fixation, and on age of the specimen. This chapter aims to describe in detail the methods of oligonucleotide aCGH, SNP aCGH and MLPA, with special attention for the use of DNA obtained from FFPE samples. These techniques have primarily been used in cancer research; however, they are also suitable for the analysis of DNA copy number aberrations in human genetic disorders.

1.2 Methods and approaches

1.2.1 Oligonucleotide aCGH

The first whole-genome microarray contained 2400 large-insert genomic clones, primarily bacterial artificial chromosomes (BACs) [8]. With the total human genome covering about 3000 Mb the resolution of this array is on average close to 1 Mb, which is about one order of magnitude higher than that obtained with classical CGH [1]. For a full coverage resolution, about 30 000 BACs have been arrayed [9], increasing the resolution by another order of magnitude. However, producing such large numbers of BACs for aCGH is expensive and time consuming; and due to the large size of the BACs, the resolution limits for BAC aCGH resolution were reached.

Several laboratories used cDNA arrays, initially designed for expression profiling, as an alternative for measuring chromosomal copy number changes [10]. Even though this approach has certainly yielded valuable information, it cannot compete with the oligonucleotide platforms in terms of its maximal achievable resolution. Oligonucleotides allow a sheer infinite resolution, great flexibility and are cost effective [6]. They also enable the generation of microarrays for any organism for which the genome has been sequenced. Using the same oligonucleotide array for CGH and expression profiling allows direct comparison of mRNA expression and DNA copy number ratios. In addition, oligonucleotide arrays are being used, designed and accepted for expression profiling, and thus are widely available.

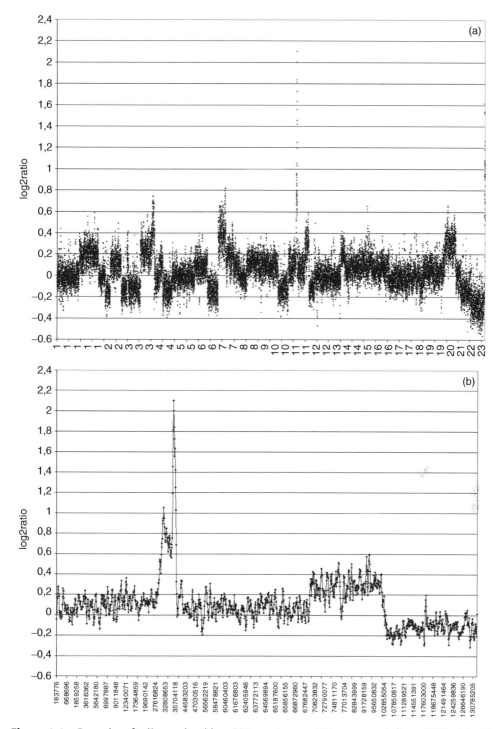

Figure 1.1 Example of oligonucleotide aCGH on tumor DNA extracted from FFPE. (a) All oligonucleotides ordered according to their chromosomal localization. (b) Oligonucleotides from chromosome 11 only. A marked amplification (indicated by an arrow) can be observed.

Commercial oligonucleotide aCGH platforms include Illumina (60 mer), Operon (70 mer), Affymetrix (25 mer), Agilent (60 mer) and NimbleGen (45–85 bp), the latter with now up to 2.1 million oligonucleotides on the array [11]. The quality of the oligonucleotide aCGH platforms is rapidly improving, with single oligonucleotides rapidly reaching the sensitivity of single BAC clones. Not all of the current oligonucleotide aCGH platforms can make a definite call for loss or gain using a single oligonucleotide, but in some cases three to five adjacent oligonucleotides are necessary for a reliable call [6, 11]. Moreover, owing to improvements in protocols, DNA isolated from FFPE tumor samples now works comparable to DNA from fresh material (Protocols 1.1–1.4) on long (>50 bp) oligonucleotide arrays (see Figure 1.1).

The principle of oligonucleotide aCGH is the same as all aCGH variants: Labeled tumor DNA (Protocol 1.5) competes with differentially labeled normal DNA for hybridizing to an array of oligonucleotides (Protocol 1.6). Using a specialized scanner and digital image processing software, the ratio of the two is measured per spot on the array. Deviations from the normal ratio of 1.0, or the $\log_2$ ratio of 0.0, represent a copy number aberration of genetic material in the tumor. The final result is DNA copy number information for all the oligonucleotides on the array, which can be ordered according to the chromosomal localization (Protocol 1.7). Graphics may represent all spots at once or only those belonging to one chromosome (see Figure 1.1). The sensitivity of CGH depends on the purity of the tumor sample and of the quality of the DNA obtained.

The oligonucleotide aCGH protocol presented provides a highly sensitive and reproducible platform applicable to DNA isolated from both fresh and FFPE tissue. We do not present protocols on the preparation of the array slides, since these can be purchased commercially.

PROTOCOL 1.1   DNA Extraction from Fresh or Frozen Tissue

Equipment and reagents

- Wizard Genomic DNA Purification Kit (Promega, A1120), containing:
  - EDTA/nuclei lysis solution
  - proteinase K (20 mg/ml)
  - RNase solution (100 mg/ml)
  - protein precipitation solution
- Phase lock gel (PLG, Eppendorf)
- Phenol solution (e.g. Sigma–Aldrich, P-4557)
- Chloroform
- Isopropanol
- Phenol/chloroform: 50% (v/v) phenol, 50% (v) chloroform
- TE: 10 mM Tris–HCl, pH 8.0, 1 mM EDTA
- Ethanol (70% (v/v) and 100% (v/v), both ice-cold)
- Sodium acetate (3 M, pH 5.2).

Method

1 To a 1.5 ml microcentrifuge tube add:

- 0.5–1 cm^3 of tissue;

- 600 μl of EDTA/nuclei lysis solution;

- 17.5 μl of proteinase K.

2 Incubate overnight at 55 °C with gentle shaking, or vortex the sample several times during the incubation.

3 Add 3 μl of RNase solution to the nuclear lysate and mix the sample by inverting the tube two to five times.

4 Incubate the mixture for 15–30 min at 37 °C.

5 Add 200 μl of protein precipitation solution to the sample and vortex vigorously. Chill the sample on ice for 10 min.

6 Centrifuge at 20 000g for 15 min at room temperature to pellet the precipitated protein.

7 Carefully transfer the supernatant containing the DNA to a fresh 1.5 ml microcentrifuge tube.

8 Add 600 μl of isopropanol (at room temperature).

9 Mix the solution by gently inverting until the white thread-like strands of DNA form a visible mass.

10 Centrifuge at 20 000g for 1 min at room temperature. The DNA will be visible as a small white pellet. Carefully aspirate supernatant by decanting the liquid. Air dry until no ethanol is visible.

11 Add 200 μl of TE to resuspend the DNA.

12 Pellet 2 ml of PLG light by centrifuging at 12 000–16 000g for 20–30 s at room temperature.

13 Add the 200 μl of DNA-containing TE to the 2 ml PLG light tube, followed by 200 μl of phenol–chloroform.

14 Mix the organic and the aqueous phases thoroughly by inversion.[a]

15 Centrifuge at 12 000–16 000g for 5 min at room temperature to separate the phases. Transfer the upper layer/supernatant to a new 1.5 ml microcentrifuge tube.

16 Add 200 μl of chloroform directly to the above new 1.5 ml microcentrifuge tube.

17 Mix thoroughly by inversion.[a]

18 Centrifuge at 12 000–16 000g for 5 min at room temperature to separate the phases.

19 Transfer the aqueous solution (above the gel) to a new 1.5 ml microcentrifuge tube.

20 Add 20 μl of 3 M sodium acetate and mix by inversion.

21 Add 2.5× the total volume of 100% (v/v) ethanol (ice cold).[b]

22 Centrifuge at 12 000–16 000g for 15 min at room temperature.

23 Discard the supernatant and add 500 µl of 70% (v/v) ethanol (ice cold). Vortex the sample and centrifuge at 20 000g for 10–15 min at 4 °C.

24 Discard the supernatant and allow the pellet to air dry until no ethanol is visible.

25 Resuspend the pellet in 100 µl of TE or water.

Notes

[a]Do not vortex.

[b]After mixing, the DNA should come out of solution.

PROTOCOL 1.2   DNA Extraction from FFPE Tissue

Equipment and reagents

- Xylene (e.g. Merck – VEL, 90380)

- Methanol

- Ethanol (100% (v/v), 96% (v/v), 70%(v/v))

- QIAamp DNA Mini Kit 250 (Qiagen, 51306), or QIAamp DNA Micro Kit 50 (Qiagen, 56304) if the amount of tissue is limited (i.e. a biopsy)

- NaSCN (e.g. Sigma), 1 M

- Proteinase K (e.g. Roche), 20 mg/ml

- RNase A (e.g. Roche), 100 mg/ml

- Phosphate-buffered saline (PBS).

Method[c]

1 Transfer two or three 50 µm FFPE tissue sections into a microcentrifuge tube.

2 Incubate with 1 ml of xylene for 7 min at room temperature, mixing a few times by vortexing.

3 Centrifuge at 14 000g for 5 min at room temperature and discard the supernatant.

4 Repeat steps 2 and 3 twice.

5 Incubate with 1 ml of methanol for 5 min at room temperature, mixing a few times by vortexing.

6 Centrifuge at 14 000g for 5 min at room temperature and discard the supernatant.

7 Repeat steps 5 and 6 once

8　Add 1 ml of PBS, mixing a few times by vortexing.

9　Centrifuge at 14 000g for 5 min at room temperature and discard the supernatant.

10　Repeat steps 8 and 9 once.

11　Incubate with 1 ml of 1 M NaSCN overnight at 38–40 °C, mixing a few times by vortexing.

12　Centrifuge at 14 000g for 5 min at room temperature and discard the supernatant.

13　Wash the pellet three times with 1 ml of PBS as in steps 11 and 12.

14　Add 200 μl of Buffer ATL (QIAamp kit) and 20 μl of proteinase K, mixing a few times by vortexing.

15　Incubate at 50–60 °C for 60 h, adding an extra 20 μl of proteinase K every 12 h.

16　Incubate with 40 μl of RNase A for 2 min at room temperature, mixing a few times by vortexing.

17　Incubate with 400 μl of Buffer AL (QIAamp kit) for 10 min at 65–75 °C, mixing a few times by vortexing.

18　Add 420 μl of 100% (v/v) ethanol and mix by vortexing thoroughly.

19　Transfer 600 μl of the solution to a QIAamp centrifuge column.

20　Centrifuge at 2000g for 1 min at room temperature and discard the flow through.

21　Repeat steps 19 and 20 until all the sample has been applied to the column.

22　Add 500 μl of Buffer AW1 (QIAamp kit) to the column.

23　Centrifuge at 2000g for 1 min at room temperature and discard the flow through.

24　Add 500 μl of Buffer AW2 (QIAamp kit) to the column.

25　Centrifuge at 14 000g for 3 min at room temperature and discard the flow through.

26　Transfer the column to a fresh microcentrifuge tube (with a lid).

27　Elute the DNA from the column by adding 75 μl of Buffer AE (QIAamp kit), preheated to 65–75 °C.

28　Leave at room temperature for 1 min.

29　Centrifuge at 2000g for 1 min at room temperature.

30　Discard the column and store the DNA at 2–8 °C.

Notes

[c]The protocol described is for the isolation of DNA from about 1 cm^2 or larger size tissue sections using the QIAmp Mini kit. In the case of small-sized tissue sections (i.e. less than 0.5 cm^2), extract DNA using the QIAmp Micro kit. The proteinase K volumes and incubation times may need to be adjusted, and the RNase treatment omitted. The quality of the DNA extracted from FFPE tissue may differ considerably. In general, older paraffin blocks yield DNA of worse quality. An important factor for preservation of DNA in FFPE tissue is the use of pH 7.0 buffered formalin fixative before embedding in paraffin wax.

PROTOCOL 1.3   DNA Concentration Measurement Using Picogreen

Equipment and reagents

- TE: 10 mM Tris–HCl, pH 8.0, 0.1 mM EDTA

- PicoGreen dsDNA reagent(Molecular Probes)[d]

- Lambda DNA standards

- Recommended microtiter plates (immunoassay microplates – flat bottom; Dynex Immulux[TM])

- Fluorescence plate reader

- Centrifuge for microtiter plates.

[e]Method

1 Prepare a series of 100 µl/well lambda DNA standards, in duplicate, within a clean, 96-well plate as follows:

Lamba DNA (2 µg/µl)	TE (µl)	Final concentration
100 µl	0	100 ng/µl
75 µl	25	750 pg/µl
50 µl	50	500 pg/µl
25 µl	75	250 pg/µl
10 µl	90	100 pg/µl
5 µl	95	50 pg/µl
0 µl	100	0 pg/µl

2 For each DNA sample, prepare duplicate dilutions of 2 µl of DNA with 98 µl of TE.

3 For each DNA sample dilution, prepare 100 µl of Picogreen[d] reagent by diluting Picogreen 200-fold in TE.

4 Add 100 µl of diluted Picogreen reagent to each diluted DNA sample and mix by pipetting up and down.

5 Centrifuge the microtitre plate at 250g for 1 min to remove possible bubbles.

6 Read in a plate reader (excitation 485 nm, emission 538 nm).

7 Calculate concentrations from the standard curve using the plate reader software package.

Notes

[d]Avoid excess exposure to light since the dye is light sensitive.

[e]For quantitation of DNA from FFPE tissue for SNP arrays, the use of Picogreen gives more reliable estimates than measurement of $A_{260 nm}$ using a spectrophotometer [12].

PROTOCOL 1.4 DNA Quality Control PCR

Equipment and reagents

- Primer stocks (100 μM):

  — RS2032018 (150 bp)

    ○ primer 1: 5'-GTGTCTCCCTTCCCACTCAA-3'
    ○ primer 2: 5'-AGCCCACCTACCTTGGAAAG-3'

  — AP000555 (PRKM1, 255 bp)

    ○ primer 3: 5'-TGGCTGATCTATGTCCCTGA-3'
    ○ primer 4: 5'-GCTCAGTTGTTTTGTGGGTAAG-3'

  — AC008575 (APC, 511 bp)

    ○ primer 5 GCTCAGACACCCAAAAGTCC
    ○ primer 6: CATTCCCATTGTCATTTTCC

- Polymerase chain reactions (PCRs) reaction buffer II without MgCl$_2$ (Applera)
- Amplitaq gold DNA polymerase (5 units/μl Applera).

Method

1 Prepare a primer working solution mix containing 20 pm/μl of primers 1 and 2, and 10 pm/μl for primers 3–6.^{f}

2 Prepare a 10 μl reaction by mixing the following:

  - 0.25 μl of primer mix
  - 1.0 μl of 10× PCR reaction buffer II without MgCl$_2$
  - 0.2 μl of 4× 10 mM dNTPs
  - 1.0 μl of 25 mM MgCl$_2$
  - 2.0 μl of template DNA (1–250 ng)^{g,h}
  - 0.1 μl of Amplitaq gold DNA polymerase
  - 5.45 μl of water.

3 Thermocycle as follows:

  - 96 °C, 10 min
  - (94 °C, 30 s; 55 °C, 30 s; 72 °C, 1 min) ×35
  - 72 °C, 5 min.

4 Analyze each PCR product by 2% (w/v) Tris–acetate–EDTA agarose gel electrophoresis. Compare the sizes of amplified products against molecular weight marker standards.^{i}

Notes

[f]The multiplex PCR amplifies three amplicons, one each of of 150 bp, 255 bp and 511 bp. This method is comparable to the van Beers method [38].

[g]Use 10 ng of genomic DNA prepared from freshly frozen tissue as a control.

[h]If the DNA concentration (see Protocol 1.9) is higher than 5 ng/µl it can be diluted in water.

[i]The 150 and 255 bp amplicons have to amplify for a DNA template considered to be suitable for aCGH.

PROTOCOL 1.5   Labeling of DNA for Oligonucleotide aCGH

Equipment and reagents

- BioPrime DNA labeling system (Invitrogen, 18094-011), containing:
  - 2.5× random primers solution
  - Klenow fragment of DNA polymerase I (40 U/µl); keep on ice at all times, or preferably use a −20 °C labcooler when taking in and out of the freezer.
- Cy3-labeled dCTP (e.g. Amersham Biosciences/Perkin Elmer)
- Cy5-labeled dCTP (e.g. Amersham Biosciences/Perkin Elmer)
- ProbeQuant G-50 Micro Columns (Amersham Biosciences)
- dNTP mixture; for 200 µl mix:
  - 4 µl of 100 mM dATP
  - 4 µl of 100 mM dGTP
  - 4 µl of 100 mM dTTP
  - 1 µl of 100 mM dCTP
  - 2 µl of 1 M Tris–HCl, pH 7.6
  - 0.4 µl of 0.5 M EDTA, pH 8.0
  - 184.6 µl of water.

Method

1 In a PCR tube, mix 300 ng[j] of genomic DNA and 20 µl of 2.5× Random Primers solution. Adjust the volume to 42 µl with water.

2 Denature the DNA mixture in a PCR machine at 100 °C for 10 min and immediately transfer to an ice/water bath for 2–5 min. Briefly centrifuge and put back on ice.

3 While maintaining on ice, add 5 µl of dNTP mixture, 2 µl of Cy3 (test) or Cy5 (ref) labeled dCTP[k] and 1 µl of Klenow DNA polymerase.

4 Mix well and incubate at 37 °C (in PCR machine) for 14 h, and then maintain at 4 °C.

5 Prepare a Probe-Quant G-50 column for removal of uncoupled dye material as follows:

- resuspend the resin in the column by vortexing;

- loosen the cap one-fourth turn and snap off the bottom closure;

- place the column in a 1.5 ml microcentrifuge tube and centrifuge at 735g for 1 min.[l]

6 Place the column into a fresh 1.5 ml tube and slowly apply 50 μl of the sample to the top center of the resin, being careful not to disturb the resin bed.

7 Centrifuge the column at 735g for 2 min. The purified sample is collected at the bottom of the support tube.

8 Discard the column and store the purified and labeled sample[m] in the dark until use on the same day, or alternatively store at −20 °C for a maximum of 10 days.

Notes

[j]For paraffin-embedded tissue, 600 ng of test and reference DNA samples should be used. We experienced that reference DNA prepared from either blood or FFPE 'normal tissue' can give equally good results.

[k]Test and reference DNA can be labeled with either Cy3 or Cy5.

[l]Start the timer and microcentrifuge simultaneously to ensure that the total centrifugation time does not exceed 1 min.

[m]It is not necessary to exactly quantify the labeled DNA or the degree of Cy5/Cy3-dCTP incorporation, because in the data analysis a normalization of the Cy5/Cy3 channels takes place.

PROTOCOL 1.6 Hybridization

Equipment and reagents

- Blocking solution:[n] 0.1 M Tris, 50 mM ethanolamine, pH 9.0: dissolve 6.055 g of Trizma base and 7.88 g of Trizma-HCl in 900 ml of water. Add 3 ml of ethanolamine (Sigma–Aldrich Chemie B.V. Zwijndrecht, Netherlands) and mix thoroughly. Adjust the pH to 9.0 using 6 N HCl. Adjust the final volume to 1 l with water.

- 20× SSC, pH 7.0 (e.g. Sigma) and dilutions in water (0.2×, 0.1× and 0.01× SSC).

- 20% (w/v) SDS solution: for preparation of 100 ml, dissolve 20 g of sodium dodecyl sulfate in 90 ml of water. Adjust the final volume to 100 ml.

- Wash solution: 4× SSC, 0.1% (w/v) SDS: 200 ml of 20× SSC, 10 ml of 10% (w/v) SDS, adjust the final volume to 1 l with water.

- Human Cot-1 DNA, 1 μg/μl (e.g. Invitrogen).

- Yeast tRNA, 100 μg/μl (e.g. Invitrogen).

- Master mix – 14.3% (w/v) dextran sulfate, 50% (v/v) formamide, 2.9× SSC, pH 7.0: Combine 1 g of dextran sulfate (USB), 3.5 ml of redistilled formamide (Invitrogen; store at −20 °C), 2.5 ml of water and 1 ml of 20× SSC. Gently shake for several hours to dissolve the dextran sulfate and store aliquoted at −20 °C.

- Washing buffer: 50% (v/v) formamide, 2× SSC, pH 7.0.

- PN buffer: 0.1 M Na_2HPO_4/NaH_2PO_4, pH 8.0, 0.1% (v/v) Igepal CA630 (e.g. Sigma).

- GeneTAC/HybArray12 hybstation (Genomic Solutions/Perkin Elmer).

Method

1. Add 0.01 volume of 10% (w/v) SDS to the blocking solution (final concentration of 0.1% (w/v) SDS) and pre-warm at 50 °C.

2. Place an oligonucleotide microarray slide in a slide rack and block residual reactive groups by incubating in pre-warmed blocking solution at 50 °C for 15 min.[o]

3. Rinse the slide twice with water.

4. Wash the slide with wash solution (pre-warmed to 50 °C) for 15–60 min.[p]

5. Rinse briefly with water, but do not allow the slide to dry prior to centrifugation.

6. Place the slide in a 50 ml tube and centrifuge at 200g for 3 min to dry.

7. Use a slide for hybridization within 1 week.

8. In a 1.5 ml tube, mix: 50 μl of Cy3-labeled test[q] DNA, 50 μl of Cy5-labeled reference DNA and 10 μl of Cot-1 DNA.[r]

9. Add 11 μl of 3 M sodium acetate, pH 5.2 (0.1 volume) and 300 μl of ice-cold 100% (v/v) ethanol, mix the solution by inversion and collect the DNA by centrifugation at 20 000g for 30 min at 4 °C.

10. Remove the supernatant with a pipette and air-dry the pellet for 5–10 min until no ethanol is visible. Carefully[s] dissolve the pellet in 13 μl of yeast tRNA and 26 μl of 20% (w/v) SDS. Leave at room temperature for at least 15 min.

11. Add 91 μl of master mix and mix gently.

12. Denature the hybridization solution at 73 °C for 10 min, and incubate at 37 °C for 60 min to allow the Cot-1 DNA to block repetitive sequences.

13. Store the following program[s] named 'CGH.hyb' on the hybstation[t]:

    (a) introduce hybridization solution, temperature 37 °C

    (b) set slide temperature: temperature: 37 °C; time: 38 hours : 00 minutes : 00 seconds, agitate: Yes

    (c) wash slides (washing buffer): six cycles, source 1, waste 2 at 36 °C, flow for 10 s, hold for 20 s

    (d) wash slides (PN buffer): two cycles, source 2, waste 1 at 25 °C, flow for 10 s, hold for 20 s

(e) wash slides (0.2× SSC): two cycles, source 3, waste 1 at 25 °C, flow for 10 s, hold for 20 s

(f) wash slides (0.1× SSC): two cycles, source 4, waste 1 at 25 °C, flow for 10 s, hold for 20 s.

14 Assemble up to six hybridization units (two slides per unit): insert rubber O-rings in the covers and put the slides on the black bottom plate. Make sure the slides are in the proper orientation with the printed side up.

15 Introduce the unit into the hybstation, press unit down with one hand while tightening the screw with the other.

16 Insert plugs into the sample ports and the waste tubes into the corresponding wash bottles.

17 On the touch screen subsequently press: start a run, from floppy, CGH.hyb, load, the positions of the slides you want to use, start, continue (the hybstation starts to warm up the slides).

18 When the hybstation is ready (visible on screen by indication of the module you have to start) apply hyb mix:

(a) press Probe to add the hyb mix for the selected slide

(b) check if a mark on the screen appears

(c) take the plug out and inject the hyb mix by pipetting it slowly into the port using a 200 μl pipette

(d) press the Finished control (check mark) and replace the plug

(e) repeat this for the next slide

(f) press the Finished control for the selected slide

(g) press the Finished control for the module

(h) repeat this for the selected module(s).

19 Take slides out after 38 h and put them in 0.01× SSC.

20 Place each slide in a 50 ml tube and centrifuge at 200*g* for 3 min to dry.

21 Immediately scan slides in a microarray scanner.

22 Cleaning the hybstation:[a]

23 Reassemble all used hybridization units with dummy slides and introduce them into the hybstation.

24 Insert plugs into all sample ports and place all tubes in a bottle of water.

25 On the touch screen subsequently press: maintenance, Machine Cleaning Cycle, the positions of the slides you used, continue.

26 When cleaning is finished, take out the hybridization units, rinse with water (never use ethanol) especially the sample port and dry the unit with the air pistol.

Notes

[n]This blocking solution is specific to the blocking of CodeLink™ slides (SurModics Inc) on which amino linker-containing oligonucleotides (dissolved at 10 μM in 50 mM sodium phosphate buffer pH 8.5) are spotted.

[o]Extend to 30 min if the blocking solution is not pre-warmed, but do not exceed 1 h.

[p]Use at least 10 ml of wash solution per slide.

[q]Test and reference DNA can be labeled with either Cy3 or Cy5.

[r]If many experiments are planned, we recommend ordering a large batch of Cot-1 DNA from the same lot.

[s]Take care to prevent foam formation due to the SDS.

[t]Alternative manual protocol for hybridization and washing: cut off the large end of a 200 μl pipette tip to fit on a 5 ml syringe and fill the syringe with rubber cement (Ross). Apply the rubber cement closely around an array. Apply a second or third layer of rubber cement thickly. Apply the hybridization mixture to the array and incubate the slide assembly in a closed incubation chamber over two nights at 37 °C on a rocking table. Following hybridization, disassemble the slide assembly and rinse the hybridization solution from the slide in a room temperature stream of PN buffer. Wash the array in wash buffer for 10–15 min at 45 °C, followed by a 10–15 min room temperature wash in PN buffer. Carefully remove the rubber cement (do not let the array dry) with tweezers/forceps, wash the array sequentially with 0.2× SSC and 0.1× SSC and centrifuge dry (250g, 3 min).

[u]Cleaning the hybstation after each hybridization is essential to maintain proper functioning of the equipment.

PROTOCOL 1.7   Scanning and Creation of a Copy Number Profile

Equipment and reagents

- High-resolution laser scanner, or imager equipped to detect Cy3 and Cy5 dyes, including software to acquire images (e.g. Microarray Scanner G2505B, Agilent Technologies)

- Feature-extraction software (e.g. Bluefuse 3.2 (BF), BlueGnome Ltd, UK)

- Gene array list (GAL-file, or equivalent) – created by the microarray printer software using the oligonucleotide plate content lists provided by the supplier of the oligo library

- Position list: a file, containing the relative positions of the oligonucleotides in the genome under investigation, provided by the supplier of the oligo library, or created by mapping the oligonucleotide sequences onto the genome concerned

- Software which calculates ratios, links the genomic position of the oligonucleotide to the experimental ratios and draws a profile (e.g. Microsoft Excel, or dedicated software such as BF).

Method

1 Allow scanner lasers to warm up for 5 min before starting.

2 Scan the microarray at 10 μm scanning resolution according to the manufacturer's protocol.

3 Store scans from both channels as separate TIFF images

4 Perform automated spot finding, using the information from the GAL-file to position the array grid over each image.

5 Perform automated spot exclusion.[v]

6 Perform automated linking of the spot ratios to the genomic positions of the corresponding oligonucleotides (using the information from the position file).

7 Perform global mode normalization.[w]

8 Draw the genomic profile (automated in BF): order normalized ratios by chromosomal mapping and display in a graph.[x]

Notes

[v]We suggest excluding spots that have a 'confidence value' lower than 0.1, or a 'quality flag' lower than 1, which will further diminish outliers. These confidence values are calculated in a proprietary manner by the BF feature extraction software.

[w]Avoid block normalization (normalization per printed block of spots on the array slide, which can be performed by either median, or intensity-dependent lowess), because this may compress the profile. Mode normalization is used to set the 'normal' level and is preferred over mean or median normalization, as it is more accurate since it ignores the ratios generated by gains, amplifications and deletions. Block normalization is sometimes used to suppress noise, although its suitability may depend on the type of sample analyzed; that is, for samples showing few aberrations, block normalization may help to suppress noise, but is not recommended for samples with multiple chromosomal aberrations (e.g. tumor samples).

[x]For more sophisticated analysis procedures and to 'call' the actual gains, losses and amplifications, we recommend the use of more dedicated software, such as the freeware CGH call [14].

1.2.2 SNP aCGH

The recently developed high-density SNP microarrays were originally developed for high-throughput genotyping for linkage analysis and association studies. These arrays have additionally proven useful to measure both genomic copy-number variations and loss of heterozygosity (LOH); that is, SNP aCGH. The ability of SNP aCGH, unlike conventional CGH, to detect copy-neutral genetic anomalies offers the benefit of detecting copy-neutral LOH [15]. Moreover, the combination of copy number abnormalities and LOH status with the parental origin of the aberrant allele can possibly be associated with the predisposition to hereditary cancer. Successful use of SNP aCGH has been reported for several cancers, such as breast, colorectal and lung cancer [16–19]. While the current high-density SNP arrays can interrogate more than a million SNPs, these arrays cannot (yet) be used reliably with DNA from FFPE tissue. This is due to the fragmented nature of the DNA isolated from FFPE tissue. For this purpose, current arrays are restricted to 6000–10 000 features.

Different methodologies and types of commercially available SNP arrays exist. These consist of either locus-specific arrays of oligonucleotides (Genechips) or of arrays with universal capture oligonucleotide on beads that are randomly assembled on arrays and subsequently decoded (Beadarrays). Genechips can detect up to 250 000 SNPs on a single chip. For each SNP, a set of locus-specific oligonucleotides is synthesized on the array. The sample is prepared according to a whole-genome sampling assay [20]. After restriction enzyme digestion of high molecular weight genomic DNA and ligation of a common adaptor, the DNA is amplified in a single-primer PCR and hybridized to a locus-specific array [21]. For Infinium arrays, genomic DNA is whole-genome amplified, subsequently fragmented, and denatured DNA is hybridized to a locus-specific array. An allele-specific primer extension assay on the array is followed by staining and scanning of the arrays using standard immunohistochemical detection methods. Currently these arrays can detect over a million SNPs on a single array [22]. Goldengate genotyping makes use of a multiplex mixture of probes for 96, 384, 768 or 1536 SNPs per array [23]. For each SNP, a combination of allele-specific and locus-specific primers is annealed to the SNP locus, the primers are tailed with common forward and reverse primers and a complementary universal capture probe to the locus-specific primer. Subsequent allele-specific primer-extension, followed by ligation, generates an allele-specific artificial PCR template. This template is then PCR-amplified and labeled. After hybridization to an array of universal-capture probes, the array is scanned in two colors, representing the two alleles of an SNP. Molecular-inversion probe (MIP) genotyping utilizes a pool of circularizable locus-specific probes with a multiplexing degree of over 10 000 SNPs per array. The 5' and 3' ends of each probe anneal upstream and downstream of the SNP. The 1 bp gap is filled; subsequent ligation seals the nick and generates a circular probe. Restriction digestion then releases the circularized probe and the resultant product is PCR-amplified using common primers [24]. The four nucleotide reactions are labeled in different colors and pooled. Subsequently, the pool is hybridized to an array of universal-capture probes and the four colors are read out in a scanner. Whereas the high-density Genechip and Infinium arrays are designed for use with high-quality genomic DNA, both the Goldengate and the MIP assay can be used to detect LOH and copy number changes in FFPE tissue [25, 26].

SNP aCGH collects both intensity and allelic information from a sample. To extract profiles of LOH and copy number abnormalities, different methods and algorithms have been reported [13,27–30]. For the interpretation of LOH and copy number abnormalities, specifically for the Goldengate assay, a limited or no method was available. Therefore, to interpret the Beadarray data an R-package BeadArraySNP was developed. The package deals with the normalization of the allele-specific signal intensities and the representation of the copy number and LOH profiles [25].

Here we describe the use of the GoldenGate assay and Beadarrays to generate high-resolution copy number profiles and LOH using DNA isolated from FFPE tissue (Protocol 1.8). We do not present Illumina protocols, since this is a commercial platform. The most recent version of the protocol (user card) can be obtained through www .illumina.com.

PROTOCOL 1.8   Data Analysis of Illumina SNP Beadarray Experiments

Equipment and reagents

- Illumina BeadScan software for genotyping and the Bioconductor (www.bioconductor.org) BeadarraySNP package [25]

- Quantile smoothing software [31].

Method[y]

1 Perform an Illumina GoldenGate™ assay,[y] according to the protocol supplied by Illumina, using 1 μg of activated DNA (isolated from FFPE tissue) dissolved in 60–100 μl of RS1 buffer.

2 Scan[z] the Illumina arrays using the Illumina BeadScan software,[aa] creating (by default) the following types of files for each of the samples on the array:

- locs: locations of beads on the array

- idat: summarized intensity information (binary format)

- XML: scanner settings.

3 Adapt the Settings.xml[bb] within the beadstudio directory in order to produce the additional file types.

```
<SavePerBeadFiles>true<\SavePerBeadFiles>
<SaveEIFFiles>false<\SaveEIFFiles>
<SaveTextFiles>true<\SaveTextFiles>
<CompressImages>true<\CompressImages>
<ExcludeOutliers>true<\ExcludeOutliers>
<IncludeXY>true<\IncludeXY>
```

4 Perform the genotyping of the scanned images using the GenCall software,[cc] producing the following types of files for each experiment.[dd]

5 *-OPA_LocusByDNA_*.csv: genotyping and quality scores (one sample per row):

- *-OPA_LocusByDNA_*DNA_Report.csv: summary of allele frequencies for each sample

- *-OPA_LocusByDNA_*Final.csv: genotyping and quality scores (each probe and sample appear on a separate row)

- *-OPA_LocusByDNA_*Locus_Report.csv: quality index summaries for all probes.

6 Begin copy number analysis[ee] by defining the samples in a sample sheet.[ff] Calculate the copy number values for all the samples in the experiment using the function standardNormalization().

7 Plot the raw and smoothed copy number data using the Quantile smoothing software.[gg]

8 Use the 50th percentiles (displayed on the plots as dotted lines; see Figure 1.2) to guide identification of gains and losses.[hh]

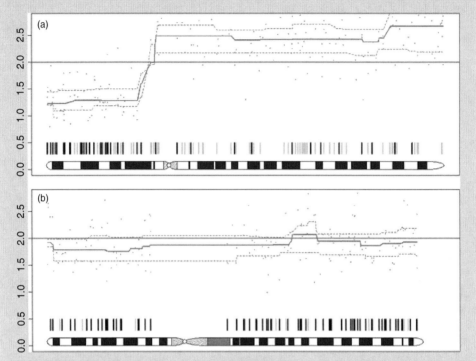

Figure 1.2 Chromosome view output of BeadArraySNP showing examples of a colorectal carcinoma. (a) Chromosome 8 with physical loss at the p-arm and gain at the q-arm. (b) An example of LOH without a copy number change on chromosome 9. Each red dot indicates the normalized individual SNP signal; the solid red lines show the smoothed copy number with the 25th and 75th percentile interval marked with the dashed red lines. The gray vertical bars above the ideogram show the heterozygous SNPs in the normal sample that retained heterozygosity in the paired tumor. Black vertical bars show LOH; that is, heterozygous in the normal and homozygous in the tumor.

Notes

[y]Twelve tumor:normal pairs (24 samples) can be profiled using four pools of approximately 1500 SNPs (collectively referred to as the linkage panel). The SNPs are hybridized to sentrix arrays, a matrix of 96 arrays, each detecting about 1500 SNPs. Thus, for each sample, four arrays of approximately 1500 SNPs each are hybridized and subsequently combined to yield around 6000 genotypes. Alternatively, a genome-wide cancer SNP panel, or custom SNP panels of 1536, 768 or 384 SNPs, can be typed.

[z]The allele-specific data obtained from Illumina SNP-arrays can be used to perform both genotyping and copy number analysis. Although the Illumina software for genotyping performs satisfactorily in most cases, we have found that there is room for improvement in performing

copy number analysis, in particular for the GoldenGate assay with fewer SNPs (6000 SNPs) compared with 100 000 or 317 000 SNP Infinium arrays.

[aa]This software can identify the individual beads and measure the intensity of each bead. In general, it is useful to organize the files created by BeadScan by putting them into separate subdirectories for each experiment.

[bb]These settings also create the following files: tiff: scanned image of sample; csv: summary values per probe – intensity, standard deviation, number of beads; txt: values for each bead – intensity, position on tif.

[cc]Illumina provides two software packages that can perform genotyping on the scanned images. Gencall was the original application for genotyping and was followed by BeadStudio, which is an integrated package for performing all kinds of data analysis for Illumina arrays. The algorithm for genotyping differs between the packages. We prefer to use the results of GenCall as the input for copy number analysis. If genotyping is performed with the Illumina Gencall software then the csv files are needed to obtain copy number values.

[dd]Beadstudio produces a report file with a specifiable number of result fields. For the copy number analysis the following fields are needed: GC Score Allele1 – AB Allele2 – AB GT Score X Raw Y Raw. The report file will contain the chosen values for each sample and probe.

[ee]This calculation consists of three stages: normalizeBetweenAlleles.SNP() Performs quantile normalization between both colors of a sample. This is allowed because the frequencies of both alleles throughout a sample are nearly identical in practice. This action also neutralizes any dye bias. normalizeWithinArrays.SNP() scales each sample using the median of the high-quality heterozygous SNPs as the normalization factor. Genomic regions that show copy number alterations are likely to show LOH, or are harder to genotype leading to a decrease quality score of the call. normalizeLoci.SNP() scales each probe using the normal samples in the experiment, assuming that these samples are diploid, and have a copy number of two.

[ff]This can be the file produced by the Illumina genotyping software, but it is useful to include more information on the individual samples here; for example, experimental groups, normal/tumor tissue. The format for these data is explained in detail in the help pages for the package. The data quality should be checked. The overview of average intensities for both channels as laid out on the physical device has proved helpful to detect technical anomalies. For the GoldenGate assay both channels should have an average intensity above 1250.

[gg]Various plots of raw or smoothed copy number data can be obtained using the Quantile smoothing software, such as: smoothed intensities from all chromosomes for a number of samples, all samples in an experiment indicating regions of gain, loss and LOH, for individual chromosomes, or a BAC-array-like plot for each sample in an experiment.

[hh]A gain is called where the 25th percentile line exceeds the 2N line and a loss is called where the 75th percentile is below the diploid line.

1.2.3 Multiple ligation-dependent probe amplification (MLPA)

The MLPA technique was first described by Schouten *et al.* [7] in 2002 and has become a rapidly growing technique used in the detection of aberrations in genes related to various diseases. MLPA is a multiplex technique for determining copy numbers of genomic DNA sequences [32] and promoter methylation status [33], as well as mRNA profiling [34]. While genomic profiling is becoming increasingly important in both the research and diagnostic setting, routine detection methods for exon deletions and duplications are still lacking.

To date, MLPA technology has shown its applicability in the research setting through several studies. These include detection of trisomies [35], Duchene and Becker muscular dystrophy [36], centrifugal muscular dystrophy and also detection of deletions and duplications of one or more exons of the *BRCA1* [37] and the *MLH1/MSH2* genes [32]. However, the protocols described in this chapter are robust and specific enough to be used as a routine diagnostic assay.

MLPA is easy to perform and relatively cheap. Up to 96 samples can be handled simultaneously and results can be obtained within 24 h. The required apparatus for MLPA (a thermocycler and sequencing type electrophoresis equipment) is present in most molecular biology laboratories. Probe target sequences are small (50–70 nucleotides), meaning that MLPA can be applied to partially degraded DNA, such as FFPE tissue DNA, or free fetal DNA obtained from maternal plasma. MLPA reactions are more sensitive to contaminants, often present in DNA samples extracted from FFPE tissues, than ordinary PCRs. Developing new MLPA probe mixes is complicated, expensive and time consuming. Each probe requires the design and preparation of a phage M13 clone, the purification of its single stranded DNA and digestion with expensive restriction endonucleases. For research purposes, it is possible to design a small number of completely synthetic probes, resulting in amplification products with lengths varying between 100 and 130 nucleotides.

MLPA is an approach based on the PCR technique, and is sufficiently sensitive, reproducible and sequence-specific to allow the relative quantification of up to 50 different targets simultaneously. In MLPA, probes and not sample nucleic acids are subject to amplification and quantification. MLPA probes consist of one short synthetic oligonucleotide and one M13-derived, long probe oligonucleotide, which hybridize to adjacent sites of the target sequence. The short probe contains a target-specific sequence (21–30 nucleotides) and a 19 nucleotide sequence at the 5' end which is identical to the sequence of a labeled PCR primer. The long MLPA probe contains 24–43 nucleotides of target-specific sequence at the 5' phosphorylated end, a 36 nucleotide sequence that contains the complement of an unlabeled PCR primer at the 3' end, and a stuffer sequence of variable length in between. This variable-length fragment gives each complete probe the necessary size difference for detection and quantification using capillary gel electrophoresis [7]. Hybridized probe oligonucleotides are then ligated by a specific ligase enzyme, but only when both probes are stably hybridized to adjacent sites of the target sequence, permitting subsequent amplification (Protocol 1.9; see Figure 1.3).

MLPA probes are identified after capillary separation by size, using a selected size standard for the size calling procedure (Protocol 1.10). The relative MLPA probe signals (fluorescent units) reflect the relative copy number of the target sequence. An indication of the DNA input in the performed MLPA reaction may be obtained by examining the dosage quotient (DQ) control fragments, fragments whose lengths always co-vary and which are present in all MLPA kits. The signals of these fragments will be prominent if the amount of sample DNA is very low. By contrast, the fifth control band of 92 nucleotides is ligation dependent and should have a signal similar to most other MLPA amplification products. Visual inspection of the peak pattern of a patient sample superimposed over a peak pattern of a reference run can be used to analyze few sample numbers. Analyzing a larger sample series, more complex diseases and MLPA runs performed with miscellaneous sample types and quality require exportation of the peak signals and reliable normalization methods (Protocol 1.11).

To date, more than 150 MLPA applications are available, and this number is still growing rapidly. MLPA kits for the most common hereditary genetic disorders are already available,

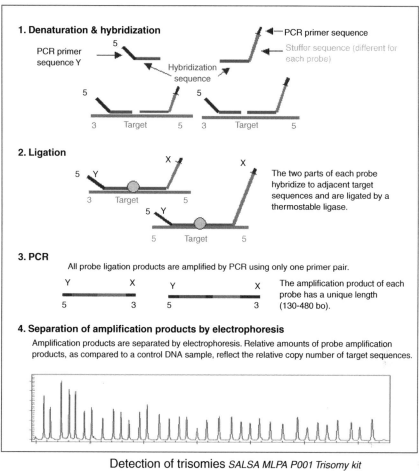

1. Denaturation & hybridization

PCR primer sequence Y

PCR primer sequence

Stuffer sequence (different for each probe)

Hybridization sequence

Target

2. Ligation

The two parts of each probe hybridize to adjacent target sequences and are ligated by a thermostable ligase.

Target

3. PCR

All probe ligation products are amplified by PCR using only one primer pair.

The amplification product of each probe has a unique length (130-480 bo).

4. Separation of amplification products by electrophoresis

Amplification products are separated by electrophoresis. Relative amounts of probe amplification products, as compared to a control DNA sample, reflect the relative copy number of target sequences.

Detection of trisomies *SALSA MLPA P001 Trisomy kit*

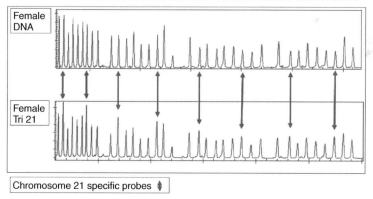

Female DNA

Female Tri 21

Chromosome 21 specific probes

Figure 1.3 Schematic of MLPA. The MLPA reaction is performed using four steps. Genomic DNA is denatured, whereafter the MLPA probes are added and incubated for 16 h, allowing complete hybridization adjacent to all target sequences. Probes completely hybridized to sequences either side of each target region are subsequently ligated to each other, enabling their exponential PCR amplification and final detection, and quantification by capillary electrophoresis. (See Plate 1.3.)

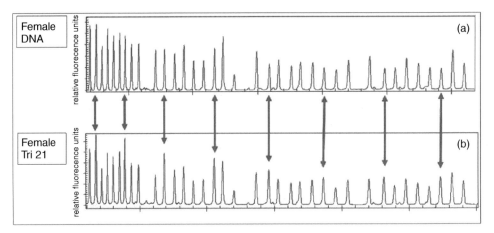

Figure 1.4 Detection of Down syndrome using the MLPA trisomy kit, including eight probes for sequences on each of chromosomes 13, 18, 21 and X, respectively, in addition to three probes for a chromosome Y target sequence. The arrows indicate the chromosome 21 probes. (a) The electropherogram of an MLPA performed on normal female DNA. (b) MLPA performed on DNA from a Down syndrome patient, clearly showing an increase of all chromosome 21 probes.

such as: Down's syndrome (see Figure 1.4), breast (*BRCA1*) [37] and colon cancer [32] (*MSH2* and *MLH1*), Duchenne [36], cystic fibrosis, centrifugal muscular atrophy, sickle cell anemia and Marfan syndrome, to name but a few. However, future MLPA kits will focus more on the detection of aberrations of genes playing roles in tumorigenesis, apoptosis, angiogenesis and pharmacogenetics.

PROTOCOL 1.9   Multiplex Ligation-Dependent Probe Amplification on DNA

Equipment and reagents

- Complete MLPA kit (MRC-Holland) containing:

    — MLPA probe mix

    — MLPA buffer

    — Ligase-65 buffer A

    — Ligase-65 buffer B

    — Ligase-65

    — ultra pure water

    — 10× SALSA PCR buffer

    — 10× SALSA PCR buffer

    — SALSA PCR forward primer (FAM or D4-labeled): 5′-GGGTTCCCTAAGGGTTGGA-3′

    — SALSA PCR reverse primer (unlabeled): 5′-GTGCCAGCAAGATCCAATCTAGA-3′

    — SALSA enzyme dilution buffer

    — SALSA polymerase

- Thermocycler (equipped with a heated lid)[ii] programmed to perform the complete MLPA program:

    — Hybridization program

        ○ 98 °C, 5 min

        ○ 25 °C, hold

        ○ 95 °C, 1 min

        ○ 60 °C, hold.

    — Ligation program

        ○ 54 °C, hold

        ○ 54 °C, 15 min

        ○ 98 °C, 5 min

        ○ 4 °C, hold.

    — PCR program

        ○ 60 °C, hold

        ○ (95 °C, 30 s; 60 °C, 30 s; 72 °C, 1 min) ×35

        ○ 72 °C, 20 min

        ○ 4 °C, hold.

Method

1 In a pre-PCR location, measure the concentrations of the stock DNA solutions (see Protocol 1.3).

2 Dilute[jj,kk] the DNA samples to working stocks of 10–20 ng/μl using 10 mM Tris-HCl, pH 8.0, 1 mM EDTA.

3 Label 0.2 ml strip tubes with sample initials and probe set number.

4 Add 5 μl of a working stock DNA solution to each tube (or water for a water blank).

5 Centrifuge the strip tubes briefly to collect the sample DNA at the bottom.

6 Put the tubes into the thermocycler and start the MLPA program (5 min at 98 °C); allow the samples to cool to 25 °C before opening the thermocycler.

7 Remove the probe mix and the MLPA buffer from the −20 °C freezer and allow to thaw. Vortex briefly.

8 Prepare a probe master mix by combining (per DNA sample) 1.5 μl of probe mix and 1.5 μl of MLPA buffer.

9 When the thermocycler reaches 25 °C, add 3 μl of probe master mix to each DNA sample. Mix well by pipetting up and down.

10 Continue the program on the thermocycler to advance from the 25 °C hold step, incubating for 1 min at 95 °C, and then for 16 h (overnight) at 60 °C.

11 In a pre-PCR location, remove the Ligase-65 buffer A and Ligase-65 buffer B from the freezer and allow to thaw. Vortex briefly.

12 Prepare[ll,mm,nn,oo] a ligase buffer mix by combining (for each reaction): 3 µl of Ligase-65 buffer A, 3 µl of Ligase-65 buffer B and 25 µl of water. Mix by vortexing.

13 Add 1 µl of Ligase-65 (per reaction) to the ligase buffer mix and mix well by vortexing.

14 Continue the program on the thermocycler to advance to the 54 °C hold step.

15 When the samples are at 54 °C, add 32 µl of the ligase buffer plus Ligase-65 mix to each reaction tube and mix well by pipetting up and down.

16 Continue the program on the thermocycler to advance to the next step (15 min incubation at 54 °C, followed by 5 min at 98 °C, and holding at 4 °C). Store[pp] the ligation products at 4 °C for up to 1 week.

17 In a pre-PCR location, remove 10× SALSA PCR buffer, SALSA PCR-primers and SALSA enzyme dilution buffer from the freezer and allow to thaw. Vortex briefly.

18 Label new tubes for PCR with the same sample initials and probe set numbers used for the hybridization and ligation reactions.

19 Prepare[ll,mm,nn,oo] a PCR buffer mix by combining (per ligation product): 4 µl of SALSA PCR buffer and 26 µl of water. Mix by vortexing.

20 Add 30 µl of the PCR buffer mix to each new tube.

21 Transfer 10 µl of each ligation product to its corresponding PCR tube.

22 Centrifuge the strip tubes briefly in order to collect the reaction mixtures at the bottom of each tube.

23 Place tubes into the thermocycler.

24 On ice, prepare a PCR master mix combining (per ligation product): 2 µl of SALSA primers, 2 µl of SALSA enzyme dilution buffer and 5.5 µl of water. Mix well by pipetting up and down.

25 Add 0.5 µl of SALSA polymerase to each reaction and carefully mix again by pipetting up and down.

26 Continue the program on the thermocycler to advance to the 60 °C hold step.

27 Add 10 µl of PCR mix to each PCR tube sitting in the thermocycler at 60 °C. Mix by pipetting up and down.

28 Continue the program, proceeding to exponential amplification.[qq]

Notes

[ii]If no heated lid is available in the PCR machine, overlay each DNA sample with 15 µl of mineral oil in order to prevent evaporation.

*jj*If the stock DNA solution concentrations are less than 10 ng/μl, use the available concentration, but note that results may be less reliable. 50–100 ng of DNA is recommended for each assay, although the acceptable range is 20–500 ng of DNA.

*kk*Both positive and negative controls (e.g. normal DNA and water respectively) can be processed in parallel.

*ll*All MLPA reaction mixtures should be prepared less than 1 h before use and stored on ice.

*mm*All reagents should be returned to the freezer after usage, reducing the loss of activity of the enzymatic solutions.

*nn*Make master mix solutions for all reactions in an experiment in order to minimize sample to sample variation. Add 10% more than required of each reagent to account for pipetting loss.

*oo*When performing MLPA on large sample numbers, multichannel pipettes are recommended.

*pp*For longer periods, storage at −20 °C is recommended.

*qq*PCR products can be stored at 4 °C for at least 1 week. As the fluorescent labels used are light sensitive, the PCR products should be stored in a dark box, or wrapped in aluminum foil.

PROTOCOL 1.10   Separation and Relative Quantification of MLPA Products

Equipment and reagents

- Capillary sequencer, or slab gel DNA sequencer, with fragment analysis software; for example

  — ABI-310 (1 capillary) – capillary: 5–47 cm, 50 μm (ABI 402839); polymer: POP-4 (ABI 4316355) or POP-6 (ABI 4306733)

  — ABI-3100 (16 capillaries), or ABI 3100 Avant (four capillaries) – capillaries: 36 cm; polymer: POP-4 (ABI 4316355)

  — ABI-3700 (96 capillaries) – capillaries: 3700 capillary array, 50 cm (ABI 4305787); polymer: POP-4, or POP-6

- Deionized formamide (ABI, 4311320)

- Labelled size standard (ROX-500 ABI GeneScan 401734; TAMRA-500 ABI GeneScan 401733).

*rr*Method

*ss,tt*ABI-310:

1 Following the PCR reactions, for each reaction mix:

  - 0.75 μl of the PCR reaction

  - 0.75 μl of water

- 0.5 μl of ROX-labeled internal size standard
- 12 μl of deionized formamide.

2 Mix by pipetting, incubate at 94 °C for 2 min and cool on ice.

3 Start the fragment separation software using the following settings:

- injection time: 5 s
- run voltage: 15 kV
- run time: 30 min
- run temperature: 60 °C
- run voltage: 15 kV
- filter: C.

ABI-3100 and ABI 3100 Avant:[ss,tt]

1 Following the PCR reactions, for each reaction mix:

- 0.5 μl of ROX-labeled internal size standard
- 8.5 μl of deionized formamide
- 1–3 μl of the PCR reaction.

2 Pipette this mixture into the injection plate.

3 Seal the injection plate with plate sealing film and incubate at 92 °C for 2 min, and hold at 4 °C for 5 min.

4 Start the fragment separation software using the following settings:

- run temperature: 60 °C
- capillary fill volume: 184 steps
- pre-run voltage: 15 kV
- pre-run time: 180 s
- injection voltage: 3.0 kV
- injection time: 10–30 s
- run voltage: 15 kV
- data delay time: 1 s
- run time: 1500 s.

ABI-3700:[ss,tt]

1 Following the PCR reactions, for each reaction mix:

- 2 μl of the PCR reaction
- 0.2 μl of ROX-labeled internal size standard
- 10 μl of deionized formamide.

2 Pipette this mixture into the injection plate.

3 Seal the injection plate with plate sealing film and incubate at 92 °C for 2 min and hold at 4 °C for 5 min.

4 Start the fragment separation software using the following settings:

- sample volume: 2.5 µl

- injection time: 10 s

- injection voltage: 10 kV

- run voltage: 7.5 kV

- run time: 4500 s

- cuvette temperature: 48 °C

- run temperature: 50 °C

- filter set: D

Notes

[rr]A description of the methods and settings for a number of commercially available capillary sequencers is provided.

[ss]The amount of the MLPA PCR required for analysis by capillary electrophoresis depends on the apparatus and fluorescent label used.

[tt]Label for MLPA: SALSA 6-FAM PCR primer-dNTP mix.

PROTOCOL 1.11 Normalization of MLPA Fragment Separations Results into Copy Number Ratios

Equipment and reagents

- Capillary sequencer, or slab gel DNA sequencer, with fragment analysis software

- The MLPA probe mix list containing the genomic map coordinates of the probes

- The size-called fragment list(s) containing the size-called fragment lengths, heights and areas of samples, and their references[uu]

- Software which calculates ratios, links the genomic position of the probes and creates charts and reports for interpretation (e.g. Microsoft Excel, or dedicated software such as MLPA-DAT, Genemarker or Seq Pilot)

Method[vv]

1 Load the probe list of your specific MLPA mix into the program.

2 Import your fragment list and define your reference data.[ww]

3 Examine the electropherogram patterns visually.[xx]

4 Separate probe signals from background signals by automatic binning and data filtering.

5 Quantify and normalize the probe signals (peaks).[yy]

6 Chart DNA copy number alterations according to genomic locations.[zz,aaa]

Notes

[uu] MLPA product fragment lists should have the following plot settings – Genescan (ABI): dye, time, peak length, peak height, peak area; Genemapper (ABI): peak dye, sample file name, peak height, peak area.

[vv] The protocol describes the creation of copy number ratio from MLPA product fragment lists using MLPA-DAT.

[ww] Reference runs are usually MLPA runs performed on normal human DNA samples within the same experiment as the samples. Multiple reference runs are recommended for probability calculation of the sample ratios found. Mean or median normalization are only recommended with larger sample numbers investigating syndromes with a low prevalence.

[xx] The DNA concentration, DNA ligation check and signal count are performed automatically.

[yy] Quantification is fully automated in MLPA-DAT and includes removal of individual run variance. The principle of normalization is twofold. First, all peak areas are converted into relative peak areas by dividing each by the sum of all peak areas. Second, a normalization factor is calculated from the ratio between the relative peak areas of all control probes of a sample and a reference DNA. Normalization settings may depend on the MLPA kit and type of sample analysed; for samples showing few aberrations, normalization can be done using all probes, but for samples with multiple chromosomal aberrations, such as tumor samples, it is recommended to use the control probes only. Control probe normalization sets the internal run normalization factor on a number of MLPA-probes known to remain constant in the expected sample types, while the amount of run variance is either determined on these probes only, or on all probes.

[zz] This is automatic in MLPA-DAT.

[aaa] A deletion of one copy of a probe target sequence will usually be apparent by a reduction in relative peak area for that probe amplification product of 35–55%. A gain in copy number from two to three copies/diploid genome will usually be apparent by an increase in relative peak area between 30 and 55%.

1.3 Troubleshooting

- **Low concentration of DNA isolated from FFPE tissue**: When a low yield of DNA is expected from FFPE tissue, glycogen can be added as a co-precipitant prior to ethanol precipitation. This has no noticeable effect on aCGH results.

- **DNA contaminants in MLPA**: MLPA reactions are more sensitive to contaminants than ordinary PCR reactions. DNA samples extracted from FFPE tissues often contain contaminants. Small remnants of phenol may act as PCR inhibitors, giving amplification products with lower average peak areas. Reducing the amount of DNA used will therefore sometimes have a beneficial effect. Normalization of these DNA samples should be performed against DNA extracted from reference DNA of the same source of tissue.

- **Low MLPA peak areas**: The quantity of a probe amplification product is primarily determined by the target sequence of the probe. MLPA probes located in guanine- and cytosine-rich areas often have difficulty reaching their target due to incomplete denaturation of the DNA. Better results may be obtained with these MLPA mixes by increasing the length of the initial denaturation step to 98 °C for 10 min. The probe signals are further influenced by the quality of the probe oligonucleotides and the amount of KCl and polymerase present during the PCR reaction. The polymerase activity may influence the relative signal strength of some probes; therefore, mixing the master mix well by pipetting up and down is strongly advised. It is also possible that the DNA input was too low. This is likely if the average signal of the four DQ (DNA quality) fragments is less than one-third of the ligation-dependent peak signal.

References

1. Kallioniemi, A., Kallioniemi, O.P., Sudar, D. *et al.* (1992) Comparative genomic hybridization for molecular cytogenetic analysis of solid tumors. *Science*, **258**, 818–821. The original publication describing CGH.

2. Kallioniemi, O.P., Kallioniemi, A., Piper, J. *et al.* (1994) Optimizing comparative genomic hybridization for analysis of DNA sequence copy number changes in solid tumors. *Genes Chromosomes Cancer*, **10**, 231–243.

3. Oostlander, A.E., Meijer, G.A. and Ylstra, B. (2004) Microarray-based comparative genomic hybridization and its applications in human genetics. *Clinical Genetics*, **66**, 488–495.

4. Solinas-Toldo, S., Lampel, S., Stilgenbauer, S. *et al.* (1997) Matrix-based comparative genomic hybridization: biochips to screen for genomic imbalances. *Genes Chromosomes Cancer*, **20**, 399–407.

5. Pinkel, D., Segraves, R., Sudar, D. *et al.* (1998) High resolution analysis of DNA copy number variation using comparative genomic hybridization to microarrays. *Nature Genetics*, **20**, 207–211. The original publication describing microarray CGH.

6. Ylstra, B., van den IJssel, P., Carvalho, B. *et al.* (2006) BAC to the future! or oligonucleotides: a perspective for micro array comparative genomic hybridization (array CGH). *Nucleic Acids Res.*, **34**, 445–450. Review of microarray CGH platforms.

7. Schouten, J.P., McElgunn, C.J., Waaijer, R. *et al.* (2002) Relative quantification of 40 nucleic acid sequences by multiplex ligation-dependent probe amplification. *Nucleic Acids Res.*, **30**, e57. The original publication describing MLPA.

8. Snijders, A.M., Nowak, N., Segraves, R. *et al.* (2001) Assembly of microarrays for genome-wide measurement of DNA copy number. *Nature Genetics*, **29**, 263–264.

9. Ishkanian, A.S., Malloff, C.A., Watson, S.K. *et al.* (2004) *Nature Genetics*, **36**, 299–303.

10. Pollack, J.R., Perou, C.M., Alizadeh, A.A. *et al.* (1999) Genome-wide analysis of DNA copy-number changes using cDNA microarrays. *Nature Genetics*, **23**, 41–46.

11. Coe, B.P., Ylstra, B., Carvalho, B. *et al.* (2007) Resolving the resolution of array CGH. *Genomics*, **89**, 647–653. Interesting paper on algorithms dealing with the resolution of diverse microarray CGH platforms.

12. Serth, J., Kuczyk, M.A., Paeslack, U. *et al.* (2000) Quantitation of DNA extracted after micro-preparation of cells from frozen and formalin-fixed tissue sections. *American Journal of Pathology*, **156**, 1189–1196.

13. Nannya, Y., Sanada, M., Nakazaki, K. *et al.* (2005) A robust algorithm for copy number detection using high-density oligonucleotide single nucleotide polymorphism genotyping arrays. *Cancer Research*, **65**, 6071–6079.

14. Van de Wiel, M.A., Kim, K.I., Vosse, S.J. *et al.* (2007) CGHcall: calling aberrations for array CGH tumor profiles. *Bioinformatics*, **23**, 892–894.

15. Bignell, G.R., Huang, J., Greshock, J. *et al.* (2004) High-resolution analysis of DNA copy number using oligonucleotide microarrays. *Genome Research*, **14**, 287–295.

16. Lindblad-Toh, K., Tanenbaum, D.M., Daly, M.J. *et al.* (2000) Loss-of-heterozygosity analysis of small-cell lung carcinomas using single-nucleotide polymorphism arrays. *Nature Biotechnology*, **18**, 1001–1005.

17. Janne, P.A., Li, C., Zhao, X. *et al.* (2004) High-resolution single-nucleotide polymorphism array and clustering analysis of loss of heterozygosity in human lung cancer cell lines. *Oncogene*, **23**, 2716–27262.

18. Zhao, X., Li, C., Paez, J.G. *et al.* (2004) An integrated view of copy number and allelic alterations in the cancer genome using single nucleotide polymorphism arrays. *Cancer Research*, **64**, 3060–3071.

19. Lips, E.H., Dierssen, J.W.F., van Eijk, R.R *et al.* (2005) Reliable high-throughput genotyping and loss-of-heterozygosity detection in formalin-fixed, paraffin-embedded tumors using single nucleotide polymorphism arrays. *Cancer Research*, **65**, 10188–10191.

20. Kennedy, G.C., Matsuzaki, H., Dong, S. *et al.* (2003) Large-scale genotyping of complex DNA. *Nature Biotechnology*, **21**, 1233–1237.

21. Matsuzaki, H., Loi, H., Dong, S. *et al.* (2004) Parallel genotyping of over 10,000 SNPs using a one-primer assay on a high-density oligonucleotide array. *Genome Research*, **14**, 414–425.

22. Gunderson, K.L., Steemers, F.J., Lee, G. *et al.* (2005) A genome-wide scalable SNP genotyping assay using microarray technology. *Nature Genetics*, **37**, 549–554.

23. Shen, R., Fan, J.B., Campbell, D. *et al.* (2005) High-throughput SNP genotyping on universal bead arrays. *Mutation Research*, **573**, 70–82.

24. Hardenbol, P., Yu, F., Belmont, J. *et al.* (2005) Highly multiplexed molecular inversion probe genotyping: over 10,000 targeted SNPs genotyped in a single tube assay. *Genome Research*, **15**, 269–275.

25. Oosting, J., Lips, E.H., van Eijk, R.R *et al.* (2007) High-resolution copy number analysis of paraffin-embedded archival tissue using SNP BeadArrays. *Genome Research*, **17**, 368–376.

26. Ji, H., Kumm, J., Zhang, M. *et al.* (2006) Molecular inversion probe analysis of gene copy alterations reveals distinct categories of colorectal carcinoma. *Cancer Research*, **66**, 7910–7919.

27. Lieberfarb, M.E., Lin, M., Lechpammer, M. *et al.* (2003) Genome-wide loss of heterozygosity analysis from laser capture microdissected prostate cancer using single nucleotide polymorphic allele (SNP) arrays and a novel bioinformatics platform dChipSNP. *Cancer Research*, **63**, 4781–4785.

28. Lin, M., Wei, L.J., Sellers, W.R. *et al.* (2004) dChipSNP: significance curve and clustering of SNP-array-based loss-of-heterozygosity data. *Bioinformatics*, **20**, 1233–1240.

29. Ishikawa, S., Komura, D., Tsuji, S. *et al.* (2005) Allelic dosage analysis with genotyping microarrays. *Biochemical and Biophysical Research Communications*, **333**, 1309–1314.

30. Herr, A., Grutzmann, R., Matthaei, A. *et al.* (2005) High-resolution analysis of chromosomal imbalances using the Affymetrix 10K SNP genotyping chip. *Genomics*, **85**, 392–400.

31. Eilers, P.H. and de Menezes, R.X. (2005) Quantile smoothing of array CGH data. *Bioinformatics*, **21**, 1146–1153.

32. Gille, J.J.P., Hogervorst, F.B.L., Pals, G. *et al.* (2002) *British Journal of Cancer*, **87**, 892–897.

33. Procter, M., Chou, L.S., Tang, W. *et al.* (2006) Molecular diagnosis of Prader–Willi and Angelman syndromes by methylation-specific melting analysis and methylation-specific multiplex ligation-dependent probe amplification. *Clinical Chemistry*, **52** (7), 1276–1283.

34. Hess, C.J., Denkers, F., Ossenkoppele, F.J. *et al.* (2004) Gene expression profiling of minimal residual disease in acute myeloid leukaemia by novel multiplex-PCR-based method. *Leukemia*, **18**, 1981–1988.

35. Diego-Alvarez, D., Ramos-Corrales, C., Garcia-Hoyas, M. *et al.* (2006) Double trisomy in spontaneous miscarriages: cytogenetic and molecular approach. *Human Reproduction*, **21**, 958–966.

36. Lalic, T.T, Vossen, R.H.A.M., Coffa, J. *et al.* (2005) Deletion and duplication screening in the DMD gene using MLPA. *European Journal of Human Genetics*, **13**, 1231–1234.

37. Hogervorst, F.B.L., Nederlof, P.M., Gille, J.J.P. *et al.* (2003) Large genomic deletions and duplications in the *BRCA1* gene identified by a novel quantitative method. *Cancer Research*, **63**, 1449–1453.

38. Van Beers, E.H., Joosse, S.A., Ligtenberg, M.J. *et al.* (2006) A multiplex PCR predictor for aCGH success of FFPE samples. *British Journal of Cancer*, **94**, 333–337. Describing a method for testing DNA quality of FFPE samples.

2
Identification of Polymorphic Markers for Genetic Mapping

Daniel C. Koboldt and Raymond D. Miller
Department of Genetics, Washington University School of Medicine, St. Louis, Missouri, USA

2.1 Introduction

Single nucleotide polymorphisms (SNPs) are the most prevalent form of DNA sequence variation in humans. At the end of 2008, public databases contained more than 12 million entries of genetic variants in humans, the vast majority of which are SNPs [1]. While the incidence of SNPs in the human genome is roughly 1 per 1000 bp on average, SNPs tend to cluster locally [2], creating regions of high SNP density with long stretches of 'SNP deserts' in between [3]. While an SNP could conceivably have four alleles (A, C, G, T), most are biallelic, with A/G the most common allele combination. Since DNA is a double helix, the opposite strand has alleles T and C. Consequently, an A/G SNP can also be described as a T/C SNP, depending upon orientation. An estimated 63% of known SNPs are A/G, 17% are A/C, 8% are C/G, 4% are A/T and the remaining 8% are single base insertions or deletions [3].

Owing to their widespread distribution across the genome, and facilitated by development of high-throughput genotyping technologies, SNPs have become important tools for genetic association studies. In an effort to understand the underlying structure of genetic variation in humans, the International HapMap Consortium [4, 5] characterized nearly 4 million SNPs in four geographically diverse human populations: the Yoruba in Ibadan, Nigeria (YRI); the CEPH population with European ancestry in Utah, USA (CEU); Han Chinese in Beijing, China (CHB); and Japanese in Tokyo, Japan. These data enabled the construction of a high-resolution haplotype map describing the block-like structure of linkage disequilibrium (LD) in the human genome. Large-scale, genome-wide association studies made possible

Genomics: Essential Methods Edited by Mike Starkey and Ramnath Elaswarapu
© 2011 John Wiley & Sons, Ltd.

by these advances have reliably identified many novel genomic locations associated with complex diseases [6].

Only rarely do the SNPs used to identify these loci prove to be the causal variants themselves. The vast majority of SNPs, in fact, are believed to be phenotypically neutral. Regions identified by association studies must be exhaustively screened to identify all DNA sequence variants before the complete genotype–phenotype relationship can be understood. Targeted DNA resequencing on capillary-based platforms (such as the ABI 3730XL) remains the gold standard for exhaustive variant discovery across a specific region of interest.

2.2 Methods and approaches

2.2.1 Repositories of known genetic variants

For humans and many other organisms, public databases contain a substantial number of SNPs and other sequence variants. The central repository for genetic variants is dbSNP [1],

Table 2.1 Organisms with more than 1000 reference SNPs in public database dbSNP (build 129) [7].

Organism	SNPs
Homo sapiens	14 708 752
Mus musculus	14 380 528
Gallus gallus	3 293 383
Oryza sativa	5 418 373
Canis familiaris	3 301 322
Pan troglodytes	1 543 208
Bos taurus	2 223 033
Monodelphis domestica	1 194 131
Anopheles gambiae	1 131 534
Apis mellifera	1 117 049
Danio rerio	662 322
Felis catus	327 037
Plasmodium falciparum	185 071
Rattus norvegicus	43 628
Saccharum hybrid cultivar	42 853
Sus scrofa	8427
Ovis aries	4181
Bos indicus × *Bos taurus*	2484
Macaca mulatta	780
Caenorhabditis elegans	1065

a database hosted by the National Center for Biotechnology Information (NCBI) in the USA. As of build 129 (April, 2008), dbSNP contained more than 1000 unique variants for at least 20 species (see Table 2.1).

Organism-specific databases, such as WormBase [8] and FlyBase [9], also contain extensive collections of known sequence variants. Extensive information about SNPs in the human genome, including their allele frequencies in different populations, can also be accessed through the University of California Santa Cruz (UCSC) Genome Browser [10, 11] and the website of the International HapMap Project [12, 13].

2.2.2 Targeted resequencing for variant discovery

The current gold standard for SNP discovery is direct resequencing of genomic DNA. In this approach, regions of interest are amplified by the polymerase chain reaction (PCR) and directly sequenced with automated DNA sequence analyzers. Traditional sequencing platforms, such as the ABI 3730XL, typically generate sequence 'reads' of 400–600 high-quality bases. Variants are identified by comparing these reads with the corresponding reference sequence.

2.2.2.1 Sample selection

Several factors should be considered when selecting samples to include for targeted resequencing. First of all, samples should be prioritized by the likelihood that they contain relevant sequence variants in the regions of interest. One approach is the 'sequencing the extremes' method, in which samples at each end of a phenotypic spectrum are chosen for variant discovery. This approach is particularly well suited for studies of quantitative phenotypes; for example, blood pressure, drug dosage, body mass index, and so on. The reasoning is that patients at one end of the spectrum are likely to have variants that confer susceptibility, whereas patients at the other end of the spectrum are likely to have variants that confer resistance.

Another important consideration during sample selection is the quality and available quantity of DNA. Sample quality and purity are especially critical, since contamination (by other tissue types, other organisms, etc.) can lead to false positives due to noisy sequence data. The quantity of available DNA from a particular sample (or patient) should also be considered. Whenever possible, the samples chosen for resequencing should have enough DNA kept in reserve for validation or further experimentation.

The number of samples to include for resequencing depends on several factors, but is typically dictated by the number of samples available and the budget for the project. While DNA resequencing can be costly, the more samples screened for sequence variants, the greater is the probability that key variants will be found.

2.2.2.2 Selecting regions of interest

The goal of targeted resequencing for mutation discovery is to choose genomic regions most likely to harbor variants of interest. There are two principal approaches to candidate gene selection. The first is knowledge based, in which the published literature and current understanding of the biology underlying a particular phenotype are used to construct lists of relevant genes. The second approach relies on experimental results (e.g. gene expression data, linkage analyses, genetic association studies) to prioritize candidate genes for mutation discovery.

Once a set of candidate genes has been compiled, the next step is to identify specific targets in those genes for resequencing. The classic approach utilized by many large-scale resequencing projects – including the pilot of the Cancer Genome Atlas project [14] – is to focus on the exons and splice regions of genes, which presumably are more likely to harbor functional variants. Evolutionarily conserved regions of promoters and untranslated regions (UTRs) may also be targeted for resequencing.

2.2.2.3 PCR Primer design

Once the genomic coordinates of sequencing targets have been identified, primers must be designed to selectively amplify those segments of DNA by PCR. Owing to the limitations of capillary-based sequencing platforms, PCR amplicons should be no longer than 600 bp in order to obtain high-quality sequences in both directions. Genomic targets larger than 500 bp should be covered by PCR amplicons using a tiling approach, ensuring that each amplicon overlaps the next by at least 100 bp (to ensure high-quality sequence coverage across the entire region). Selecting the appropriate PCR primers is not a simple procedure; ideally, primers should avoid repetitive regions, locations of known sequence variants, areas of extremely high or low GC content and potential primer–dimer interactions. Fortunately, many freely available primer design tools, such as Primer3Plus, automate amplicon tiling and primer selection while taking all of these factors into consideration (see Protocol 2.1).

PROTOCOL 2.1   PCR Primer Design with Primer3Plus

Equipment and reagents

- The gene name, transcript name or protein name for a gene of interest
- The UCSC Genome Browser website[a]
- The Primer3Plus web software.[b]

Method

1 Navigate to the UCSC Genome Browser website.[c] Click on the 'Gene Sorter' link at the top of the page.

2 Select the appropriate organism from the pull-down menu, and type the gene name or search term into the query box. Click the 'Go' button.

3 Click on the appropriate gene from the search results.[d]

4 Under 'Sequence and Links to Tools and Databases', click on the 'Genomic Sequence' link.[e]

5 Under 'Sequence Retrieval Region Options', choose the desired gene targets.[f]

6 In the same section, choose the radio button next to 'One FASTA record per region.' Specify that 500 bases upstream *and* downstream are to be included.

7 Under 'Sequence Formatting Options' choose 'Exons in upper case, everything else in lower case.'

8 Click the 'Mask Repeats' checkbox and set the radio button, 'to N.'[g]

9 Click the 'Submit' button.

10 In a separate internet browser window, navigate to the Primer3Plus web tool.

11 Under Task, select 'Sequencing' from the pull-down menu.

12 Paste the target gene sequence into the sequence field. Using brackets ([ ]), mark the region or regions of interest within the gene sequence.[h]

13 Click the 'General Settings' tab and set the following options: Primer Size – Min = 20 Opt = 23 Max = 26; Primer TM – Min = 54 Opt = 55 Max = 56; Primer GC% – Min = 20.0 Max = 50.0. Select the appropriate Mispriming/Repeat Library.[i]

14 Adjust the Advanced Settings. Set Product Size Min = 200, Opt = 400, Max = 600. Under the 'Sequencing' section of the Advanced Settings tab, set: Spacing = 500, Interval = 400, Lead = 50, Pick Reverse Primers = Checked.

15 Click the Pick Primers button.[j]

16 If primer design succeeds, the primer selections will be highlighted in your sequence. Scroll down to see the primer sequences. Check the box next to the desired primers to and click the 'Send to Primer3Manager' button.

17 Check the primers with the Primer3Manager. For each primer, verify that TM is between 54 and 56 °C and that GC content is between 20 and 50%.

Notes

[a]Available from http://genome.ucsc.edu/ [10].

[b]Available from http://www.bioinformatics.nl/cgi-bin/primer3plus/primer3plus.cgi [15].

[c]If the sequence for PCR primer design is already available then proceed to step 10.

[d]A list of genes should appear in table form. Click on the required gene description to bring up the summary page.

[e]This should bring the 'Get Genomic Sequence Near Gene' page.

[f]Generally, these should include 5′ UTR exons, CDS exons and 3′ UTR exons.

[g]This will prevent the design of primers in repetitive regions, and also make the target (exon) sequence easy to identify.

[h]This can also be performed by highlighting the desired sequence with the mouse and then clicking the brackets button ([ ]) next to 'Mark selected region.'

[i]Repeat/Mispriming libraries are species specific. Select the library for the species of interest.

[j]The next page may take a few minutes to load.

2.2.2.4 Targeted resequencing of genomic DNA

Once samples have been chosen, targets identified, and requisite PCR primers designed, it is time to proceed to the actual DNA sequencing. As the sequencing reads will be utilized for variant discovery, obtaining the highest quality data is crucial. Ideally, PCR amplicons should be sequenced in both directions to generate a 'read pair' – two reads across the same genomic region, but on opposite strands. Many downstream SNP discovery tools use read pair information to assess the confidence of variant predictions. The disadvantage of sequencing in both directions, of course, is the added expense of universal primer sequences. However, this expense can be avoided by using a '10-to-1' protocol for sequencing, in which one PCR primer serves as the sequencing primer when it is added at $10\times$ concentration (see Protocol 2.2).

PROTOCOL 2.2   PCR and Sequencing with the 10-to-1 Protocol

Equipment and reagents

- $10\times$ PCR buffer (0.5 M KCl, 0.1 M Tris-HCl (pH 8.3), 35 mM $MgCl_2$)

- $10\times$ 4 dNTPs (4×1 mM)

- $10\times$ Primer $1^{k,l}$ ($10 \times \mu$M)

- $10\times$ Primer 2^l (1 μM)

- Hot-start Taq DNA polymerase (JumpStart Taq, Sigma–Aldrich or Platinum Taq, Invitrogen Life Technologies)

- BigDye version 3 mix (Applied Biosystems)

- $5\times$ sequencing buffer (Applied Biosystems)

- 96- or 384-well purification plates (Princeton Separation or Genetix)

- Column cleanup kit (Qiagen or GE Healthcare).

Method

PCR

1 Set up PCR amplifications in a 96- or 384-well format. For each target, set up a 10 µl reaction containing $1 \times$ PCR buffer, $1 \times$ 4 dNTPs, $1 \times$ Primer 1, $1 \times$ Primer 2, 4 ng of DNA and 0.15 U of hot start Taq DNA polymerase.

2 Thermocycle as follows: 95 °C, 2 min; (92 °C, 10 s; 58 °C, 20 s; 68 °C, 30 s) $\times$ 35; 68 °C, 10 min.

3 Verify a successful PCR by gel electrophoresis before sequencing.[m]

Sequencing

4 For each PCR product, set up a 12 µl sequencing reaction containing 2.5 µl of PCR reaction mixture, 2 µl of BigDye version 3 mix and 1.0 µl of $5\times$ sequencing buffer.

5 Thermocycle as follows: 96 °C, 2 min; (96 °C, 15 s; 50 °C, 1 s; 60 °C, 4 min) $\times$ 25.

6 Remove unincorporated dye terminators.[n] Load reactions in 96- or 384-well format onto an ABI 3730XL for capillary electrophoresis (Applied Biosystems).

Notes

[k]Primer 1 is also the sequencing primer.

[l]Order primers in 96-well format (Integrated DNA Technologies or Qiagen Operon), and adjust the primer stock concentrations to 40 µM.

[m]For best results, perform enzymatic cleanup of each PCR product by adding commercially available shrimp alkaline phosphatase and exonuclease (SAP-EXO; Princeton Separation or Genetix).

[n]For superior results, use a commercial kit from Qiagen (QiaQuick) or GE Healthcare (Illustra AutoSeq G-50).

2.2.2.5 Extracting read sequences and quality scores

Typically, capillary sequencers generate one file per well containing the sequence read from one sample across one amplicon. These machine-readable binary files are usually in one of two formats: ABI or SCF. For most downstream applications, it is necessary to process traces with a basecalling program that interprets the chromatogram to call bases and assign quality values to each. The most commonly used basecaller, *Phred* [16], was developed at the University of Washington (see Protocol 2.3).

PROTOCOL 2.3   Basecalling with the Phred Program

Equipment and reagents

- Sequencing traces in ABI or SCF format
- UNIX/LINUX operating system
- The *Phred* program[o]

Method

1 Create a project directory with the following subdirectories: *chromat_dir*, *edit_dir*, *phd_dir* and *poly_dir*.

2 Copy all binary trace files (.abi or .scf) into the *chromat_dir* sub-directory.

3 Run *Phred* to perform basecalling and extract sequences/qualities. Instruct *Phred* to trim off low-quality bases using the following options:[p]

   (a) `phred -trim_alt ' ' trim_cutoff 0.05 -trim_trim_qual`

   (b) To append all read sequences to a single file and all read qualities to a single file:
   `phred -id chromat_dir/ -s -sa myFile.fa -q -qa myFile.qual`

   (c) To save read sequences and qualities as files (one per trace) in output subfolders:
   `phred -id chromat_dir/ -s -sd fasta_dir/ -q -qd qual_dir/`

   (d) To convert traces from ABI to SCF format: `phred -id chromat_dir/ -c -cd trace_dir/`

Table 2.2 Common command-line options for the *Phred* program.

Parameter	Argument	Default	Description
-help	None	None	Display helpful information
-if	<filename>	None	Read input filenames from file
-id	<dirname>	None	Read input files from <dirname>
-s	None	Nofile	Write × .seq sequence file(s)
-sa	<filename>	None	Append sequence files to <filename>
-sd	<dirname>	Nofile	Write × .seq file(s) to <dirname>
-q	None	Nofile	Write × .qual quality file(s)
-qa	<filename>	None	Append quality files to <filename>
-qd	<dirname>	Nofile	Write × .qual file(s) to <dirname>
-c	None	Nofile	Write × phred SCF file(s)
-cd	<dirname>	Nofile	Write × SCF file(s) to <dirname>
-d	None	Nofile	Write × .poly poly file(s)
-dd	<dirname>	Nofile	Write × .poly file(s) to <dirname>
-trim_alt	<enzyme seq>	Notrim	Enable alternate auto trim
-trim_cutoff	<n>	0.05	Trim_alt error probability
-trim_fasta	None	None	Trim FASTA bases and qual. values
-trim_scf	None	None	Trim SCF bases and qual. values
-trim_phd	None	None	Trim base call data in phd files
-process_nomatch	None	None	Process chromats with unmatchable primerID string

Notes

[o] Available from http://www.phrap.org/phredphrap/phred.html [17]. For common command-line options, see Table 2.2.

[p] For a complete listing of command-line options, invoke Phred with the –help option.

2.2.2.6 *Sequence alignment and assembly*

To perform variant discovery with resequencing data, it is necessary to anchor sequence reads on a reference sequence for making comparisons (alignment), and it is usually advisable to layer multiple overlapping reads on the reference to generate a consensus call at each base (assembly). The *cross_match* and *Phrap* [18] programs created by the University of Washington are commonly used to perform such tasks with capillary-based resequencing data. The reference sequence input to *Phrap* must be in trace file (SCF) format and can be created with the *sudophred* utility (see Protocol 2.4). Subsequently, the real resequencing traces can be assembled together with the reference using the *phrap* command (see Protocol 2.5). Typically, assemblies are written to ACE (.ace) files, a format utilized by many downstream analysis tools.

PROTOCOL 2.4 Creating Traces from the Reference Sequence with the *sudophred* Utility

Equipment and reagents

- Reference sequence for the target region in FASTA format.
- The *sudophred* utility from the *Polyphred* package.[q]

Method

1 Create or verify the existence of the *chromat_dir* and *phd_dir* subdirectories.

2 Run the *sudophred* tool as follows: sudophred [reference.fasta] −r

3 To specify a minimum base quality value for the reference, use the −q parameter: sudophred [reference.fasta] −r −q 20

4 For further options, invoke the command: sudophred −h

5 Verify that reference trace files were created in the *chromat_dir* and *phd_dir* subdirectories.

Notes
[q]Available from http://droog.gs.washington.edu/polyphred [19].

PROTOCOL 2.5 Assembling Reads on the Reference Sequence with the *Phrap* program

Equipment and reagents

- Reference sequence traces generated by *sudophred* (see Protocol 2.4)
- Clone and subclone vector sequences in FASTA format for vector screening
- Sequence files in trace or PHD format
- The *Phred/Phrap/cross_match* programs[r]
- The *phredPhrap* script.[s]

Method

1 Create the four directories expected by the *phredPhrap* program.

2 Move or copy the trace files into the *chromat_dir* subdirectory.

3 Move or copy the *phredPhrap* script into the *edit_dir* subdirectory.

4 Run the *phredPhrap* script, which automates several steps:

(a) *Phred* for basecalling and conversion of reads to PHD format

(b) *Phd2fasta*, to convert the PHD files from *Phred* to FASTA format

(c) *Cross_match*, to mask vector sequence

(d) *Phrap*, to assemble the reads and build an ACE file[t]

(e) Additional ancillary programs to aid visualization in *Consed*.

Notes

[r]Available from http://www.phrap.org. [20].

[s]Available from http://droog.gs.washington.edu/polyphred [19].

[t]Typically, assemblies are written to ACE (.ace) files, a format utilized by many downstream analysis tools.

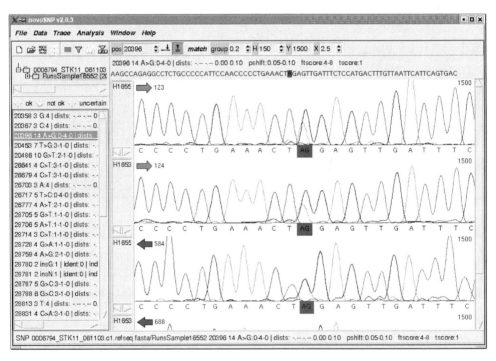

Figure 2.1   SNP detection in ABI 3730 sequence data with NovoSNP. SNP discovery efforts can be organized by projects, each with its own reference sequence (top left). Raw sequence files are basecalled and aligned to the reference sequence, after which a list of candidate sequence changes (left) is generated. Each prediction can be manually reviewed by visualizing the traces as they align to the reference sequence (right). (See Plate 2.1.)

2.2.2.7 Variant detection

Once reads are assembled on a reference sequence, variants can be detected by pairwise comparisons between the reads and the reference sequence. If an assembly was generated using the *phredPhrap* script (see Protocol 2.5), then *Polyphred* [19] can be used to detect variants (see Protocol 2.6). This program, also developed at the University of Washington, is one of the most widely used tools for mutation detection. It detects heterozygous and homozygous SNPs as well as indels, scores each variant, and provides relevant genotypes from the read basecalls. Both the *phredPhrap* suite and *Polyphred*, while intuitive, require basic knowledge of Linux/UNIX. Alternatively, the NovoSNP tool [21] (Figure 2.1) can perform variant discovery and visualization in a more user-friendly interface (see Protocol 2.7). While less scalable than *Polyphred*, NovoSNP runs on Windows, Mac or Linux computers and requires only a reference sequence (FASTA file) and sequence traces (in binary format).

PROTOCOL 2.6   Variant detection in *Phrap* assemblies with *Polyphred*

Equipment and reagents

- Directory structures, *Phred* output and assemblies for sequence traces and reference sequence from running *phredPhrap* (see Protocol 2.*[u]*)

- The *Polyphred* program.*[v]*

Method

1   Run the *phredPhrap* script from the *edit_dir* subdirectory if you have not already done so (see Protocol 2.5).

2   Run the *Polyphred* program*[w]* from the *edit_dir* subdirectory:

  (a) cd edit_dir/

  (b) polyphred −ace [ace_file] −refcomp [refseq_id] [options]*[x]*

3   Review the *Polyphred* output.*[y]*

Notes

*[u]*If the *phredPhrap* script ran successfully, there should be four subdirectories: *chromat_dir, edit_dir, phd_dir* and *poly_dir*. There should be one file per trace in the *chromat_dir, phd_dir* and *poly_dir* subdirectories. There should also be an assembly (.ace) file in the *edit_dir* folder.

*[v]*Available from http://droog.gs.washington.edu/polyphred [19].

*[w]*Recommended options for SNP detection include:

- -t genotype: specifies output with Consed-compatible tags and SNP genotypes.

- -quality 25: specifies the quality threshold to use bases for variant calling.

- -score 25: specifies the score threshold for variant calling.

[x][refseq_id] is the name of the assembly created by phredPhrap. The [refseq_id] is the extension of the reference sequence trace if different from the sudophred default (.REF).

[y]To learn about how to interpret Polyphred output, see the Polyphred documentation at http://droog.gs.washington.edu/polyphred/poly_doclist.html.

PROTOCOL 2.7   Graphical Variant Discovery with the NovoSNP program

Equipment and reagents

- The *novoSNP* program[z],[aa]
- Reference sequence(s) in FASTA format
- Sequence trace files in binary (ABI or SCF) format.

Method

1 Create a new project in novoSNP. Under the File menu, select 'New Project.' In the pop-up window, navigate to where the project will be stored and give it a name; for example, MyNovoSNP.proj.

2 Add a reference sequence to the project. Under the Data menu, select 'Add refseq.' In the pop-up window, navigate to the folder that contains the reference sequence and select the FASTA file.

3 Add sequence traces to the project. Under the Data menu, select 'Add Runs.' Navigate to the folder that contains the binary trace files, and click the 'Select' button. In the pop-up window that appears, select the following options:

   (a) Simple base qualities: Always

   (b) Quality clipping: On

   (c) Cutoff: 0.02.

4 Click on 'Add all' or 'Add selected' to add traces.[bb]

5 View the read alignments. Under the Window menu, select 'Alignment.'[cc]

6 Manually review the variants. The variants detected by novoSNP appear in a panel on the left-hand side of the program. Click a variant to display the traces across that position in the right-hand window. Left-clicking and dragging on the trace window will scroll the trace display right or left. If desired, edit the settings in the novoSNP toolbar to adjust the trace display.

7 Automatically filter the variants. Under the Window menu, select 'Filter.'[dd]

8 Generate SNP or trace reports.[ee] Under the Analysis menu, select 'Reports.' Choose the type of report required and name it.

2.3 Troubleshooting

2.3.1 Primer design

Primer design typically fails for one of the following reasons:

- **Primer GC content**: In regions with unusually high or low G + C content, primer design fails because the percentage G + C content falls outside the desirable range (20–50%), or because potential primers have melting temperatures that fall outside the permitted range. Try adjusting the permitted maximum melting temperature first and then the allowable percentage G + C.

- **Repetitive sequence**: Often a region of interest is flanked by repetitive sequence such that unique primers cannot be designed on both sides. Such regions are typically resistant to most primer-based assays. However, occasionally PCR will succeed if one (and only one) primer is designed within a repetitive region.

- **PCR product size**: If candidates are available for both forward and reverse primers, but no suitable assay can be designed, try adjusting the PCR product size from 80–400 bp to 80–1000 bp.

2.3.2 PCR amplification

A number of variables can contribute to PCR failure. Some of the most common issues are:

- **Variable primer annealing sites (SNP-in-Primer)**: When PCR primers are designed across a sequence variant such as an SNP, PCR failure or allele dropout (loss of a single allele during PCR) may occur, particularly if the variant is near the 3′ end of the primer. To avoid such issues, use an SNP-marked reference sequence when designing PCR primers. Variable positions marked with Ns will not be considered for primer design.

- **Poor sample quality**: Degraded, contaminated or variable-concentration DNA samples amplify poorly. Samples whose quality and concentration are unknown should be quantified by Picogreen assay [24] and assessed for integrity by agarose gel electrophoresis prior to PCR.

2.3.3 Working with binary trace files

Often *Phred* and other programs have difficulty reading binary files from non-native operating systems (such as MacOS). In UNIX/Linux systems, the file type of a binary sequence trace file can be displayed using the command:

```
file trace.b1.ab1
```

where trace.b1.ab1 is the name of a trace file. The file type should be 'data.' If it is 'compressed data,' then the files can be uncompressed with the *gunzip* command. If another file type is detected, then the trace is likely corrupt (or empty) and should be either removed or replaced from the original file location.

2.3.4 *Phred/Phrap*

Many issues with *Phred*, especially those related to parameters, can be addressed by viewing the usage information:

```
phred -doc
```

Alternatively, complete documentation for *Phred* is available at http://www.phrap.org/phredphrap/phred.html.

References

1. Sherry, S.T., Ward, M.H., Kholodov, M. *et al.* (2001) dbSNP: the NCBI database of genetic variation. *Nucleic Acids Research*, **29** (1), 308–311.

2. Koboldt, D.C., Miller, R.D. and Kwok, P.Y. (2006) Distribution of human SNPs and its effect on high-throughput genotyping. *Human Mutation*, **27** (3), 249–225.

3. Miller, R.D., Taillon-Miller, P. and Kwok, P.Y. (2001) Regions of low single-nucleotide polymorphism incidence in human and orangutan xq: deserts and recent coalescences. *Genomics*, **71** (1), 78–88.

4. The International HapMap Consortium (2005) A haplotype map of the human genome. *Nature*, **437** (7063), 1299–1320.

5. The International HapMap Consortium (2007) A second generation human haplotype map of over 3.1 million SNPs. *Nature*, **449** (7164), 851–861.

6. The Wellcome Trust Case Control Consortium (2007) Genome-wide association study of 14,000 cases of seven common diseases and 3,000 shared controls. *Nature*, **447** (7145), 661–678.

7. dbSNP http://www.ncbi.nlm.nih.gov/projects/SNP/ (last accessed May 2010). The NCBI database of sequence variation. This is the central resource for searching databases of known SNPs and indels.

8. WormBase http://www.wormbase.org/ (last accessed May 2010). This is a central community resource for researchers that work with *Caenorhabditis elegans* and other nematodes. It serves as the central repository for sequence variation, mutants, phenotypes and other information related to nematode species.

9. FlyBase http://flybase.org/ (last accessed May 2010). This is the central community resource for researchers that work with *Drosophila melanogaster* and other fly species. It contains extensive information, including a database of known sequence variants, specific to fly model organisms.

10. UCSC http://genome.ucsc.edu/ (last accessed May 2010). The UCSC Genome Browser. This widely used visualization tool makes it possible to browse genomes at any resolution (individual bases to entire chromosomes), along with tracks showing annotated genes, conserved sequences, regulatory elements and known sequence variations.

11. Karolchik, D., Baertsch, R., Diekhans, M. *et al.* (2003) The UCSC genome browser database. *Nucleic Acids Research*, **31** (1), 51–54.

12. HapMap http://www.hapmap.org (last accessed May 2010). The website of the International HapMap Project provides allele frequencies and genotype data for millions of SNPs characterized in several human population.

13. Thorisson, G.A., Smith, A.V., Krishnan, L. and Stein, L.D. *et al.* (2005) The International HapMap Project Web site. *Genome Research*, **15** (11), 1592–1593.

14. Cancer Genome Atlas Consortium (2008) Comprehensive genomic characterization defines human glioblastoma genes and core pathways. *Nature*, **455** (7216), 1061–1068.

15. Primer3Plus http://www.bioinformatics.nl/cgi-bin/primer3plus/primer3plus.cgi (last accessed May 2010). This online tool for provides interactive, highly customizable design of PCR assays based on the *primer3* design algorithm.

16. Ewing, B. and Green, P. (1998) Base-calling of automated sequencer traces using phred. II. Error probabilities. *Genome Research*, **8** (3), 186–194.

17. *Phred* http://www.phrap.org/phredphrap/phred.html (last accessed May 2010). This is the site for downloading *Phred*, the widely used basecaller for capillary-based sequencing that provides a numeric quality score for each base position.

18. De la Bastide, M. and McCombie, W.R. (2007) Assembling genomic DNA sequences with PHRAP. *Current Protocols in Bioinformatics*, **11**: 14.

19. *Polyphred* http://droog.gs.washington.edu/polyphred (last accessed May 2010). This is the site for downloading *Polyphred*, a suite of programs from the University of Washington for basecalling, alignment, assembly and SNP/indel discovery in capillary-based sequencing data.

20. *Phrap* http://www.phrap.org (last accessed May 2010). This is the site for downloading *Phrap*, a widely used assembly program, and cross_match, a sequence alignment algorithm.

21. Weckx, S., Del-Favero, J., Rademakers, R. *et al.* (2005) novoSNP, a novel computational tool for sequence variation discovery. *Genome Research*, **15** (3), 436–442.

22. NovoSNP http://www.molgen.ua.ac.be/bioinfo/novosnp (last accessed May 2010). This is the site for downloading novoSNP, a Java-based, platform-independent, graphical tool for visualizing ABI 3730 sequence traces and detecting variations in them.

23. Rijk, P.D. and Del-Favero, J. (2007) novoSNP3: variant detection and sequence annotation in rese-quencing projects. *Comparative Genomics*, volume 2 (ed. N.H. Bergman). Methods in Molecular Biology, volume 396. Humana Press: 331–344.

24. Ahn, S.J., Costa, J. and Emanuel, J.R. (1996) PicoGreen quantitation of DNA: effective evaluation of samples pre- or post-PCR. *Nucleic Acids Research*, **24** (13), 2623–2625.

3

Genotyping and LOH Analysis on Archival Tissue Using SNP Arrays

Ronald van Eijk, Anneke Middeldorp, Esther H. Lips, Marjo van Puijenbroek, Hans Morreau, Jan Oosting and Tom van Wezel
Department of Pathology, Leiden University Medical Center, Leiden, The Netherlands

3.1 Introduction

Genotyping is instrumental in the elucidation of the genetics underlying the cause of human diseases, such as cancer. Linkage analysis and association studies can improve our understanding of the relation between genetic information and disease phenotype and identify genomic loci that harbor disease-causing genes. Ultimately, this could result in the identification of the underlying genes. Linkage analysis, for example, led to the successful identification of genes that cause breast cancer, *BRCA1* and *2* [1–3], and colorectal cancer, *MLH1* and *MSH2* [4–7]. Over the last few decades genotyping has evolved from time-consuming and low-throughput methods – such as the determination of restriction fragment length polymorphisms (RFLPs, [8]) or simple sequence length polymorphisms (SSLPs [9]) – to the high-throughput genotyping using single nucleotide polymorphism (SNP) arrays.

   In this chapter we will discuss methods and SNP array platforms for high-throughput genotyping with a focus on the use of DNA from formalin-fixed paraffin embedded (FFPE) tissue and the detection of loss of heterozygosity (LOH).

Genomics: Essential Methods   Edited by Mike Starkey and Ramnath Elaswarapu
© 2011 John Wiley & Sons, Ltd.

3.2 Methods and approaches

3.2.1 Arrays

Different methodologies of SNP typing and types of commercially available SNP arrays have been developed. Basically, two types of array exist: locus-specific arrays of oligonucleotides and arrays with universal capture oligonucleotides. The SNP typing assays include methodologies such as allele-specific primer extension and whole genome sampling. We will briefly discuss four different methods. Two different genotyping methods, molecular-inversion probe (MIP) genotyping and Goldengate genotyping, are based on high level multiplex polymerase chain reaction (PCR) with universal primers in combination with universal arrays. Goldengate genotyping makes use of a multiplex mixture of probes for 96, 384, 768 or 1536 SNPs per array [10]. For each SNP, a combination of allele-specific and locus-specific primers is annealed to the SNP locus. These primers are tailed with common forward and reverse primers, and a universal capture probe complementary to the locus-specific primer. The small gap between the allele and locus-specific probes is filled by subsequent allele-specific primer extension and sealed by a ligase, resulting in an allele-specific artificial PCR template. This template is then PCR amplified using fluorescently labeled universal PCR primers. The resulting probe is hybridized to an array of universal-capture probes and the array is scanned in a special reader, generating two fluorescent signals representing the two different alleles of an SNP. MIP genotyping utilizes a pool of locus-specific probes with a multiplexing degree of over 10 000 SNPs per array. The 5' and 3' ends of each circularizable probe anneal upstream and downstream of the SNP. The 1 bp gap is filled in a different reaction for each nucleotide. The probes are subsequently circularized using ligase to seal the remaining nick; non-annealed and non-circular probes are removed by exonuclease treatment. Restriction and digestion then releases the circularized probe and the resulting template is PCR amplified using common primers [11]. The four nucleotide reactions are labeled in different colors and pooled. Subsequently, the pool is hybridized to an array of universal-capture probes and the four colors are read out. Genechips can detect up to a million SNPs on a single chip. For each SNP, a set of locus-specific 25-mer oligonucleotides is present on the array. The sample is prepared according to the whole-genome sampling assay [12], a method in which the genomic complexity is reduced through restriction enzyme (RE) treatment of high-quality genomic DNA and ligation of a common adaptor to the digested DNA. The subsequent single primer PCR step results in the reduction of the genomic complexity through efficient size-selection in the PCR reaction. The product is hybridized to a locus-specific array. The SNPs on the array are selected from the DNA that is represented after the complexity reduction PCR [13]. Finally, Infinium arrays are locus-specific arrays with allele-specific capture probes. In the assay, genomic DNA is whole-genome amplified and subsequently fragmented. The resulting probes are then denatured and hybridized to the array. An 'on the array'-allele-specific primer extension assay is followed by staining and read out using standard immunohistochemical detection methods. Currently, these arrays can detect over a million SNPs on a single array [14].

3.2.2 Genotyping

After scanning of the SNP arrays, the signal intensities have to be converted into genotype calls. SNP calling software is available for each platform: Beadstudio for Goldengate and

Infinium, GTYPE and Genotyping Console for Genechips, and GTGS for the MIP assay. All programs are essentially similar, in that three clusters are automatically computed for each SNP; the heterozygous AB and homozygous AA, or BB. The clusters are based upon the allele-specific signal intensities. Genotyping errors and no-calls will hamper linkage and association studies, and reliable SNP calls are essential for these applications. Therefore, additional genotyping algorithms are becoming available to improve the quality of the genotypes from SNP arrays. Examples of these methods are SNIPer [15] AccuTyping [16], SNPchip [17] or RLMM [18].

3.2.3 Linkage and association analysis

The development of relatively inexpensive, high-throughput SNP arrays has transformed genetic and genomic research. Their availability enables whole-genome association studies for the unraveling of the complex genetics of disease. Success was reported for inflammatory bowel disease [19] and type 2 diabetes mellitus [20] and recently for breast, prostate and colorectal cancers [21–24]. An increasing number of reports of successful linkage are appearing with a variety of phenotypes, such as Bardet–Biedl syndrome [25], neonatal diabetes [26] or alcoholism [27]. In the study of colorectal cancer, arrays with over 10 000 SNPs were successfully used to identify a susceptibility locus on chromosome 3q21–q24 [28] and a locus on chromosome 10q23 that is associated with hereditary mixed polyposis syndrome [29]. Linkage analysis using genotypes from SNP arrays can be performed using freely available packages such as Mendel [30], Merlin [31] or Allegro [32]. In order to facilitate the analysis and processing of genotypes of vast size, several freely available tools have been developed; for example, Alohomora [33], SNPlink [34], CompareLinkage [35] and Easylinkage [36]. The tools convert genotypes into the proper format for linkage programs, perform different levels of quality control and error removal, and graphically present the linkage analysis data. SNPlink also automatically removes SNPs that are in high linkage disequilibrium (LD) with nearby SNPs, since LD can falsely inflate linkage statistics.

3.2.4 Formalin-fixed, paraffin-embedded tissue

One increasingly important application is genotyping and genomic profiling of FFPE samples. Across the world, large collections of FFPE samples with clinical follow up exist in the archives of pathology departments. Ideally, the DNA extracted from these tissues (see Protocols 3.1–3.3) might serve to perform linkage and association studies, or to generate tumor profiles of LOH and genomic abnormalities. The availability of clinical follow up for these samples will certainly strengthen the research. For linkage analysis, the use of FFPE tissue for genotyping also allows incorporation into analyzes of individuals for which leukocyte DNA is unavailable. A major drawback of the use of DNA from FFPE samples is DNA quality (see Protocol 3.4). As a consequence of formalin fixation of the tissue, extracted DNAs show varying levels of degradation. The degree of degradation depends largely on the length and the method of fixation, and the age of the specimen. The implication for SNP arrays is that not all the methods are suitable for use (or will not perform consistently) with FFPE DNAs. Whereas the high-density Genechip and Infinium arrays are designed for use with high-quality genomic DNA, both the Goldengate and the MIP assay can be used for genotyping and for detection of LOH and copy number changes in FFPE tissue [37–39].

PROTOCOL 3.1   Non-Column-Based DNA Isolation from FFPE Tissue[a]

Equipment and reagents

- Xylene

- Ethanol (100% (v/v), 70% (v/v))

- PK1: 10 mM Tris-HCl, pH 8.3, 50 mM KCl, 2.5 mM MgCl$_2$, 0.45% (v/v) NP-40, 0.45% (v/v) Tween-20, 0.01% gelatin.

- Proteinase-K (10 mg/ml)

- PPS: protein precipitation solution (e.g. Promega, A7951)

- TE: 10 mM Tris-HCl, pH 8.0, 0.1 mM EDTA

- Tissue arrayer (e.g. Beecher Instruments)

[b]Method

1 Identify histologically normal and tumor areas from the FFPE tumor. Using a tissue arrayer, collect three punches (0.6 mm) from normal tissue and three from tumor tissue in separate tubes.

2 Add 1 ml of xylene to a tube with three FFPE tissue punches.

3 Vortex and mix on a rotating wheel for 15 min at room temperature.

4 Centrifuge at 13 000g for 3 min at room temperature.

5 Carefully remove the xylene from the tissue cores.

6 Repeat steps 2–5.

7 Add 1 ml of 100% (v/v) ethanol.

8 Vortex and mix on a rotating wheel for 15 min at room temperature.

9 Centrifuge at 13 000g for 3 min at room temperature.

10 Carefully remove ethanol from the tissue pellet.

11 Repeat steps 7–10.

12 Air dry for 5 min.

13 Add 150 μl of PK1 buffer to the dried tissue pellet.

14 Add 5 μl of proteinase-K.

15 Mix and pulse centrifuge.

16 Incubate overnight in a heat block at 56 °C.

17 Heat-inactivate the proteinase-K at 100 °C for 10 min.

18 Centrifuge at 13 000g for 10 min at room temperature.

19 Transfer the supernatant containing the DNA to new tube.

20 Add 50 μl of PPS to the supernatant.

21 Vortex and chill on ice for 5 min.

22 ^{c}Centrifuge at 13 000g for 5 min at room temperature.

23 Carefully transfer the supernatant containing the DNA to a new tube.

24 ^{d}Add 150 μl of isopropanol (at room temperature) to the supernatant and mix by inverting the tube.

25 Precipitate the DNA by centrifuging at 13 000g for 5 min at room temperature.

26 Carefully remove the supernatant.

27 ^{e}Wash the pellet with 200 μl of 70% (v/v) ethanol.

28 Repeat steps 26 and 27.

29 Air dry the DNA^f pellet and dissolve in 100 μl of TE.

Notes

^{a}For the Goldengate assay, we favor the use of DNA isolated using a relative simple precipitation procedure after proteinase K digestion. Although column, or bead based methods, are likely to yield cleaner DNA, smaller fragments from the already fragmented FFPE tissue DNA are better preserved using a precipitation method.

^{b}Genomic DNA can be isolated from fresh frozen tumors or from normal blood leukocytes, using the Wizard Genomic DNA Purification Kit (Promega).

^{c}The precipitated protein will form a white pellet.

^{d}When a low yield is expected from the tissue, glycogen can be added prior to precipitation.

^{e}The pellet might not be visible and so mark the outside edge of the tube facing outwards in the centrifuge, or ensure that the hinge of the lid is facing outwards during centrifugation.

^{f}DNA size and quality can be assessed by electrophoresis through a 1.5 % agarose gel and by multiplex PCR (see Protocol 3.5) or a similar method as described by van Beers [40].

PROTOCOL 3.2 Column-Based DNA Isolation from FFPE Tissue^g

Equipment and reagents

• NucleoSpin Tissue XS Genomic DNA Purification from Tissue kit (Machery-Nagel)

• Xylene

• Ethanol [100% (v/v)]

• PK1: 10 mM Tris-HCl, pH 8.3, 50 mM KCl, 2.5 mM MgCl$_2$, 0.45% (v/v) NP-40, 0.45% (v/v) Tween-20, 0.01% gelatin.

- Proteinase-K (10 mg/ml)
- PPS: protein precipitation solution (e.g. Promega, A7951)
- TE: 10 mM Tris-HCl, pH 8.0, 0.1 mM EDTA
- Tissue arrayer (e.g. Beecher Instruments)

Method

1 Identify histologically normal and tumor areas from the FFPE tumor. Using a tissue arrayer, collect three punches (0.6 mm) from normal tissue and three from tumor tissue in separate tubes.

2 Add 1 ml of xylene to a tube with three FFPE tissue punches.

3 Vortex and mix on a rotating wheel for 15 min at room temperature.

4 Centrifuge at 13 000g for 3 min at room temperature.

5 Carefully remove the xylene from the tissue cores.

6 Repeat steps 2–5.

7 Add 1 ml of 100% (v/v) ethanol.

8 Vortex and mix on a rotating wheel for 15 min at room temperature.

9 Centrifuge at 13 000g for 3 min at room temperature.

10 Carefully remove ethanol from the tissue pellet.

11 Repeat steps 7–10.

12 Air dry for 5 min.

13 Add 150 μl of PK1 buffer to the dried tissue pellet.

14 Add 5 μl of proteinase-K.

15 Mix and pulse centrifuge

16 Incubate overnight in a heat block at 56 °C.

17 Heat-inactivate the proteinase-K at 100 °C for 10 min.

18 Centrifuge at 13 000g for 10 min at room temperature.

19 Transfer the supernatant containing the DNA to new tube.

20 Add 80 μl of NucleoSpin buffer B3.

21 Vortex twice for 5 s each, and incubate at 70 °C for 5 min. Vortex briefly at the end of the incubation.

22 Allow the lysate to cool down to ambient temperature.

23 Add 80 μl of 100% (v/v) ethanol to the lysate.

24 Vortex twice for 5 s each.

25 Centrifuge at 13 000g for 10 s to collect all the liquid in the bottom of the tube.

26 Place a NucleoSpin® Tissue XS column into a 2 ml collecting tube, and apply the sample to the column.

27 Centrifuge at 13 000g for 1 min.

28 Discard the flow-through.

29 Place the column into a new 2 ml collecting tube.

30 Add 50 µl of buffer B5 to the membrane.

31 Centrifuge for 1 minute at 11 000g.[h]

32 Add 50 µl of buffer B5 directly onto the membrane.

33 Centrifuge at 13 000g for 2 min.

34 Place the column into a 1.5 ml microcentrifuge tube.

35 Apply 50 µl of buffer TE directly onto the center of the silica membrane of the column.

36 [i]Centrifuge at 13 000g for 1 min.

Notes

[g]This protocol can be used instead of the non-column based method described in Protocol 3.1. Protocol 3.2 is more expensive than Protocol 3.1 and the yield of DNA is slightly lower. However, the purity of the DNA isolated is higher.

[h]It is not necessary to discard the flow-through. Reuse the collecting tube.

[i]For a higher DNA yield, repeat step 35 by pipetting the eluate from step 36. back onto the column.

PROTOCOL 3.3   DNA Concentration Measurement Using Picogreen

Equipment and reagents

- TE: 10 mM Tris-HCl, pH 8.0, 0.1 mM EDTA
- Picogreen reagent (Molecular Probes)[j]
- Lambda DNA standards
- Microtiter plates (Dynex Immulux™)
- Fluorescence plate reader
- Centrifuge for microtiter plates

^{k}Method

1 Prepare a series of 100 μl/well lambda DNA standards, in duplicate, within a clean, 96-well plate as follows:

Lamba DNA (2 μg/μl)	TE (μl)	Final concentration
100 μl	0	100 ng/μl
75 μl	25	750 pg/μl
50 μl	50	500 pg/μl
25 μl	75	250 pg/μl
10 μl	90	100 pg/μl
5 μl	95	50 pg/μl
0 μl	100	0 pg/μl

2 For each DNA sample, prepare duplicate dilutions of 2 μl of DNA with 98 μl of TE.

3 For each DNA sample dilution, prepare 100 μl of Picogreen^j reagent by diluting Picogreen 200-fold in TE.

4 Add 100 μl of diluted Picogreen reagent to each diluted DNA sample and mix by pipetting up and down.

5 Centrifuge the microtiter plate at 250g for 1 min to remove possible bubbles.

6 Read in a plate reader (excitation 485 nm, emission 538 nm).

7 Calculate concentrations from the standard curve using the plate reader software package.

Notes

^{j}Avoid excess exposure to light since the dye is light sensitive.

^{k}Alternatively, the DNA concentration can be calculated from measuring the $A_{260 nm}$ using a spectrophotometer (e.g. Nanodrop). However, for quantification of DNA from FFPE tissue the use of Picogreen gives more reliable estimates than measurement of $A_{260 nm}$ using a spectrophotometer because only double-stranded DNA (and no degradation products) are measured.

PROTOCOL 3.4   DNA Quality Control PCR

Equipment and reagents

- Primer stocks (100 μM):
  — RS2032018 (150 bp)
    ◦ primer 1: 5'-GTGTCTCCCTTCCCACTCAA-3'
    ◦ primer 2: 5'-AGCCCACCTACCTTGGAAAG-3'

— AP000555 (PRKM1, 255 bp)

- primer 3: 5'-TGGCTGATCTATGTCCCTGA-3'

- primer 4: 5'-GCTCAGTTGTTTTGTGGGTAAG-3'

— AC008575 (APC, 511 bp)

- primer 5 GCTCAGACACCCAAAAGTCC

- primer 6: CATTCCCATTGTCATTTTCC

- PCR reaction buffer II without $MgCl_2$ (Applera)

- Amplitaq Gold DNA Polymerase (5 units/μl Applera)

Method

1 Prepare a primer working solution mix containing 20 pm/μl of primers 1 and 2, and 10 pm/μl for primers 3–6.[l]

2 Prepare a 10 μl reaction by mixing the following:

- 0.25 μl of primer mix

- 1.0 μl of 10× PCR reaction buffer II without $MgCl_2$

- 0.2 μl of 4× 10 mM dNTPs

- 1.0 μl of 25 mM $MgCl_2$

- 2.0 μl of template DNA (5 ng/μl)[m]

- 0.1 μl of Amplitaq Gold DNA polymerase

- 5.45 μl of water.

3 Thermocycle as follows:

- 96 °C, 10 min

- (94 °C, 30 s; 55 °C, 30 s; 72 °C, 1 min) × 35

- 72 °C, 5 min.

4 Analyze each PCR product by 2% (w/v) TAE agarose gel electrophoresis.[n] Compare the sizes of amplified products against molecular weight marker standards.

Notes

[l]The multiplex PCR amplifies three amplicons, of 150, 255 and 511 bp. This method is comparable to the van Beers method [40]. See Figure 3.1.

[m]Use 10 ng of high molecular genomic DNA (e.g. prepared from freshly frozen tissue) as a control.

[n]DNA is considered to be of high quality if all three amplicons are visible on an agarose gel. However, DNA samples showing both the 150 and 255 bp amplicons can be used for further investigations.

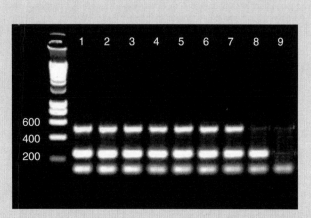

Figure 3.1 Agarose gel electrophoresis of quality control PCR amplification of DNA prepared from FFPE tissues. Lanes 1–9 contain DNA samples prepared from different FFPE tumor samples. Lanes 1–8 exemplify 'acceptable' DNA and lane 9 'unacceptable' DNA. The left lane contains a molecular weight marker (Smartladder; Eurogentec). All the samples are analyzed on a 2 % (w/v) agarose gel in Tris-acetate-EDTA (TAE) buffer.

3.2.5 Loss of heterozygosity

LOH of chromosomal regions is a common characteristic of cancer [41], and their detection provides insight into the genomic regions that might harbor potential tumor-suppressor genes. LOH was originally analyzed using simple sequence-length polymorphisms and is found through the genotyping of tumor and normal DNA from the same patient. LOH is identified as the regions where the heterozygous markers in the normal DNA become homozygous in the tumor. LOH detected at the genotype level can be a result of allelic imbalances (AIs) that can be caused by either physical loss or gain of alleles. In addition, mitotic recombination leads to LOH. With the advent of the SNP arrays, high-throughput analysis of LOH in cancer has become possible; this was first described in a study on small-cell lung carcinomas [42]. Moreover, the identification of the haplotypes that are lost in a tumor could be deployed in the study of hereditary cancer. An explosion of studies on LOH in various cancers has been observed. The methods to identify and visualize LOH are being developed and are improving. Since paired normal samples for tumors and cell lines are not always available, the efforts to infer LOH from tumor samples only are noteworthy [43, 44]. The application of high-throughput SNP arrays for genotyping and LOH analysis in FFPE samples is a major advance and opens up the tissue archives for these studies.

We do not present the Illumina protocol for genotyping and LOH analysis, using the Illumina Goldengate assay and Sentrix arrays, as it is a commercial platform (the most recent version of the protocol can be obtained through www.illumina.com). Alternatively, we discuss data management, and the analysis of genotypes and LOH, including the use of a chromosome visualization tool [45] in Spotfire DecisionSite (see Protocols 3.5 and 3.6).

PROTOCOL 3.5 Visualization of Genotyping and LOH Data in Spotfire

Equipment and reagents

- A PC with the following specification:

  — Intel Pentium 2.0 GHz or above

  — memory size ≥ 2 GB

  — hard drive ≥ 100 GB

  — video display 1280×1024 (recommended)

  — 17 inch LCD monitor (recommended)

  — operating system Windows XP-SP2 (32 bits) or Windows XP-SP2 (64 bits)

- Microsoft NET framework 1.1 (or above)

- Microsoft Access 2003

- Spotfire DecisionSite 9.1 for Functional Genomics (Spotfire, Somerville, MA, USA)

Method

1 Open Spotfire DecisionSite for Functional Genomics.

2 Paste the data from the final query containing all sample information (Protocol 3.5, step 14) from the Access database into Spotfire.

3 Spotfire automatically generates a scatter plot.[o] However, some settings have to be changed to visualize SNP and LOH information for every tumor sample (see Figure 3.2).

   (a) Select Start Position (bp) on the x-axis. This shows the base pair position of the SNPs from p-ter to q-ter.

   (b) Select the ABnormRationorm for one of the tumors on the y-axis to visualize the quality scores relating to possible LOH.

   (c) Go to Edit Properties. Select the Trellis tab and Add variable Chromosome Name (Bind To Columns). Set Columns to 1 and Rows, for example, to 5 to generate a visualization for sets of five chromosomes per sample.

   (d) Select the Markers tab and set Color By the ABnormRationorm for the selected tumor and Categorical. Customize the Shape to Diamonds. Set Labels to Locus_Name.

4 In the Query Devices pane, select the Heterozygous SNPs for the paired Normal of the tumor sample under analysis by deselecting the AA and BB genotypes. This confines the analysis to the SNPs that are heterozygous in the normal.

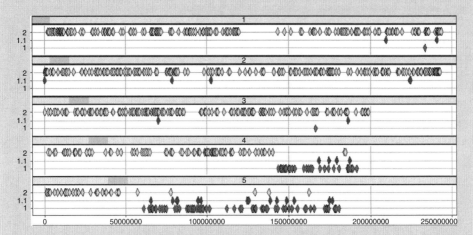

Figure 3.2 Spotfire Genotype and LOH visualization of a single tumor sample relative to a paired normal sample. Five panels (1–5) for chromosomes 1–5 are shown. For each panel, on the *x*-axis the position of each SNP is depicted in base pairs from the p-telomere to the q-telomere of the chromosome. An SNP that is heterozygous in both the normal sample and paired tumor is represented in yellow diamonds on the 2-line on the *y*-axis. SNPs that are heterozygous in the normal sample but homozygous in the tumor sample are represented in red diamonds on the 1-line, while SNPs that are heterozygous in the normal but with a quality ratio below 0.8 in the paired tumor sample are represented in blue diamonds on the 1.1 line. LOH is called in regions (relative to their base pair positions on the *x*-axis), marked by more red and blue markers (SNPs) than yellow markers. (See Plate 3.2.)

5 Analyze the scatter plot visualization to determine patterns of potential LOH as described in Figure 3.2.

6 Repeat step 3b–5 to analyze every tumor : normal pair.

Notes

[o]See [45] for more details.

PROTOCOL 3.6   Illumina Data Storage, LOH and Genotype Analysis

Equipment and reagents

- A PC with the following specification:
  - Intel Pentium 2.0 GHz or above
  - memory Size ≥ 2 GB
  - hard drive ≥ 100 GB
  - video display 1280 × 1024 (recommended)

— 17 inch LCD monitor (recommended)

— operating system Windows XP-SP2 (32 bits) or Windows XP-SP2 (64 bits)

• Microsoft NET framework 1.1 (or above)

• Illumina Beadstudio version 2 and Gencall version 6

• Microsoft Access.

Method

1 Create a 'general' Access database to keep track of all samples.[p]

2 Create a separate folder for every new Illumina experiment.[q,r]

3 Create a specific Access database for each experiment in the 09_Database folder.

4 For each experiment, import the 'final Genotyping report' and the 'Locus-By-DNA' reports into the database (New, Import Table function in Access).[s]

5 Import the genome information table as provided by Illumina in this database.[t]

6 Link the experiment database to the general sample sheet database (created in step 1) by using the 'Link Table' wizard in Access.

7 Make an Access query (e.g. with the Simple Query Wizard) containing the 'final Genotyping report' and add an extra colum n where the array (SentrixID) is defined (SentrixID:Left$(tbl_LocusByDNA_final!DNA_Name;7). Save the query as qryGenotypingfinal.

8 Make an Access query (e.g. with the Simple Query Wizard) containing the qryGenotypingfinal, the 'Locus-By-DNA' report and the Samplesheet table.

9 To combine the qryGenotypingfinal and the 'Locus-By-DNA' table, join the Locus_Name and SentrixID columns (use: view join properties). Link the samplesheet table to the Locus-By-DNA table by joining the DNA_Name to the Sentrix_Position.

10 Select all columns for visualization in the query and add the following calculated columns to the query:

   (a) AB: [Allele1] and [Allele2] which combines the SNP calls into one column.

   (b) ABnorm: IIf([AB] = 'AB';'2';'1') which renames heterozygous SNPs to 2 and homozygous SNPs to 1.

   (c) Ratio: [GC_Score]/[Gentrain_Score] which calculates GCS/gene train score (GTS) ratios for quality assignment and LOH analysis.[u]

   (d) ABnormRationorm: IIf([ABnorm] = '2'; (IIf([ratio] > 0.8; [ABnorm]; '1.1'));[ABnorm]).

   (e) This column sets each heterozygous SNP with a ratio <0.8 to 1.1 instead of 2.[v]

11 Save the query with a descriptive name; for example, qry_finalresults. This query contains all results for all samples on the array.

12 Use the step 11 query to make two different Crosstab Queries (use the Access Crosstab Query Wizard) making use of the query from step 11. For both, Locus_Name is set to Row Heading and Sample_Name to Column Heading.

   (a) In the first the AB (see step 10a) is used for Values. Save this Crosstab Query as ctqry_ABresults.

   (b) In the second the ABnormRationorm (see step 10d) is used for Values. Save this Crosstab Query as ctqry_ABNRresults.

13 Make a final query where the ctqry_ABresults is combined with the ctqry_ABNRresults and the genomic information table (step 5) by joining the SNP (Locus) names and selecting all columns in the query. Save this query as final_query.

14 For further analysis and visualization, import or paste the data from the final_query out of step 13 into Spotfire (see Protocol 3.6).

Notes

[p]Record sample names, sample positions on arrays, sample sheet information and so on in this database (see Figure 3.3b).

Figure 3.3 panel (a) — Illumina Directory Structure:

- 01_Samplesheet
- 02_OPA_Manifest
- 03_Raw_data
- 04_Cluster_files
- 05_Familydata
- 06_Workspace
- 07_Outputfiles
- 08_Reports
- 09_Database
- 10_DNA_samples
- 11_analyseR

Figure 3.3 panel (b):

Field Name	Data Type
SampleID	AutoNumber
Sample	Text
Sample_Code	Text
Sample_Name	Text
Sample_Well	Text
Sample_Plate	Text
Folderdate	Text
Sample_Group	Text
Pool_ID	Text
Sentrix_ID	Text
NorTum	Text
Isolate	Text
Row	Number
Col	Number
OPA	Text
RowC	Text
ColC	Text
HybNR	Text
remarks	Text
decode	Text
date	Date/Time
selection1	Yes/No
selection2	Yes/No
selection3	Yes/No
Sentrix_Position	Text

Figure 3.3 Illumina data storage. In panel A an overview is shown of the Illumina directory structure with 11 subfolders containing data for the different steps in the analysis flow. In panel B an overview is presented of the different columns present in the Illumina Sample Sheet Table.

[q]Name the folder exp_year_month_day_project; for example, Exp081125_myproject.

ʳAn experiment is defined as the results of one Illumina array and the genotyping and LOH analysis of all the DNA samples typed on the array. Every experiment folder contains 11 subfolders containing data for different steps in the analysis flow (see Figure 3.3a).

ˢThe final Genotyping report is generated in Beadstudio and contains genotyping and quality information. The Locus-By-DNA report is exported from Gencall and contains information about the SNPs, the linkage panel and the array used for the experiment.

ᵗGenomic information about all the SNPs is provided by Illumina in the oligonucleotide pool assay manifest, a small CD provided in the Sentrix Array box.

ᵘLOH detection can be impaired by a low tumor percentage; normal cells and (tumor infiltrating) leukocytes that are present in the isolate will result in diminished LOH calling. Ideally, pure tumor material should be used for LOH detection, although in practice tumors will be tested with up to 40 % of non-tumor cells. In these cases, most SNPs will still be called heterozygous; however, their genotype quality will be reduced. We therefore use the quality scores that are automatically assigned to each genotype by the gene calling program Beadstudio. Each SNP gets two different quality scores; the sample-specific gene call score (GCS) and the SNP-specific GTS. Since reduced quality scores in tumor tissue might indicate LOH, we calculate the GCS/GTS ratio for each SNP and sample. For normal samples, we expect the ratio to be high (between 0.8 and 1). Therefore, only the high-quality heterozygous SNPs in the normal sample, with a GCS/GTS ratio between 0.8 and 1.0, are included in LOH analysis. LOH is called for these SNPs when the tumor genotype call is homozygous, or is heterozygous and has a GCS/GTS ratio below 0.8.

ᵛFor a heterozygous SNP with a GCS/GTS ratio <0.8 the SNP genotyping quality is defined as 'low.'

3.3 Troubleshooting

- **Frequently asked questions:** Illumina has generated a list of 'general problematic issues' and suggestions on how to resolve the problems. This information can be found on www.illumina.com under faqs (frequently asked questions) for DNA analysis.

- **Experimental design:** A typical experiment is the whole genome-genotyping and LOH analysis of 12 tumors and paired normal DNAs. These 24 samples are genotyped using four pools of approximately 1500 SNPs, the linkage panels. The SNPs are hybridized to Sentrix arrays, a matrix of 96 arrays, each detecting 1500 SNPs. Thus, for each sample, four arrays of approximately 1500 SNPs each are hybridized and subsequently combined to approximately 6000 genotypes. These together form the linkage panel. Alternatively, a genome-wide cancer SNP panel, or custom SNP panels of 1536, 768 or 384 SNPs, could be tested.

- **Genotyping of FFPE tissue DNA using the Illumina Goldengate assay and Sentrix arrays:** For each array, the Goldengate protocol requires 250 ng of activated (biotinylated) DNA in the single-use DNA activation (final volume 10 μl) or 2 μg of DNA in the multiple-use DNA activation (final volume of 100 μl). Fan *et al.* [46] describe that less DNA can be used. For the use of FFPE DNA, we routinely use 1 μg of DNA as input in the multiple-use DNA activation and dissolve the resulting biotinylated DNA in 60 μl of RS1 buffer.

References

1. Hall, J.M., Lee, M.K., Newman, B. *et al.* (1990) Linkage of early-onset familial breast cancer to chromosome 17q21. *Science*, **250**, 1684–1689.

2. Castilla, L.H., Couch, F.J., Erdos, M.R. *et al.* (1994) Mutations in the *BRCA1* gene in families with early-onset breast and ovarian cancer. *Nature Genetics*, **8**, 387–391.

3. Wooster, R., Bignell, G., Lancaster, J. *et al.* (1995) Identification of the breast cancer susceptibility gene BRCA2. *Nature*, **378**, 789–792.

4. Leach, F.S., Nicolaides, N.C., Papadopoulos, N. *et al.* (1993) Mutations of a mutS homolog in hereditary nonpolyposis colorectal cancer. *Cell*, **75**, 1215–1225.

5. Peltomaki, P., Aaltonen, L.A., Sistonen, P. *et al.* (1993) Genetic mapping of a locus predisposing to human colorectal cancer. *Science*, **260**, 810–812.

6. Lindblom, A., Tannergard, P., Werelius, B. and Nordenskjold, M. (1993) Genetic mapping of a second locus predisposing to hereditary non-polyposis colon cancer. *Nature Genetics*, **5**, 279–282.

7. Papadopoulos, N., Nicolaides, N.C., Wei, Y.F. *et al.* (1994) Mutation of a mutL homolog in hereditary colon cancer. *Science*, **263**, 1625–1629.

8. Wyman, A.R. and White, R. (1980) A highly polymorphic locus in human DNA. *Proceedings of the National Academy of Sciences of the United States of America*, **77**, 6754–6758.

9. Tautz, D. (1989) Hypervariability of simple sequences as a general source for polymorphic DNA markers. *Nucleic Acids Research*, **17**, 6463–6471.

10. Shen, R., Fan, J.B., Campbell, D. *et al.* (2005) High-throughput SNP genotyping on universal bead arrays. *Mutation Research*, **573**, 70–82. Describes the development of a highly multiplexed SNP genotyping assay for high-throughput genetic analysis of large populations on a bead array platform.

11. Hardenbol, P., Yu, F., Belmont, J. *et al.* (2005) Highly multiplexed molecular inversion probe genotyping: over 10,000 targeted SNPs genotyped in a single tube assay. *Genome Research*, **15**, 269–275.

12. Kennedy, G.C., Matsuzaki, H., Dong, S. *et al.* (2003) Large-scale genotyping of complex DNA. *Nature Biotechnology*, **21**, 1233–1237. Demonstrates that oligonucleotide arrays designed for CGH provide a robust and precise platform for detecting chromosomal alterations throughout a genome.

13. Matsuzaki, H., Loi, H., Dong, S. *et al.* (2004) Parallel genotyping of over 10,000 SNPs using a one-primer assay on a high-density oligonucleotide array. *Genome Research*, **14**, 414–425.

14. Gunderson, K.L., Steemers, F.J., Lee, G. *et al.* (2005) A genome-wide scalable SNP genotyping assay using microarray technology. *Nature Genetics*, **37**, 549–554. A whole-genome genotyping assay that combines specific hybridization of WGA DNA to arrayed probes with allele-specific primer extension and signal amplification.

15. Hua, J., Craig, D.W., Brun, M. *et al.* (2007) SNiPer-HD: improved genotype calling accuracy by an expectation-maximization algorithm for high-density SNP arrays. *Bioinformatics*, **23**, 57–63.

16. Hu, G., Wang, H.Y., Greenawalt, D.M. *et al.* (2006) AccuTyping: new algorithms for automated analysis of data from high-throughput genotyping with oligonucleotide microarrays. *Nucleic Acids Research*, **34**, e116.

17. Scharpf, R.B., Ting, J.C., Pevsner, J. and Ruczinski, I. (2007) *SNPchip*: R classes and methods for SNP array data. *Bioinformatics*, **23**, 627–628.

18. Rabbee, N. and Speed, T.P. (2006) A genotype calling algorithm for Affymetrix SNP arrays. *Bioinformatics*, **22**, 7–12.

19. Duerr, R.H., Taylor, K.D., Brant, S.R. *et al.* (2006) A genome-wide association study identifies IL23R as an inflammatory bowel disease gene. *Science*, **314**, 1461–1463.

20. Sladek, R., Rocheleau, G., Rung, J. *et al.* (2007) A genome-wide association study identifies novel risk loci for type 2 diabetes. *Nature*, **445**, 881–885.

21. Easton, D.F., Pooley, K.A., Dunning, A.M. *et al.* (2007) Genome-wide association study identifies novel breast cancer susceptibility loci. *Nature*, **447**, 1087–1093.

22. Haiman, C.A., Le Marchand, L., Yamamato, J. *et al.* (2007) A common genetic risk factor for colorectal and prostate cancer. *Nature Genetics*, **39**, 954–956.

23. Schumacher, F.R., Feigelson, H.S., Cox, D.G. *et al.* (2007) a common 8q24 variant in prostate and breast cancer from a large nested case-control study. *Cancer Research*, **67**, 2951–2956.

24. Tomlinson, I., Webb, E., Carvajal-Carmona, L. *et al.* (2007) A genome-wide association scan of tag SNPs identifies a susceptibility variant for colorectal cancer at 8q24.21. *Nature Genetics*, **39**, 984–988.

25. White, D.R., Ganesh, A., Nishimura, D. *et al.* (2007) Autozygosity mapping of Bardet–Biedl syndrome to 12q21.2 and confirmation of *FLJ23560* as *BBS10*. *European Journal of Human Genetics*, **15**, 173–178.

26. Sellick, G.S., Garrett, C. and Houlston, R.S. (2003) A novel gene for neonatal diabetes maps to chromosome 10p12.1-p13. *Diabetes*, **52**, 2636–2638.

27. Zhang, C., Cawley, S., Liu, G. *et al.* (2005) A genome-wide linkage analysis of alcoholism on microsatellite and single-nucleotide polymorphism data, using alcohol dependence phenotypes and electroencephalogram measures. *BMC Genetics*, **6** (Suppl 1), S17.

28. Kemp, Z., Carvajal-Carmona, L., Spain, S. *et al.* (2006) Evidence for a colorectal cancer susceptibility locus on chromosome 3q21–q24 from a high-density SNP genome-wide linkage scan. *Human Molecular Genetics*, **15**, 2903–2910.

29. Cao, X., Eu, K.W., Kumarasinghe, M.P. *et al.* (2006) Mapping of hereditary mixed polyposis syndrome (HMPS) to chromosome 10q23 by genomewide high-density single nucleotide polymorphism (SNP) scan and identification of *BMPR1A* loss of function. *Journal of Medical Genetics*, **43**, e13.

30. Lange, K., Weeks, D. and Boehnke, M. (1988) Programs for pedigree analysis: MENDEL, FISHER, and dGENE. *Genetic Epidemiology*, **5**, 471–472.

31. Abecasis, G.R., Cherny, S.S., Cookson, W.O. and Cardon, L.R. (2002) Merlin – rapid analysis of dense genetic maps using sparse gene flow trees. *Nature Genetics*, **30**, 97–101.

32. Gudbjartsson, D.F., Jonasson, K., Frigge, M.L. and Kong, A. (2000) Allegro, a new computer program for multipoint linkage analysis. *Nature Genetics*, **25**, 12–13.

33. Ruschendorf, F. and Nurnberg, P. (2005) ALOHOMORA: a tool for linkage analysis using 10K SNP array data. *Bioinformatics*, **21**, 2123–2125.

34. Webb, E.L., Sellick, G.S. and Houlston, R.S. (2005) SNPLINK: multipoint linkage analysis of densely distributed SNP data incorporating automated linkage disequilibrium removal. *Bioinformatics*, **21**, 3060–3061. Describing the detection of linkage and removal of LD from high density SNP data.

35. Leykin, I., Hao, K., Cheng, J. *et al.* (2005) Comparative linkage analysis and visualization of high-density oligonucleotide SNP array data. *BMC Genetics*, **6**, 7.

36. Lindner, T.H. and Hoffmann, K. (2005) easyLINKAGE: a PERL script for easy and automated two-/multi-point linkage analyses. *Bioinformatics*, **21**, 405–407.

37. Lips, E.H., Dierssen, J.W.F., van Eijk, R. *et al.* (2005) Reliable high-throughput genotyping and loss-of-heterozygosity detection in formalin-fixed, paraffin-embedded tumors using single nucleotide polymorphism arrays. *Cancer Research*, **65**, 10188–10191. The first paper that describes genotyping and LOH analysis using Illumina Beadarrays.

38. Oosting, J., Lips, E.H., van Eijk, R. *et al.* (2007) High-resolution copy number analysis of paraffin-embedded archival tissue using SNP BeadArrays. *Genome Research*, **17**, 368–376. The design and validation of copy number measurements using BeadArray's and the application to FFPE tissue.

39. Ji, H., Kumm, J., Zhang, M. *et al.* (2006) Molecular inversion probe analysis of gene copy alterations reveals distinct categories of colorectal carcinoma. *Cancer Research*, **66**, 7910–7919.

40. Van Beers, E.H., Joosse, S.A., Ligtenberg, M.J. *et al.* (2006) A multiplex PCR predictor for aCGH success of FFPE samples. *British Journal of Cancer*, **94**, 333–337. Demonstrates that WGA is capable of increasing the yield of starting DNA material with identical genetic sequence.

41. Rajagopalan, H. and Lengauer, C. (2004) Aneuploidy and cancer. *Nature*, **432**, 338–341.

42. Lindblad-Toh, K., Tanenbaum, D.M., Daly, M.J. *et al.* (2000) Loss-of-heterozygosity analysis of small-cell lung carcinomas using single-nucleotide polymorphism arrays. *Nature Biotechnology*, **18**, 1001–1005. First paper describing the use of SNP arrays for the detection of LOH in cancer.

43. LaFramboise, T., Harrington, D. and Weir, B.A. (2006) PLASQ: a generalized linear model-based procedure to determine allelic dosage in cancer cells from SNP array data. *Biostatistics*, **8**, 323–336.

44. Beroukhim, R., Lin, M., Park, Y. *et al.* (2006) Inferring loss-of-heterozygosity from unpaired tumors using high-density oligonucleotide SNP arrays. *PLoS Computational Biology*, **2**, e41.

45. Van Eijk, R., Oosting, J., Sieben, N., *et al.* (2004) Visualization of regional gene expression biases by microarray data sorting. *Biotechniques*, **36**, 592–594, 596.

46. Fan, J.B., Oliphant, A., Shen, R. *et al.* (2003) Highly parallel SNP genotyping. *Cold Spring Harbor Symposia on Quantitative Biology*, **68**, 69–78. Technical note on the properties of the beadarray platform for SNP genotyping.

4
Genetic Mapping of Complex Traits

Nancy L. Saccone
Department of Genetics, Division of Human Genetics, Washington University School of Medicine, St Louis, Missouri, USA

4.1 Introduction

This chapter surveys current major approaches to mapping genes for multifactorial human diseases using either family-based or population-based samples. Complex, multifactorial diseases are generally hypothesized to be attributable to the effects of multiple genes with incomplete penetrance. Gene–gene and gene–environment interactions are also expected to play a role in the disease etiology.

Disease gene mapping strategies may be divided into two categories that are not mutually exclusive: linkage based and association based. Linkage mapping relies on the co-segregation of marker genotypes with phenotype within families, thereby detecting genetic loci that affect disease risk. Association mapping methods are designed to detect loci for which genotype or allelic status is correlated with phenotypic status on a population level, and can be carried out using either family samples or unrelated individuals. Association-based mapping has gained popularity in recent years due to technological advances that allow large-scale genotyping of single nucleotide polymorphism (SNP) markers, enabling genome-wide association studies (GWAS) of human diseases. This popularity also stems from earlier observations of Risch and Merikangas [1], who noted that association designs promise greater power to detect genes for complex traits, compared with sib-pair linkage approaches.

This chapter will begin with an emphasis on association mapping, and in particular, large-scale, GWAS, as several high-profile association-based studies are currently making headlines. Large-scale association studies of unrelated cases and controls have reported and replicated gene discoveries for important diseases such as diabetes [2–4], breast cancer [5], smoking [6–9] and lung cancer [8, 10–12].

Genomics: Essential Methods   Edited by Mike Starkey and Ramnath Elaswarapu
© 2011 John Wiley & Sons, Ltd.

Linkage methods, both parametric and non-parametric, will be outlined next. Although large-scale association studies are now becoming more prevalent as a means for discovering novel disease genes, and some researchers are questioning what the role of classical family-based linkage analysis will be in this new era, new linkage findings continue to be reported and will inform and influence the design and interpretation of future association studies. Furthermore, active research into combined use of linkage and association information in family samples [13] continues to highlight the relevance of linkage analysis for today's researchers.

Investigators will wish to know what study design is optimal for detecting genes involved in their disease of interest. It is not always easy to determine a clear, simple answer to this question. However, this chapter will outline strengths and weaknesses of different methods and describe some of the implications of particular design choices.

4.2 Methods and approaches

4.2.1 Association methods: unrelated case–control samples

We will first discuss the setting in which the samples are unrelated case and control individuals. The advantages of this study design include the relative ease of ascertaining unrelated participants compared with families. There are some special issues that require close attention when using unrelated samples; for example, the cases and controls should be selected to minimize the possibility of cryptic population structure (population stratification) which can increase Type I error (false positives). Family-based samples can be more difficult to recruit and the resulting samples are less efficient in terms of the sample size needed for a given level of power, compared with a case–control sample [14]. However, family-based designs are sometimes chosen because they are robust to population stratification.

Owing to the popularity of SNP genotyping for large-scale association mapping projects, most of our discussion will presume that SNP markers with exactly two alleles (bi-allelic) are being used. However, we will comment on some instances in which methods discussed are also appropriate for multi-allelic markers (such as microsatellites).

Protocols for carrying out the analyses discussed will depend in great part on the analytic software chosen by the investigator. Rather than outlining multiple software-specific protocols, we discuss the conceptual basis of methods, cite available software packages and refer interested investigators to the documentation provided with those particular programs. For illustration, we will also present two general protocols in the context of a large-scale association study of unrelated case–control samples: a protocol for study design and data quality control, and a protocol for analysis based on a particular analysis method (logistic regression).

4.2.1.1 Association in unrelated case–control samples: study design and data quality

Here we discuss some design issues, and also methods for preprocessing the data before carrying out statistical tests of association with disease. These include data cleaning checks and tests for population substructure. Then we will discuss the major statistical methods used for testing association between genetic markers and disease status in unrelated case–control samples.

Initial study design

Broadly speaking, there are two important dimensions to designing a case–control study: choosing which subjects to genotype and choosing which genetic marker loci (usually SNPs) to genotype for disease association testing.

Study subjects For a case–control study, definition of the case group is usually relatively clear to the investigators, who will have decided upon a disease of interest, though details may be dictated by the resources and patient access of the study. Ideally, investigators will be able to decide on a consistent and precise definition of case. However, depending on the disease and study design, one may weigh the importance of determining a definitive clinical diagnosis versus using less precise definitions, or measures which are merely correlated with clinical diagnoses but may be useful for capturing underlying biological features of the disease. These phenotypic decisions will depend on the disease and the goals of the study. For example, if the goal is to collect a very large sample through the collaboration of different investigators with different patient pools or data sources, a consistent definition may be impractical, in which case potential 'site' differences may be modeled as covariates in the final analysis.

For the control group, investigators may choose 'clean' controls who do not manifest the disease and who, furthermore, may be past the age of risk for disease onset. Alternatively, they may choose population-based controls, for which a certain proportion will likely have the disease (though specific phenotype information may not be available). The former plan is more efficient, in that a cleaner control sample should have greater power than a random population-based control sample of the same size. However, the latter plan may be more feasible and less expensive because those controls do not need to be diagnosed or clinically assessed. Costbenefit analyses indicate that unscreened population-based controls may be a cost-effective choice for a range of situations, especially for disorders with lower population prevalence [15]. Recently, the Wellcome Trust Case–Control Consortium (WTCCC) used a single large sample of population-based subjects as a common control dataset for genome-wide analyses of multiple diseases for which distinct sets of cases were recruited [16]; their successful identification of both known and novel variants associated with disease risk underscores the effectiveness of this approach.

Selecting SNPs Whether the overall design is a genome-wide association scan or a targeted study of candidate genes, it is important to decide what markers to genotype. SNPs are becoming the marker of choice for association due to low error rates and available technology for high-throughput genotyping, so the question becomes, 'Which SNPs should be genotyped?' Fixed platforms that include hundreds of thousands up to millions of SNP markers across the genome are commercially available from companies such as Illumina (www.illumina.com) and Affymetrix (www.affymetrix.com). These platforms, or GWAS genotyping arrays, have become popular for genome-wide association scans and have led to recent findings across many complex diseases [16].

For a candidate gene study, custom selection of SNPs is typically carried out. To reduce costs, often only a subset of known SNPs in and near the targeted genes will be genotyped. The selected SNPs may be chosen to 'tag' additional variation that can be represented with acceptable power by the assayed SNPs. Such tag SNP selection methods, which use correlation among the alleles of SNP loci, may be loosely classed into two categories: haplotype based and linkage disequilibrium (LD) based. Of course, the two kinds of approaches are related, since regions of high LD exhibit reduced haplotype diversity and vice versa. The availability of extensive LD data on multiple human populations via the International HapMap Consortium (www.hapmap.org) has been a critical resource for designing association studies [17, 18].

LD occurs when alleles at two loci co-occur at frequencies different than expected under independent assortment. The traditional measures of pairwise LD include the LD coefficient $D = h_{11} - p_{A1} p_{B1}$, where p_{A1} (or p_{B1}) is the allele frequency of allele 1 at locus A (or locus B), and h_{11} is the frequency of the haplotype consisting of allele 1 at locus A and allele 1 at locus B (that is the $1-1$ haplotype). Because the range of D varies depending on the allele frequencies at loci A and B, D is often normalized to range from -1 to 1, giving the normalized disequilibrium coefficient $D' = D/|D|_{max}$, where $|D|_{max}$ is the maximum possible (absolute) value for D given the allele frequencies of the two loci. High values of $|D'|$ are often used to recognize genomic regions of reduced recombination. The correlation coefficient is $r = D/(p_{A1} p_{A2} p_{B1} p_{B2})^{0.5}$, and r^2 is commonly used for determining whether one locus may serve as a proxy for another due to strong correlation between the loci, as discussed further below.

A typical SNP tagging method in the haplotype-based category is to choose 'haplotype tag SNPs' to represent contiguous blocks of SNPs that show reduced haplotype diversity [19–21]. These tag SNPs then allow all haplotypes, or all common haplotypes, to be differentiated from each other based on genotypes only at the tags.

In contrast, popular LD-based methods do not necessarily take explicit note of the haplotypes that may be inferred or observed across multiple SNPs. Rather, they use pairwise LD measures (which for genotype data usually require estimation of two-locus haplotype frequencies) to determine a group of markers which can serve as proxies for unassayed markers based on the pairwise correlations. Carlson *et al.* [22] developed a greedy algorithm to define 'bins' of markers, not necessarily contiguous, where at least one marker in the bin has sufficiently high r^2 with all other markers in that bin; such a marker may then be chosen as a tag for that bin.

The r^2 bin method has gained popularity due to ease of use and interpretability, especially considering the following useful relationship between the power to detect disease association and the strength of LD between the tags and bin members. For a given value of r^2 between a disease locus and SNP marker, the sample size needed to have equivalent power to detect disease association with alleles at the marker, rather than at the true locus, is increased by a factor of approximately $1/r^2$ [23]. More precise relationships can also be computed [24]. A popular threshold to define bin tags is to require that the tag has $r^2 \geq 0.8$ with all bin members. It is useful to keep in mind that allele frequencies must be similar for two SNPs to have a high value of r^2. Thus, it is typical for an LD block defined by D' to be partitioned into multiple r^2 bins, each of which may consist of non-consecutive SNPs, according to the underlying allele frequencies of the markers in the block.

For researchers wishing to select tag SNPs for particular genes or regions of interest, the HapMap website offers a browser interface allowing selection of tag SNPs from a selection of methods, including the r^2 bin method. Figure 4.1 illustrates the use of the HapMap browser (http://www.hapmap.org/cgi-perl/gbrowse/hapmap_B35/) to select tags for the gene *CHRNA5*. Other software tools for tag SNP selection include SNPtagger [20], Snagger [25] and Haploview [19] for haplotype tagging, and LDselect [22] and Tagger [26] for r^2 bin tagging. The SNP Annotation and Proxy Search (SNAP) website [27] is a user-friendly tool that will look up all SNPs tagged by a provided list of SNPs.

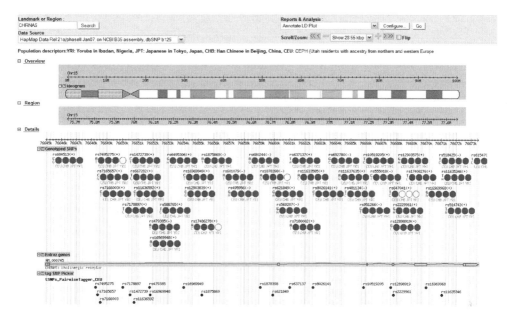

Figure 4.1 An example of the browser display from www.hapmap.org, accessed by clicking on the 'HapMap Genome Browser (B35 – full data set)' link. The gene *CHRNA5* was entered into the 'Landmark or Region' field. The 'Scroll/Zoom' box indicates that the display is showing 28.55 kbp. The 'Overview' panel, or track, indicates the full chromosome on which this gene lies, and the chromosomal region blown up under the 'Region' panel. The 'Details' panel shows the SNPs genotyped by HapMap in the selected region, and also displays a pie chart of the allele frequencies for each SNP in each of the four HapMap population samples: CEU (Centre de Polymorphisme Humaine, CEPH; Utah residents with ancestry from northern and western Europe), YRI (Yoruba in Ibadan, Nigeria), CHB (Han Chinese in Beijing, China) and JPT (Japanese in Tokyo, Japan). The last two tracks display the gene and also tag SNPs which have been selected according to the default settings: tags represent r^2 bins where all bin members satisfy $r^2 \geq 0.8$ with at least one tag in the CEU population. The parameters for tag SNP selection may be modified using the 'Reports and Analysis' drop-down menu, which is currently set on 'Annotate LD Plot': click on the arrow to the right, select 'Annotate tag SNP Picker' and choose the desired parameters. (See Plate 4.1.)

The tagging methods described above are concerned with selecting SNPs that, because of SNP–SNP correlations or haplotype patterns, serve as good representatives of the variation observed at additional loci in the region of interest. In addition to that key consideration, however, it is useful to consider genotyping additional SNPs that are of particular interest; for example, because of a previously reported association with that SNP or because it is likely to be of functional significance in a gene (e.g. a non-synonymous coding SNP). To facilitate such design choices, the HapMap browser allows the user to include a predetermined list of SNPs in the 'tagging' set. Key functional information for SNPs (such as all coding SNPs within a gene) can be obtained from the 'gene' and 'SNP' databases of the National Center for Biotechnology Information (NCBI, www.ncbi.nlm.nih.gov).

An interesting and important design question for a GWAS is the 'genomic coverage' provided by a given commercial GWAS platform. Recent studies have evaluated coverage and cost efficiency of several commercial arrays by calculating the fraction of common (minor allele frequency ≥ 0.05) SNPs that are tagged by SNPs on the chip with an r^2 surpassing a range of thresholds [28, 29]. A GWAS design may also benefit from taking into account coverage of specific groups of genes for which there are strong prior hypotheses for involvement with the disease of study [30, 31]. Investigators using a commercial genotyping array may then wish to genotype supplementary SNPs to ensure good coverage of important target genes, appropriate to the specific disease or class of diseases. For example, recent studies to assess coverage and develop resources to help supplement current genotyping arrays have been carried out for cardiovascular [31] and addiction [30] diseases.

In addition to selecting markers which are to be tested for disease association, case–control studies should genotype markers that may be used to investigate the sample for potential population stratification, which can lead to false positive association between markers and disease. This issue is discussed further in the next section.

Data quality control

Before analyzing the genotype data for association with disease status, it is important to examine the genotype data for possible problems and either resolve discrepancies or remove problem observations.

For unrelated cases and controls, it is difficult to test for genotyping errors because there is no family data to permit detection of deviation from Mendelian inheritance patterns. Nevertheless, there are some useful tools to help clean the genotype data. In addition, depending on the genotyping platform used, there may be additional quality control measures specific to that technology that the investigator should take advantage of; discussion of these platform-specific issues is beyond the scope of this chapter, and interested investigators can follow up with individual companies or manufacturers for more details. Such platform-dependent considerations can include techniques for assessing the genotyping clusters formed by the data when making genotype calls; often these assessments are made by experienced laboratory personnel in conjunction with statistical tests for deviation from Hardy–Weinberg equilibrium (HWE) (discussed in more detail below). In addition, as an example, Illumina can provide information about the likely performance of particular SNPs on their platforms *prior* to genotyping, which can be important in helping investigators select SNPs at the design stage.

One oft-used quality control technique is to check each genotyped locus for consistency with HWE. Deviations from HWE may be attributable to several different reasons, including inbreeding, selection or even association with the disease under study. However, when HWE is severely violated in the control sample, it is common for such loci to be viewed with suspicion and removed from further analysis. To test for significant deviation from HWE, one may perform a chi-square test, or preferably Fisher's exact test when expected genotype counts are low, using software such as PEDSTATS [32] or PLINK [33]. For a genome-wide study covering hundreds of thousands of SNPs, however, it is expected that several SNPs will have significant HWE p-values simply because of the number of SNPs assayed, so the HWE information is sometimes used to flag potential problem SNPs for re-examination of the raw genotype cluster plots, without necessarily removing them from analysis. After association testing, however, it is worthwhile to check top associated SNPs for consistency with HWE.

For each genotyped marker, the call rate (that is, the rate of successfully called genotypes among all genotypes attempted) should be examined. Low call rates may be indicative of genotyping problems for that marker locus, and the resulting genotypes may not be reliable. Of special note, if the 'missingness' – that is, the partial absence of genotype calls – is dependent on the true genotype or on the phenotype, the resulting bias in the data can cause problems for analysis and interpretation. For example, if no-calls tend to occur for heterozygotes only, or for homozygotes only, this can lead to deviation from HWE [34]. PLINK [33] provides useful tests that are able to detect some patterns of non-random missing data. In addition, if, at a given marker, the call rate in cases is significantly different than that in controls, then this could lead to a spurious allele frequency difference; therefore, results should be examined closely for such patterns before conclusions are drawn from them.

Case–control samples have the advantage that they are generally easier to collect than family-based samples. This advantage has boosted the popularity of this study design. In addition, this design is more efficient than the popular parent–child trio design discussed in Section 4.2.2 [14]. However, this design has one oft-cited potential disadvantage: it is susceptible to false positive (spurious) associations if there is so-called cryptic population structure, or, in other words, if population stratification is present. Consider the following example. Suppose the case and control samples are drawn from an overall population consisting of two unrecognized subpopulations that happen to have very different rates of the disease under study, possibly due to culture, environment, differing standards of health care or other (non-genetic) factors specific to these populations. Then any genetic marker having very different allele frequencies in the two populations may appear associated in the overall case–control sample, even if there is no association with disease in each subpopulation when considered separately. Figure 4.2 illustrates this phenomenon, which is an example of what statisticians sometimes call 'Simpson's paradox.' Figure 4.2 depicts two populations having different prevalences of disease from each other and also different allele frequencies at a marker with two alleles. Each population, individually, shows no association between allele and disease status. For example, in population 1, 75% of the '1' alleles are from cases and 75% of the '2' alleles are from cases; or viewed another way, 1/6 of the control alleles are '1' and 1/6 of the case (affected) alleles are also '1.' However, in the combined population there appears to be significant association between allele and disease, where allele 1 appears to increase risk for disease.

A well-designed study must, therefore, protect against this confound. The prevailing approach is to use a chosen set of genotype markers to test for possible substructure in the case–control data. Such markers should be unlinked (or unlikely to be linked) to disease-causing loci and are often chosen to be distributed across the genome and known to have differing frequencies in different populations. Methods for assessing these markers include the program STRUCTURE [35], which allows one to infer the presence of distinct subpopulations using genotypes at a modest number of appropriately selected genetic markers. In a typical application, the data will be clustered into a predetermined number of clusters, so that each individual's probability of cluster membership is estimated. These clusters are examined to determine, first, if there is strong evidence that the sample members may be divided into two or more subpopulations (e.g. with some individuals very likely in one cluster and others very likely in another). Second, it is important to assess whether these subpopulations are represented in significantly different proportions in the case and control groups.

Analysis is simplified if there is no evidence for substructure. However, if substructure does exist, then it is possible to take this into account by appropriately adjusting for the detected ancestry differences in the analysis. The genomic control method [36] corrects

Population 1: allele 1 more prevalent ($p_1 = 0.8$) and disease more prevalent ($K = 0.75$). No association between allele status and disease status within this population: $\chi^2 = 0$.

Population 2: allele 2 more prevalent ($p_1 = 0.2$) and disease less prevalent ($K = 0.25$) No association between allele status and disease status within this population: $\chi^2 = 0$.

Population 1 + Population 2.

Apparent association between allele status and disease: $\chi^2 = 5.33$; p-value = 0.021 (1 degree of freedom). There is a higher proportion of disease (dark shading) among the "1"s than among the "2"s.

Figure 4.2 An example of spurious marker–disease association due to population stratification. The two possible alleles (1 or 2) at the marker are denoted by the corresponding numeral. When the allele is from an affected case, the numeral is shaded black; when the allele is from an unaffected control, the numeral is unshaded (white). *K* is the population prevalence of the disease (proportion of the population that is affected with the disease).

for the effects of stratification by computing an overall 'inflation factor' to adjust resulting association statistics. The more recent EIGENSTRAT method [37] uses principal components analysis to adjust for stratification and addresses the limitations of using a uniform correction despite possible differences in allele frequencies across ancestral populations. EIGENSTRAT is also amenable to using large numbers of markers such as would be available in a GWAS. Various studies have identified ancestry informative markers (AIMs) that are expected to be especially useful to test for substructure [38–42]. Some commercial SNP genotyping arrays, such as Illumina's DNA Test Panel (360 SNPs), also can provide SNPs that are informative for assessing potential population stratification.

PROTOCOL 4.1   Generic Design and Quality Control for a Disease Association Study Using Unrelated Case and Control Subjects[a]

Equipment and tools

- Association analysis software; for example PLINK [33].[b]

Method

Study and Experiment Design

1 If a large-scale, genome-wide study is planned,[c] carefully select the most appropriate genotyping platform.

2 If a modest-sized (e.g. candidate gene) study is planned,[d] select SNPs that cover the regions of interest (e.g. by tagging all common SNPs at a chosen LD threshold).

3 Include duplicate samples[e] and individuals with known genotype (e.g. Centre de Polymorphisme Humaine, CEPH controls) on genotyping plates.

4 Include both case samples and control samples (randomly selected and distributed among the wells) in each genotyping plate to prevent spurious association signals driven by potential plate effects (e.g. if differences in plate handling result in systematic biases in genotype calls on different plates).

Quality Control (Cleaning) for Genotype Data[f]

5 Calculate genotyping call rates (percent of called (non-missing) genotypes) per sample and per SNP. (In PLINK, use the '--missing' option.)

6 Among all samples, check for duplicated samples (samples that share a very high proportion of identical genotypes),[g] without referring to sample status or the labeling of planned duplicates. Confirm that planned duplicates do have matching genotypes. Keep only one member of any duplicate pair (or multiple).[h] (In PLINK, use the '-- genome' option.)

7 Among all samples, check for (unexpected) relatives (samples that share a proportion of shared alleles consistent with a specific family relationship). (In PLINK, use the '-- genome' option.)

8 If X chromosome genotypes are available, check all samples to ensure that the reported sex matches the X chromosome genotype calls (e.g. no heterozygote X chromosome genotypes for females).[i] (In PLINK, use the '--check-sex' option.)

9 Check for effects of genotyping plate/batch on SNP call rates and allele frequencies.[j]

10 After removing the problem data/samples in steps 2–5, again calculate call rates (percentage of called (non-missing) genotypes) per sample and per SNP. (In PLINK, use the '--missing' option.)

11 Filter samples, then SNPs, to keep only those with call rates above an acceptable threshold.

12 Check self-reported race using a program such as EIGENSTRAT or STRUCTURE.[k]

13 Compute HWE (by race/ethnicity for studies of more than one racial/ethnic group) for each SNP.[l] (In PLINK, use the '--hardy' option.)

14 Compute allele frequencies by race; compare with known allele frequencies from dbSNP when available.

Notes

[a]The emphasis of the protocol is large-scale genome-wide designs.

[b]For added illustration, appropriate PLINK [33] commands are included for some of the steps.

[c]In general, special population stratification SNPs do not need to be added for a GWAS.

[d]SNPs for population stratification testing must also be included for genotyping.

[e]These should be placed so that a plate rotation can be detected; for example for a 96-well plate of 8 rows and 12 columns, a duplicate of the sample in row A, column 1 can be usefully placed anywhere except row H, column 12.

[f]If the association study is family-based, this cleaning protocol still provides a useful guideline. Additional checks for Mendelian errors should also be carried out (e.g. using PLINK, the ' – mendel' option). In addition, certain tests, such as Hardy-Weinberg checks, should be carried out on unrelated individuals (e.g. founders from each family), rather than on the full sample including relatives.

[g]This 'blinded' approach will both confirm correctness of known, planned duplicates, and also flag unexpected duplicates.

[h]For example, it is useful to keep the duplicate with the highest genotyping call rate.

[i]Problems should be investigated carefully, as they may indicate inadvertent sample swaps. As an example of further investigation, problem samples should be checked to see whether they come from the same plate (which may indicate an inadvertent plate rotation, or other mishandling that may affect all samples on the plate, and not just the samples that exhibit an obvious sex discrepancy).

[j]Plates or batches exhibiting an excess of poor calls may need to be redone.

[k]As in step 5, problems should be investigated carefully, as they may indicate inadvertent sample swaps.

[l]SNPs with very low (significant) p-values may need to be dropped. Examination of genotyping cluster plots for these SNPs may indicate poor clustering and unreliable genotype calls, confirming that they should be removed prior to the genetic association analyses.

4.2.1.2 Association in unrelated case–control samples: analytic methods

Testing single markers for disease association

The principle behind case–control association mapping of disease genes is relatively straightforward. After genotyping case and control samples, one is interested in detecting significant differences in allele or genotype frequencies between cases and controls. Analysis methods for testing for these differences can be as simple as chi-square tests for the corresponding 2×2 or 3×2 tables in the case of diallelic SNP markers (or $N \times 2$ and $N(N + 1)/2 \times 2$ for a marker with N alleles). Cochran's test of trend is often favored, as it is equivalent to a test for allele frequency differences when HWE holds, but does not require the data to be consistent with HWE [43].

The trend test is simple and often useful. However, investigators may wish to account for covariates that may be important, such as sex or age. Or, it may be that a study has recruited participants from two different recruiting sites, and the rates of cases and controls collected by the two sites are different. Thus, the recruitment site is predictive of case–control status in the subjects, and we may wish to control for this effect before testing for genetic effects.

An alternative analysis method that allows covariates to be included would be logistic regression. Genotype status at each marker is coded as an ordinal variable representing the number of copies of a fixed allele. The non-genetic base model then predicts case status as a function of the k covariates (e.g. site in the example stated above): $\ln[P/(1-P)] = \alpha + \beta_1 x_1 + \ldots + \beta_k x_k$, where P is the probability of being a case, and $x_i, i = 1, \ldots, k$, are the covariate variables included. This base model is compared with a full model $\ln[P/(1-P)] = \alpha + \beta_1 x_1 + \ldots + \beta_k x_k + \beta g$, where g is genotype at the analyzed marker, coded ordinally (that is, coded as the number of copies of a fixed allele, often the minor allele in the overall sample). The likelihood ratio chi-square statistic may be computed to obtain the p-value for the effect of the genetic marker. Note that both the base model and the full model must be computed for the same set of data; therefore, if there are individuals having missing genotype at this marker or missing covariate variable values, then they must be removed from the data prior to the logistic regression analysis. Other codings of the genotype may be tested similarly Protocol 4.2 describes an analysis based on logistic regression and the likelihood ratio chi-square. The user-friendly software package PLINK [33] also implements logistic-regression-based tests of association.

Multilocus association analysis

The first pass analysis of a large-scale case–control association study is usually single-SNP analysis of the kind described in the previous section. Additional information can be gained by multilocus analysis, looking at groups of markers in various ways. However, there is often a penalty to be paid in the form of correction for the number of tests (see below). For example, given 100 000 genotyped, bi-allelic SNPs, pairwise analyses of all possible two-way interactions of (not necessarily contiguous) loci leads to $(100\,000) \times (99\,999)/2 = 4\,999\,950\,000$ tests.

One way to group multiple SNP loci for analysis is to consider haplotypes; that is, the combination of alleles occurring at linked loci along a chromosomal strand. Haplotype analyses have some biological justification. A commonly cited motivation for haplotype analysis is that a disease-causing but un-genotyped variant may lie on a particular haplotype background, and analysis of that haplotype will reveal the association. Alternatively, it may be that the combined allelic state across the haplotype is biologically functional and causes the disease. However, when testing haplotypes, the number of tests, and the degrees of freedom for a given test, can rapidly increase. A typical approach is to use a 'sliding window' of N SNPs for a range of values for N. For each fixed window, the investigator must decide whether to include all observed haplotypes in the analysis or either to ignore or pool together rare haplotypes. An analysis of H haplotype categories may then be carried out in various ways, from a traditional chi-square or exact test of the resulting $2 \times H$ table of case–control status and estimated haplotype frequencies, to haplotype trend regression analysis [44] and analysis of weighted haplotype frequency differences [45].

Haplotype analysis is often used to stand in for direct analysis of an untyped locus for which the susceptibility allele lies on a specific haplotype background. An alternative analysis approach with a similar intent is genetic imputation, which infers 'missing' geno-types at untyped loci. Imputation typically relies on available LD data from a 'reference population' (e.g. HapMap) between typed and untyped loci. These 'in silico' genotypes can then be tested for association with phenotype. The Wellcome Trust GWAS of multiple diseases mentioned above used a method of Marchini *et al.* [46]. This imputation method uses HapMap LD and estimates of fine-scale recombination across the genome to obtain probabilities for each possible genotype call at an untyped locus and takes the uncertainty of imputed genotypes into account in tests for association at the locus. Several other imputation programs are also now available [47–50]. Whatever software is used, it is important to choose an appropriate external reference population as the source of LD information. Recent research suggests that, in European-descent populations, common SNPs can be reliably imputed using the single CEU HapMap reference panel, while mixtures of at least two HapMap panels produce the highest imputation accuracy for most other populations [51]. Accurate imputation at rare SNPs will likely require larger reference samples beyond what is currently available from HapMap.

Gene–gene interaction analysis is also of great importance for complex disease studies. It is widely hypothesized that complex diseases arise in part because of such epistatic effects. Specific interactions between SNP loci may be readily tested within the logistic regression framework described above, using a standard product term for the interaction. Other methods for testing interactions include the multidimensional reduction (MDR) method [52–54], an extension of MDR to allow covariates [55] and the recursive partitioning method (RPM) [56, 57].

Multiple testing

Multiple testing and potential inflation of false positive rates are not new concerns for statistical genetics and gene mapping. Nevertheless, with the advent of large-scale, GWAS designs, the problem of potential false positives can seem especially pressing. The traditional statistical significance level of 0.05 is certainly too permissive if applied to each test without correcting for the number of tests, and yet, in contrast, a 0.05 experiment-wide error rate may seem overly conservative when investigators have invested considerable resources into generating the data.

The experiment-wide error rate may be estimated using permutation to generate an empirical p-value; similarly, Bonferroni-style corrections that account for correlations (LD) between SNPs [58, 59] may be applied to obtain an experiment-wide significance level. An important alternative approach to determine which SNPs to label as 'findings worth following up' is based on controlling the false discovery rate (FDR). Unlike the significance level, which is the proportion of results which would be declared positive if the null hypothesis were true, the FDR is the proportion of false positive results among all the declared positives. The FDR is thus arguably more relevant than the significance level for studies in which resources will be invested in following up the declared positive results. A low FDR would thus improve expectations that most of that follow-up effort will 'pay off.' Publicly available programs to calculate FDRs include QVALUE (http://faculty.washington.edu/jstorey/qvalue) [60].

Power

An integral part of planning a genetic mapping study is determining (estimating) the required sample size to be able to detect a given effect. Power of an experimental test at a given significance level is defined as the probability of rejecting the null hypothesis at the specified significance level α when that null hypothesis is false.

For the simple chi-square test for allele frequency differences between cases and controls, power (or necessary sample sizes for a desired level of power) may be calculated using non-central chi-square distributions. The necessary calculations have been implemented in the Genetic Power Calculator, a user-friendly web-based calculator at http://pngu.mgh.harvard.edu/~purcell/gpc/ [61]; power for case–control studies of discrete traits is one of several options. For this calculation, the user must specify a disease model parameterized in terms of the population prevalence of the disease, the disease gene frequency and the genotypic relative risks that determine the disease penetrances for having one or two copies of the disease allele. Power to detect this underlying disease locus can then be calculated for a genotyped biallelic marker (SNP) having a given allele frequency and strength of LD with the disease locus, the latter measured by $|D'|$. Other user-friendly power calculators for genetic association studies include those from Menashe *et al.* [62] and Gordon *et al.* [63].

The LD measure r^2 has a very useful property with regard to the power of an association mapping study: for a given value of r^2 between a disease locus and marker, regardless of phase, the sample size needed to have equivalent power to detect disease association with alleles at the marker, rather than at the true locus, is increased by a factor of approximately $1/r^2$ [23]. Though this formula is approximate, and it has been pointed out that it can overestimate the power of a tag set in certain circumstances [24], it provides a useful guideline for estimating the sample size needed if r^2 bin tagging SNPs are genotyped as proxies for other SNPs in a bin defined by a given r^2 value.

Summary

Association analysis of unrelated case control samples is currently a popular approach to disease gene mapping. Technologies that are now enabling GWAS are continuing to increase genotyping capacity and reduce costs. More discussion of methods and issues for case–control association studies may be found in the excellent review article by Balding [64, 65].

PROTOCOL 4.2 Logistic Regression for Association Analysis of Case–Control SNP Genotype Data

Equipment and tools

- SAS Software (Cary, IN, USA)

Method

Following is a description of single SNP analysis using the likelihood ratio chi-square statistic from logistic regression models (discussed in Section 4.2.1.2).[m]

1 Create the dataset. The SAS dataset named 'fulldata' contains one observation for each subject in the study. The variable 'ctrl' has a value of 1 if the subject is a control and a value of 0 if the subject is a case. The variable 'cov1' is a binary covariate; for example, this could be gender. The variable 'loc1' contains the genotype data at the marker to be analyzed, coded as the number of copies of the minor allele; missing values are coded by the usual SAS missing indicator for numeric variables.

2 Run the following SAS program to compute the likelihood-ratio chi-square statistic comparing the full logistic regression model to the reduced model in a one degree-of-freedom test for the effect of genotype:[m]

```
data subset;
set fulldata;
if loc1 = . then delete;
***analyze only the subset for which loc1 is non-missing;
run;
*** 1st proc logistic: base model, 1 covariate included;
proc logistic data=subset outest=base_loc1;
model ctrl = cov1;
title "1st proc logistic: base model";
run;
*** 2nd proc logistic: full model, 1 covariate + genotype;
*** genotype variable, loc1, is coded as number of copies of allele 1;
proc logistic data=subset outest=full_loc1;
model ctrl = cov1 loc1;
title "2nd proc logistic: full model";
run;
        *** get likelihoods for each model;
data like_base;
set base_loc1
(rename = (_LNLIKE_ = lnlike_base));
keep lnlike_base;
data like_full;
set full_loc1
(rename (_LNLIKE_ = lnlike_full));
keep lnlike_full;
run;
data mrg;
merge like_base like_full;
chisq_1df = -2 * (lnlike_base - lnlike_full);
pvalue = 1 - probchi(chisq_1df,1);
run;
```

3 Repeat the analysis for each SNP to be analyzed.[o]

4 Logistic-regression-based association analysis can also be carried out in PLINK using the '--logistic' option.

Notes

[m]The protocol described here is based on one used for the analysis of a large-scale GWAS and candidate gene study of nicotine dependence [6, 7]. A single 'proc logistic' step would suffice to obtain the p-value for the one degree of freedom test of the single genotype term; however the code shown here, with two 'proc logistic' steps, readily generalizes to carry out tests with more degrees of freedom (e.g. if the genotype is coded with two degrees of freedom, or if interactions between covariate and genotype are to be jointly tested with the genotype term).

[n]The program subsets the dataset fulldata so that only observations for which loc1 is not missing are used; this ensures that both logistic regression runs are carried out on the same set of observations. The variable 'pvalue' thus indicates the strength of evidence for association between case/control status and the single SNP being analyzed.

[o]In practice, step 2 would be programmed into a loop across all SNPs for analysis. A "by" statement to merge by the SNP name should be added to the merge step.

4.2.2 Association methods: family-based samples

As explained in the previous section, a case–control association study design is susceptible to false positives attributable to population admixture rather than due to proximity of the associated marker to a trait-causing mutation. This problem can be minimized by using family-based designs. Use of so-called family-based controls was popularized by the transmission disequilibrium test (TDT) [66]. The premise of the TDT is that, if a locus has an allele associated with disease, a parent who is heterozygous at the locus will transmit the associated allele to affected offspring more often than the proportion of 0.5 expected under the null of no association. In effect, the transmitted 'case' alleles and the non-transmitted 'control' alleles are contributed by a single person (the parent) and thus are expected to be well matched, given the ancestry of the parent. Thus, the potentially confounding effect of cryptic population stratification is reduced or eliminated.

The original TDT was geared for analysis of trios consisting of two parents and an affected offspring. Extensions have allowed the analysis of more extended pedigrees [67]. A unifying framework for family-based association tests (FBATs) was introduced by Rabinowitz and Laird [68] and Laird et al. [69]. The key feature of their approach is that it computes the null distribution of marker alleles in offspring conditional on appropriate features of the data, so that this conditional distribution follows Mendelian segregation regardless of the phenotype configuration in the family. When parental genotypes are known, the distribution is computed conditional on all phenotypes and parental genotypes. When parental genotypes are missing, the distribution is conditional also on the offspring genotype configuration. The point is that resulting tests of association that compare the observed pattern of allele segregation with that expected under the conditional null distribution have a correct Type I error rate and are protected from biases caused by population admixture. A software toolkit for carrying out a broad range of FBAT-based tests is available at http://www.biostat.harvard.edu/~fbat/default.html, and new tests based on the core conditional framework continue to be developed [70–72]. The PLINK package [33] also carries out basic family-based association tests.

4.2.2.1 Quality control for family data

In addition to some of the safeguards mentioned in the discussion of unrelated case–control data, additional measures are important for family data (whether for association analysis or for linkage designs described below). The availability of reported familial relationships, and the reliance of analysis methods on these relationships, means that, first, these relationships should be validated by evaluating the compatibility of the genotype data with these relationships. The family structures also provide an added means of detecting some genotyping errors. Several user-friendly and well-documented programs for these tasks exist, such as PEDSTATS [32] and PREST [74, 75].

4.2.3 Linkage methods: parametric LOD score analysis

The logarithmic odds (LOD) score method was introduced by Newton Morton in a land-mark article [76]. The method is now sometimes described as a 'parametric' method, to indicate that it requires specification of a disease model (mode of inheritance and values for parameters such as disease allele frequency). The method was formulated in the context of the single trait gene setting, and its suitability for simple Mendelian traits is clear. Specifying a disease model for a complex trait is more problematic. Nevertheless, some complex diseases exhibit subtypes that follow Mendelian inheritance, and discovery of genetic loci responsible for these subtypes may also give insights into the underlying biology of the more general disease.

4.2.3.1 Principles of classical parametric LOD score linkage analysis

Suppose we have a family with an observed configuration of phenotypes (affected/unaffected by the disease under study) and genotypes at a genetic marker locus M. We begin by assuming a 'single major locus' disease model that can be described in terms of four 'known' (or estimated) parameters: penetrances f_{AA}, f_{Aa} and f_{aa} for the genotype classes AA, Aa and aa at the trait locus, and gene frequency p of the trait locus. Thus, f_{AA} is the probability that a person with genotype AA is affected by the disease. The goal is to estimate the recombination fraction θ between M and the trait locus (that is, the probability θ that a recombination would occur between these two loci) and determine if it is significantly different from the value of $\theta = 0.5$ expected under the null hypothesis of no linkage. The LOD curve $Z(\theta)$ as a function of $f_{AA}, f_{Aa}, f_{aa}, p$ and θ is then defined by

$$Z(\theta) = \frac{L(\text{data} \mid \theta, f_{AA}, f_{Aa}, f_{aa}, p)}{L\left(\text{data} \mid \theta = \frac{1}{2}, f_{AA}, f_{Aa}, f_{aa}, p\right)}$$

where L denotes the likelihood function and 'data' denotes the observed phenotypes and genotypes at M for the family. The maximum of $Z(\theta)$ is called the maximum LOD score and denoted Z. The LOD score may be converted to a p-value by noting that the likelihood ratio chi-square statistic is -2 times the difference between the natural log (base e) of the likelihood at $\theta = 0.5$ and the natural log of the maximum likelihood value. That is, $2(\log_e 10)Z = 4.6Z \approx \chi_1^2$. Thus, a LOD score of, say 3, is equivalent to a chi square of 13.8 with one degree of freedom, corresponding to a p-value of 0.0001 for a one-sided test of $\theta = 0.5$.

An LOD $= 3$ is classically considered a threshold for declaring significant linkage; however, this value arose from observations of Morton [76] that corresponded to a sequential test for linkage, in which families would be added one at a time to the sample, the LOD score would be computed and linkage would be declared if at a point in this sequence the LOD rose above 3. In practice, the LOD score of 3 criterion has been used even when the analysis is not sequential. As noted above, this would correspond to $\alpha = 0.0001$ for a single marker. Given the density of marker coverage in current genome-wide linkage screens, Lander and Kruglyak [77] have recommended that to ensure a genome-wide error rate of 0.05, the more appropriate LOD threshold is 3.6, assuming an infinitely dense map.

In the simplest case of data for which the number of recombinant (k) and non-recombinant $(n - k)$ meioses can be counted out of the total of n meioses, the maximum of $Z(\theta)$ will occur when $\theta = k/n$. However, because we typically consider θ to range between 0 and 0.5, the maximum likelihood estimate for θ is taken to be

$$
\hat{\theta} = \begin{cases} \dfrac{k}{n} & \text{if } k \leq \dfrac{n}{2} \\[2ex] \dfrac{1}{2} & \text{if } k > \dfrac{n}{2} \end{cases}
$$

and the maximum LOD score is then $\log_{10} L(\hat{\theta})/L(0.5)$. For more complex data, however, software is used to evaluate the likelihood functions and compute LOD scores.

The early computer programs for two-point (i.e. single marker) LOD score linkage analysis, such as LIPED [78], were revelatory in their time and allowed for rapid analysis of complex pedigrees. User-friendly software packages in popular use today for rapid multipoint parametric linkage analysis include MERLIN [79] and Genehunter [80]; these programs also implement non-parametric linkage analysis, discussed next.

4.2.4 Linkage methods: non-parametric methods

Parametric methods have been successful in identifying genes for rare Mendelian disorders for which the mode of inheritance is typically recognizable. However, multifactorial diseases such as type II diabetes and psychiatric diseases often do not lend themselves to specification of an unequivocal disease model. This has led to the popularity of 'non-parametric' methods that do not require a disease model to specified, including affected sib pair (ASP) analysis, affected pedigree member (APM) analysis and variance components (VCs) analysis.

Despite the labeling of these methods as 'model-free' or 'non-parametric,' it is important to be aware that equivalencies between ASP and parametric tests have been demonstrated in particular cases [81]. Furthermore, implicit model assumptions exist for ASP tests, and the power of such tests is thus influenced by the appropriateness of these assumptions and the true underlying model [82].

ASP and APM methods evaluate whether affected relatives share more than the expected number of marker alleles identical by descent (IBD), where two alleles are IBD if they are inherited from the same ancestral source. Alleles that appear to have the same value but are not known to be from the same ancestor are called identical by state (IBS). For a sibling pair, the expected IBD proportion is 0.5; this is equivalent to one out of the two possible alleles being shared IBD. At a particular marker or genetic map position, the evidence for linkage is typically measured by the maximum LOD score (MLS). That is, the LOD score is maximized over the parameters used to parameterize the single major locus model; for example, over the IBD sharing probabilities at the putative trait locus.

Modern multipoint relative-pair methods have their roots in an early idea of Penrose [82]. Penrose proposed tabulating the proportion of sib-pairs that are alike or not alike for two observable states, or 'phenotypes,' which nowadays are in fact typically a disease trait and genotype status at a marker. The test for linkage in a more focused ASP design consists of comparing the phenotypic similarity of each sib pair with its allelic similarity, the latter measured by IBD status. Because with an ASP design the sib pair members are both affected with the disease, they are necessarily phenotypically similar and we expect that if a marker is linked to a disease locus (or is itself the disease locus) then the IBD status at that marker will be elevated above the 0.5 proportion expected under the null hypothesis of no linkage. Several formal statistical tests to compare these null and alternative hypotheses are available; among the best known are the mean test and the proportion test. The former has been shown to perform best for dominant traits and the latter is best for recessive traits [83]. This general idea of comparing IBD status and phenotypic status also has been extended to analysis of quantitative traits [84].

VC linkage methods, implemented in programs such as SOLAR [86], ACT [87] and SEGPATH [88], are well suited to analysis of quantitative traits but may also be applied to dichotomous or polychotomous traits via a threshold liability model. The typical approach uses the multivariate normal distribution to model the likelihood of a pedigree, and thus departure of the data from this assumption of multivariate normality can be a concern [89]. One alternative is to use the multivariate t distribution, an option implemented in SOLAR.

Because modern techniques have enabled dense genotyping of markers, non-parametric sib-pair studies typically generate data at multiple markers across each chromosome and analyze the data using multipoint methods. The programs MERLIN and Genehunter have already been mentioned above; these implement both parametric and non-parametric multi-point linkage analyses, and are now more widely used than the earlier landmark programs MAPMAKER/SIBS [90] and Neil Risch's ASPEX (http://aspex.sourceforge.net/) for multipoint sib-pair analysis.

Many of the programs mentioned above have historically required familiarity with UNIX operating systems and with formats for the input and output of the specific programs. The recently developed package easyLINKAGE-Plus [91] provides a user-friendly, automated way to set up and run analyses using any of several popular programs for both parametric and non-parametric linkage, and it may be applied to either SNP or microsatellite markers.

4.2.5 Summary and conclusions

We have described association- and linkage-based approaches to mapping genes influencing complex traits. Recent successes using genome-wide association mapping designs have brought attention both to specific genes involved in key biological processes and to this general approach as a feasible and effective way to identify genes that harbor common variants involved in common complex diseases. However, a limitation of the genome-wide association approach is that it is well-powered to detect association only for common alleles. Once such a variant is identified, the surrounding gene(s) or region may certainly be investigated further for additional rare variants that may also influence disease outcomes. However, to detect a locus at which multiple rare variants and no common alleles cause the disease, we may need still to rely on classical linkage analysis. In any case, no single approach should be expected to fit all settings, and the range of strategies discussed here will remain relevant even as the field of human genetics evolves.

4.3 Troubleshooting

4.3.1 Combining datasets

Sometimes a study will need to combine two datasets that were genotyped and cleaned separately; for example, two distinct samples that have been genotyped on two overlapping sets of SNPs. A merged dataset may increase the sample size, the number of SNPs, or both, leading to improved power and/or genomic coverage. However, the merging process must be performed with care. We outline some key steps that need attention when creating a merged dataset for two distinct genotyped samples.

First, one must determine which SNPs have been genotyped in common in two samples, taking into account possible changes in rs (reference SNP) numbers for SNPs. To do this accurately, the lists of genotyped SNPs may be submitted to dbSNP in 'batch' mode (http://www.ncbi.nlm.nih.gov/projects/SNP/). This process will return information on whether any rs numbers have been changed or consolidated to a single rs number, according to the latest dbSNP build.

Second, in each sample, the coding of the alleles must be compared to see whether the allele coding matches, or whether one sample is using the complementary nucleotides of the other. This task is straightforward for SNPs other than self-complementary 'A/T' or 'C/G' SNPs; for example, if both samples call the alleles 'C' and 'T', the allele coding matches, whereas if one sample uses 'C' and 'T' and the other uses 'G' and 'A', alleles should be flipped, in one sample, to the complementary nucleotides to be consistent across both samples before merging. Aligning the allele coding is more difficult for self-complementary 'A/T' or 'C/G' SNPs, as it can be unclear whether the allele called ' A' in one sample corresponds to the 'A' or the 'T' allele in the second sample. If information about the 'strand' of the call is available, or if flanking primer sequence is available, this can be used to check for matching. Checking allele frequencies separately in both groups, and comparing with frequencies recorded at dbSNP, can help to align alleles as long as the two samples have similar population histories (e.g. both of European descent) and the allele frequencies are far from 50%.

4.3.1.1 Annotating and displaying SNPs and results

To interpret results from a large-scale gene mapping study, and to create tables and figures for publication, it is important to annotate the SNPs and their corresponding statistical results with information about gene locations and other information available from public databases such as dbSNP and HapMap. However, integrating statistical results with consistent, up-to-date annotation information on a genome-wide scale (e.g. for a GWAS) can be laborious. Fortunately, software tools are now available that aid the presentation of genome-wide association data. The publicly available program WGAviewer [92] is a user-friendly solution that annotates and provides visualization tools for GWAS results. For details on available functions and usage, please refer to the website at http://people.genome.duke.edu/~dg48/WGAViewer/.

References

1. Risch, N. and Merikangas, K. (1996) The future of genetic studies of complex human diseases. *Science (New York, NY)*, **273**, 1516–1517.

2. Saxena, R., Voight, B.F., Lyssenko, V. *et al.* (2007) Genome-wide association analysis identifies loci for type 2 diabetes and triglyceride levels. *Science (New York, NY)*, **316**, 1331–1336.

3. Scott, L.J., Mohlke, K.L., Bonnycastle, L.L. *et al.* (2007) A genome-wide association study of type 2 diabetes in Finns detects multiple susceptibility variants. *Science (New York, NY)*, **316**, 1341–1345.

4. Zeggini, E., Weedon, M.N., Lindgren, C.M. *et al.* (2007) Replication of genome-wide association signals in UK samples reveals risk loci for type 2 diabetes. *Science (New York, NY)*, **316**, 1336–1341.

5. Easton, D.F., Pooley, K.A., Dunning, A.M. *et al.* (2007) Genome-wide association study identifies novel breast cancer susceptibility loci. *Nature*, **447**, 1087–1093.

6. Bierut, L.J., Madden, P.A., Breslau, N. *et al.* (2007) Novel genes identified in a high-density genome wide association study for nicotine dependence. *Human Molecular Genetics*, **16**, 24–35.

7. Saccone, S.F., Hinrichs, A.L., Saccone, N.L. *et al.* (2007) Cholinergic nicotinic receptor genes implicated in a nicotine dependence association study targeting 348 candidate genes with 3713 SNPs. *Human Molecular Genetics*, **16**, 36–49.

8. Thorgeirsson, T.E., Geller, F., Sulem, P. *et al.* (2008) A variant associated with nicotine dependence, lung cancer and peripheral arterial disease. *Nature*, **452**, 638–642.

9. Berrettini, W., Yuan, X., Tozzi, F. *et al.* (2008) α-5/α-3 nicotinic receptor subunit alleles increase risk for heavy smoking. *Molecular Psychiatry*, **13**, 368–373.

10. Amos, C.I., Wu, X., Broderick, P. *et al.* (2008) Genome-wide association scan of tag SNPs identifies a susceptibility locus for lung cancer at 15q25.1. *Nature Genetics*, **40**, 616–622.

11. Hung, R.J., McKay, J.D., Gaborieau, V. *et al.* (2008) A susceptibility locus for lung cancer maps to nicotinic acetylcholine receptor subunit genes on 15q25. *Nature*, **452**, 633–637.

12. Liu, P., Vikis, H.G., Wang, D. *et al.* (2008) Familial aggregation of common sequence variants on 15q24-25.1 in lung cancer. *Journal of the National Cancer Institute*, **100**, 1326–1330.

13. Li, M., Boehnke, M. and Abecasis, G.R. (2005) Joint modeling of linkage and association: identifying SNPs responsible for a linkage signal. *American Journal of Human Genetics*, **76**, 934–949.

14. Morton, N.E. and Collins, A. (1998) Tests and estimates of allelic association in complex inheritance. *Proceedings of the National Academy of Sciences of the United States of America*, **95**, 11389–11393.

15. Moskvina, V., Holmans, P., Schmidt, K.M. and Craddock, N. (2005) Design of case–controls studies with unscreened controls. *Annals of Human Genetics*, **69**, 566–576.

16. Wellcome Trust Case Control Consortium (2007) Genome-wide association study of 14,000 cases of seven common diseases and 3,000 shared controls. *Nature*, **447**, 661–678. A landmark example of a large-scale case-control genome-wide association study.

17. International HapMap Consortium (2005) A haplotype map of the human genome. *Nature*, **437**, 1299–1320. Describes Phase I results from the International HapMap Project.

18. Frazer, K.A., Ballinger, D.G., Cox, D.R. *et al.* (2007) A second generation human haplotype map of over 3.1 million SNPs. *Nature*, **449**, 851–861. Describes phase II results from the international HapMap Project.

19. Barrett, J.C., Fry, B., Maller, J. and Daly, M.J. (2005) Haploview: analysis and visualization of LD and haplotype maps. *Bioinformatics (Oxford, England)*, **21**, 263–265.

20. Ke, X. and Cardon, L.R. (2003) Efficient selective screening of haplotype tag SNPs. *Bioinformatics (Oxford, England)*, **19**, 287–288.

21. Stram, D.O., Haiman, C.A., Hirschhorn, J.N. *et al.* (2003) Choosing haplotype-tagging snps based on unphased genotype data using a preliminary sample of unrelated subjects with an example from the multiethnic cohort study. *Human Heredity*, **55**, 27–36.

22. Carlson, C.S., Eberle, M.A., Rieder, M.J. *et al.* (2004) Selecting a maximally informative set of single-nucleotide polymorphisms for association analyses using linkage disequilibrium. *American Journal of Human Genetics*, **74**, 106–120.

23. Pritchard, J.K. and Przeworski, M. (2001) Linkage disequilibrium in humans: models and data. *American Journal of Human Genetics*, **69**, 1–14.

24. Moskvina, V. and O'Donovan, M.C. (2007) Detailed analysis of the relative power of direct and indirect association studies and the implications for their interpretation. *Human Heredity*, **64**, 63–73.

25. Edlund, C.K., Lee, W.H., Li, D., Van Den Berg, D.J. and Conti, D.V. (2008) Snagger: a user-friendly program for incorporating additional information for tagSNP selection. *BMC Bioinformatics*, **9**, 174.

26. De Bakker, P.I., Yelensky, R., Pe'er, I. *et al.* (2005) Efficiency and power in genetic association studies. *Nature Genetics*, **37**, 1217–1223.

27. Johnson, A.D., Handsaker, R.E., Pulit, S.L. *et al.* (2008) SNAP: a web-based tool for identification and annotation of proxy SNPs using HapMap. *Bioinformatics (Oxford, England)*, **24**, 2938–2939.

28. Li, C., Li, M., Long, J.R. *et al.* (2008) Evaluating cost efficiency of SNP chips in genome-wide association studies. *Genetic Epidemiology*, **32**, 387–395.

29. Li, M., Li, C. and Guan, W. (2008) Evaluation of coverage variation of SNP chips for genome-wide association studies. *European Journal of Human Genetics*, **16**, 635–643.

30. Saccone, S.F., Bierut, L.J., Chesler, E.J. *et al.* (2009) Supplementing high-density SNP microarrays for additional coverage of disease-related genes: addiction as a paradigm. *PLoS One*, **4**, e5225.

31. Keating, B.J., Tischfield, S., Murray, S.S. *et al.* (2008) Concept, design and implementation of a cardiovascular gene-centric 50 K SNP array for large-scale genomic association studies. *PLoS ONE*, **3**, e3583.

32. Wigginton, J.E. and Abecasis, G.R. (2005) PEDSTATS: descriptive statistics, graphics and quality assessment for gene mapping data. *Bioinformatics (Oxford, England)*, **21**, 3445–3447. Describes PEDSTATS, a user-friendly program for performing quality control checks of both family-based and unrelated samples.

33. Purcell, S., Neale, B., Todd-Brown, K. *et al.* (2007) PLINK: a tool set for whole-genome association and population-based linkage analyses. *American Journal of Human Genetics*, **81**, 559–575. Introduces and describes PLINK, an excellent and widely used software package for analysis of genome-wide SNP data.

34. Suarez, B.K., Taylor, C., Bertelsen, S. *et al.* (2005) An analysis of identical single-nucleotide polymorphisms genotyped by two different platforms. *BMC Genetics*, **6** (Suppl. 1), S152.

35. Pritchard, J.K., Stephens, M. and Donnelly, P. (2000) Inference of population structure using multilocus genotype data. *Genetics*, **155**, 945–959.

36. Devlin, B. and Roeder, K. (1999) Genomic control for association studies. *Biometrics*, **55**, 997–1004.

37. Price, A.L., Patterson, N.J., Plenge, R.M. *et al.* (2006) Principal components analysis corrects for stratification in genome-wide association studies. *Nature Genetics*, **38**, 904–909. The publication for EIGENSTRAT, an excellent and popular program for testing and correcting for population stratification.

38. Tian, C., Hinds, D.A., Shigeta, R. *et al.* (2007) A genomewide single-nucleotide–polymorphism panel for Mexican American admixture mapping. *American Journal of Human Genetics*, **80**, 1014–1023.

39. Tian, C., Hinds, D.A., Shigeta, R. *et al.* (2006) A genomewide single-nucleotide–polymorphism panel with high ancestry information for African American admixture mapping. *American Journal of Human Genetics*, **79**, 640–649.

40. Tian, C., Plenge, R.M., Ransom, M. *et al.* (2008) Analysis and application of European genetic substructure using 300 K SNP information. *PLoS Genetics*, **4**, e4.

41. Yang, B.Z., Zhao, H., Kranzler, H.R. and Gelernter, J. (2005) Practical population group assignment with selected informative markers: characteristics and properties of Bayesian clustering via STRUCTURE. *Genetic Epidemiology*, **28**, 302–312.

42. Yang, N., Li, H., Criswell, L.A. *et al.* (2005) Examination of ancestry and ethnic affiliation using highly informative diallelic DNA markers: application to diverse and admixed populations and implications for clinical epidemiology and forensic medicine. *Human Genetics*, **118**, 382–392.

43. Sasieni, P.D. (1997) From genotypes to genes: doubling the sample size. *Biometrics*, **53**, 1253–1261.

44. Zaykin, D.V., Westfall, P.H., Young, S.S. *et al.* (2002) Testing association of statistically inferred haplotypes with discrete and continuous traits in samples of unrelated individuals. *Human Heredity*, **53**, 79–91.

45. Zaitlen, N., Kang, H.M., Eskin, E. and Halperin, E. (2007) Leveraging the HapMap correlation structure in association studies. *American Journal of Human Genetics*, **80**, 683–691.

46. Marchini, J., Howie, B., Myers, S.S *et al.* (2007) A new multipoint method for genome-wide association studies by imputation of genotypes. *Nature Genetics*, **39**, 906–913.

47. Nicolae, D.L. (2006) Testing Untyped Alleles (TUNA) – applications to genome-wide association studies. *Genetic Epidemiology*, **30**, 718–727.

48. Li, Y., Willer, C.J., Sanna, S. and Abecasis, G.R. (2009) Genotype imputation. *Annual Review of Genomics and Human Genetics*, 10, 387–406.

49. Servin, B. and Stephens, M. (2007) Imputation-based analysis of association studies: candidate regions and quantitative traits. *PLoS Genetics*, **3**, e114.

50. Browning, B.L. and Browning, S.R. (2009) A unified approach to genotype imputation and haplotype-phase inference for large data sets of trios and unrelated individuals. *American Journal of Human Genetics*, **84**, 210–223.

51. Huang, L., Li, Y., Singleton, A.B. *et al.* (2009) Genotype-imputation accuracy across worldwide human populations. *American Journal of Human Genetics*, **84**, 235–250.

52. Hahn, L.W., Ritchie, M.D. and Moore, J.H. (2003) Multifactor dimensionality reduction software for detecting gene–gene and gene–environment interactions. *Bioinformatics (Oxford, England)*, **19**, 376–382.

53. Ritchie, M.D., Hahn, L.W. and Moore, J.H. (2003) Power of multifactor dimensionality reduction for detecting gene–gene interactions in the presence of genotyping error, missing data, phenocopy, and genetic heterogeneity. *Genetic Epidemiology*, **24**, 150–157.

54. Ritchie, M.D., Hahn, L.W., Roodi, N. *et al.* (2001) Multifactor-dimensionality reduction reveals high-order interactions among estrogen-metabolism genes in sporadic breast cancer. *American Journal of Human Genetics*, **69**, 138–147.

55. Lou, X.Y., Chen, G.B., Yan, L. *et al.* (2007) A generalized combinatorial approach for detecting gene-by-gene and gene-by-environment interactions with application to nicotine dependence. *American Journal of Human Genetics*, **80**, 1125–1137.

56. Culverhouse, R. (2007) The use of the restricted partition method with case–control data. *Human Heredity*, **63**, 93–100.

57. Culverhouse, R., Klein, T. and Shannon, W. (2004) Detecting epistatic interactions contributing to quantitative traits. *Genetic Epidemiology*, **27**, 141–152.

58. Cheverud, J.M. (2001) A simple correction for multiple comparisons in interval mapping genome scans. *Heredity*, **87**, 52–58.

59. Li, J. and Ji, L. (2005) Adjusting multiple testing in multilocus analyses using the eigenvalues of a correlation matrix. *Heredity*, **95**, 221–227.

60. Storey, J.D. and Tibshirani, R. (2003) Statistical significance for genomewide studies. *Proceedings of the National Academy of Sciences of the United States of America*, **100**, 9440–9445.

61. Purcell, S., Cherny, S.S. and Sham, P.C. (2003) Genetic Power Calculator: design of linkage and association genetic mapping studies of complex traits. *Bioinformatics (Oxford, England)*, **19**, 149–150.

62. Menashe, I., Rosenberg, P.S. and Chen, B.E. (2008) PGA: power calculator for case-control genetic association analyses. *BMC Genetics*, **9**, 36.

63. Gordon, D., Haynes, C., Blumenfeld, J. and Finch, S.J. (2005) PAWE-3D: visualizing power for association with error in case-control genetic studies of complex traits. *Bioinformatics (Oxford, England)*, **21**, 3935–3937.

64. Zondervan, K.T. and Cardon, L.R. (2007) Designing candidate gene and genome-wide case-control association studies. *Nature Protocols*, **2**, 2492–2501.

65. Balding, D.J. (2006) A tutorial on statistical methods for population association studies. *Nature Reviews Genetics*, **7**, 781–791. A helpful overview of methods and issues for case–control association studies.

66. Spielman, R.S., McGinnis, R.E. and Ewens, W.J. (1993) Transmission test for linkage disequilibrium: the insulin gene region and insulin-dependent diabetes mellitus (IDDM). *American Journal of Human Genetics*, **52**, 506–516.

67. Martin, E.R., Monks, S.A., Warren, L.L. and Kaplan, N.L. (2000) A test for linkage and association in general pedigrees: the pedigree disequilibrium test. *American Journal of Human Genetics*, **67**, 146–154.

68. Rabinowitz, D. and Laird, N. (2000) A unified approach to adjusting association tests for population admixture with arbitrary pedigree structure and arbitrary missing marker information. *Human Heredity*, **50**, 211–223.

69. Laird, N.M., Horvath, S. and Xu, X. (2000) Implementing a unified approach to family-based tests of association. *Genetic Epidemiology*, **19** (Suppl 1), S36–S42.

70. Lewinger, J.P. and Bull, S.B. (2006) Validity, efficiency, and robustness of a family-based test of association. *Genetic Epidemiology*, **30**, 62–76.

71. Horvath, S., Xu, X, Lake, S.L. *et al.* (2004) Family-based tests for associating haplotypes with general phenotype data: application to asthma genetics. *Genetic Epidemiology*, **26**, 61–69.

72. Xu, X., Rakovski, C. and Laird, N. (2006) An efficient family-based association test using multiple markers. *Genetic Epidemiology*, **30**, 620–626.

73. Rakovski, C.S., Weiss, S.T., Laird, N.M. and Lange, C. (2008) FBAT-SNP-PC: an approach for multiple markers and single trait in family-based association tests. *Human Heredity*, **66**, 122–126.

74. McPeek, M.S. and Sun, L. (2000) Statistical tests for detection of misspecified relationships by use of genome-screen data. *American Journal of Human Genetics*, **66**, 1076–1094.

75. Sun, L., Wilder, K. and McPeek, M.S. (2002) Enhanced pedigree error detection. *Human Heredity*, **54**, 99–110. Develops and describes PREST and related programs for performing quality control checks for family-based data.

76. Morton, N.E. (1955) Sequential tests for the detection of linkage. *American Journal of Human Genetics*, **7**, 277–318.

77. Lander, E. and Kruglyak, L. (1995) Genetic dissection of complex traits: guidelines for interpreting and reporting linkage results. *Nature Genetics*, **11**, 241–247.

78. Ott, J. (1974) Estimation of the recombination fraction in human pedigrees: efficient computation of the likelihood for human linkage studies. *American Journal of Human Genetics*, **26**, 588–597.

79. Abecasis, G.R., Cherny, S.S., Cookson, W.O. and Cardon, L.R. (2002) Merlin – rapid analysis of dense genetic maps using sparse gene flow trees. *Nature Genetics*, **30**, 97–101.

80. Kruglyak, L., Daly, M.J., Reeve-Daly, M.P. and Lander, E.S. (1996) Parametric and nonparametric linkage analysis: a unified multipoint approach. *American Journal of Human Genetics*, **58**, 1347–1363.

81. Knapp, M., Seuchter, S.A. and Baur, M.P. (1994) Linkage analysis in nuclear families. 2: Relationship between affected sib-pair tests and lod score analysis. *Human Heredity*, **44**, 44–51.

82. Whittemore, A.S. (1996) Genome scanning for linkage: an overview. *American Journal of Human Genetics*, **59**, 704–716.

83. Penrose, L.S. (1935) The detection of autosomal linkage in data which consist of pairs of brothers and sisters of unspecified parentage. *Annals of Eugenics*, **6**, 133–138.

84. Whittemore, A.S. and Tu, I.P. (1998) Simple, robust linkage tests for affected sibs. *American Journal of Human Genetics*, **62**, 1228–1242.

85. Haseman, J.K. and Elston, R.C. (1972) The investigation of linkage between a quantitative trait and a marker locus. *Behavior Genetics*, **2**, 3–19.

86. Almasy, L. and Blangero, J. (1998) Multipoint quantitative-trait linkage analysis in general pedigrees. *American Journal of Human Genetics*, **62**, 1198–1211.

87. Amos, C.I. (1994) Robust variance-components approach for assessing genetic linkage in pedigrees. *American Journal of Human Genetics*, **54**, 535–543.

88. Province, M.A., Rice, T.K., Borecki, I.B. *et al.* (2003) Multivariate and multilocus variance components method, based on structural relationships to assess quantitative trait linkage via SEGPATH. *Genetic Epidemiology*, **24**, 128–138.

89. Allison, D.B., Neale, M.C., Zannolli, R. *et al.* (1999) Testing the robustness of the likelihood-ratio test in a variance-component quantitative-trait loci-mapping procedure. *American Journal of Human Genetics*, **65**, 531–544.

90. Kruglyak, L. and Lander, E.S. (1995) Complete multipoint sib-pair analysis of qualitative and quantitative traits. *American Journal of Human Genetics*, **57**, 439–454.

91. Hoffmann, K. and Lindner, T.H. (2005) easyLINKAGE-Plus – automated linkage analyses using large-scale SNP data. *Bioinformatics (Oxford, England)*, **21**, 3565–3567. Describes easyLinkage-Plus, a user-friendly program enabling linkage analysis of either microsatellite or SNP markers.

92. Ge, D., Zhang, K., Need, A.C. *et al.* (2008) WGAViewer: software for genomic annotation of whole genome association studies. *Genome Research*, **18**, 640–643. Describes WGAViewer, a user-friendly software package that provides SNP annotation and visualization for genome-wide association results.

5

RNA Amplification Strategies: Toward Single-Cell Sensitivity

Natalie Stickle[1], Norman N. Iscove[2], Carl Virtanen[1], Mary Barbara[2], Carolyn Modi[1], Toni Di Berardino[3], Ellen Greenblatt[3], Ted Brown[4] and Neil Winegarden[1]

[1] *University Health Network Microarray Centre, Toronto, Ontario, Canada*
[2] *Ontario Cancer Institute, Princess Margaret Hospital, University Health Network, Toronto, Ontario, Canada*
[3] *Mount Sinai Hospital Centre for Fertility and Reproductive Health, Toronto, Ontario, Canada*
[4] *Samuel Lunenfeld Research Institute, Mount Sinai Hospital, Joseph and Wolf Lebovic Centre, Toronto, Ontario, Canada*

5.1 Introduction

Microarray technology has largely become commonplace in genomics research initiatives, and yet, despite the major potential of the technology, there has been debatable success from the many microarray experiments that have thus far been performed. One of the key contributors to the mixed review of microarray technology has been the fact that the limitations of the technology have largely, to date, limited the types of experiment that can be performed. In particular, owing to the fairly high sample requirements for microarray technology, researchers have resorted to the study of bulk, heterogeneous tissues which have obfuscated any meaningful signatures that might have been obtained if more pure populations of cells had been used.

5.1.1 The need for amplification

The first microarray protocols published called for in excess of 10 μg of total RNA and in many of the earliest publications, as much as 100–500 μg of total RNA [1, 2]. Given that a single mammalian cell might contain from 2 to 35 pg of total RNA (Table 5.1), these

Genomics: Essential Methods Edited by Mike Starkey and Ramnath Elaswarapu
© 2011 John Wiley & Sons, Ltd.

Table 5.1 Approximate amounts of total RNA in various cell types.

Cell type	Approximate total RNA per cell (pg)	Reference
COS-7	35	http://www1.qiagen.com/HB/RNeasyMiniKit_EN
Epithelial cells (human)	8–15	http://tools.invitrogen.com/content/sfs/manuals/ 15596018%20pps%20Trizol%20Reagent%20061207.pdf
Fibroblasts (human)	5–7	http://tools.invitrogen.com/content/sfs/manuals/ 15596018%20pps%20Trizol%20Reagent%20061207.pdf
HeLa	12–15	http://www1.qiagen.com/HB/RNeasyMiniKit_EN
		http://www.5prime.com/products/nucleic-acid-purification/ rna-purification/perfectpure-rna-cultured-cell-kit .aspx?details=1
Hep G2	13	http://www.5prime.com/products/nucleic-acid-purification/ rna-purification/perfectpure-rna-cultured-cell-kit .aspx?details=1
Leukocytes (human)	6–11	http://tools.invitrogen.com/content/sfs/manuals/ 15596018%20pps%20Trizol%20Reagent%20061207.pdf
Macrophages (human)	5–25	http://tools.invitrogen.com/content/sfs/manuals/ 15596018%20pps%20Trizol%20Reagent%20061207.pdf
Macrophages (mouse)	1.5–2	http://tools.invitrogen.com/content/sfs/manuals/ 15596018%20pps%20Trizol%20Reagent%20061207.pdf
NIH 3T3	10–19	http://www1.qiagen.com/HB/RNeasyMiniKit_EN
Saccharomyces cerevisiae	2.5	http://www1.qiagen.com/HB/RNeasyMiniKit_EN

requirements for starting material required between 1×10^6 and 3×10^8 cells. As such, many early investigations utilized cell culture in order to obtain sufficient amounts of material. When researchers turned to the use of clinical samples, or tissue from animal models, they were often forced to utilize bulk tissue isolates. As many tissues, and most tumors, are highly heterogeneous, the signals obtained via a microarray were actually an aggregate of all of the individual profiles from the various cell types which made up the heterogeneous tissue sample. It has previously been demonstrated that in mixed cell populations, the dominant cell type completely obscures the profile of any minor cell types [3]. Indeed, Szaniszlo *et al.* showed that, even for cells existing at levels as high as 25% of the total mixture, the dominant cell type would wash out the signal [3].

Additional problems stemmed from the fact that access to bulk tissue was not always practical or possible. Many clinical samples were derived by needle biopsy or fine needle aspirates (FNAs), which simply do not provide enough material to profile on a microarray without amplification. It has been shown that FNAs and core biopsies typically only produce around 2 μg of total RNA, which is at the lower limit for microarray analysis in the absence of amplification [4, 5]. Further, in one study it was shown that the yield from FNAs was highly variable, ranging from 30 000 to 2 580 000 cells [5]. Assersohn *et al.* were only able to process 15% of the total number of patient samples due to the limited RNA obtained from the majority of FNAs.

As a greater understanding of the complexity of diseases, and cancers in particular, has developed it has become increasingly clear that the study of pure cell populations is critical. More researchers are turning to methods such as laser capture microdissection (LCM), fluorescence-activated (assisted) cell sorting (FACS), micromanipulation and paramagnetic bead-based separation. While each of these technologies allows for the purification of more homogeneous cell populations, it is often difficult to obtain completely pure cell populations without resorting to single-cell analysis.

The need for technologies that can assay samples at the single-cell level is becoming increasingly apparent. During development, it has been shown that many cells may look identical morphologically, but that the cells have already started down their path of differentiation and, thus, are showing discrete gene expression profiles [6]. In order to characterize these stages of development fully, it is necessary to look at individual cells. Neurological systems are classical examples of heterogeneous systems, which benefit greatly from the study of single cells [7–9]. In addition, there is increasing interest in rare cell populations, such as stem cells, cancer stem cells, precancerous lesions and circulating tumor cells, often comprising a very small number of cells or even a single cell. Despite the large number of publications which have used microarrays to study various cancers, there have been few true breakthroughs that have resulted. One possible explanation for this is that while the studies conducted to date have contributed greatly to the understanding of cancer biology, they have in fact missed what is perhaps the most important message: that of the cancer stem cell or cancer initiating cell. With a frequency of only 1 in 100 to 1 in 100 000 cells [10–12], clearly the signatures of these cells would be lost among those of the more abundant cell types.

Thus, it is clear that in order to push toward more meaningful analysis of the biology of development, physiology and disease, improvements in the microarray technique that allow for greatly reduced amounts of input material are required. While the sensitivity of the technology itself has been increased somewhat through increased array densities, improved substrates and better detection technologies, the gains in overall sensitivity have been modest at best. The greatest advancements have come from amplification technologies that have in many ways opened the door to analysis of minute RNA samples, including single-cell analysis.

5.1.2 Amplification approaches

The restrictive nature of the relatively high sample size requirement was recognized with the initial development of microarray technology, and companies such as Affymetrix, Agilent and Illumina have all incorporated amplification as a standard part of their procedures. Despite this, the sample requirements still remain relatively high, with between 0.1 and 8 µg of total RNA (10 000–20 000 cells) being required (Table 5.2). Most of these vendors have concentrated on isothermal, linear amplification strategies based on *in vitro* transcription (IVT) reactions. While many researchers heralded these improvements, the sample requirements are still too restrictive for many important biological questions.

Table 5.2 Comparison of various amplification methodologies.

Vendor	Kit	Amp type	Amp method	Total RNA input (ng)	Time required
Affymetrix	One Cycle	Sample	IVT	1000–15 000	1 d
	Two Cycle	Sample	IVT	10–100	2 d
Agilent	Quick Amp	Sample	IVT	200	6 h
Ambion	MessageAmp™ II	Sample	IVT	0.1–20	2 d
	MessageAmp™ III	Sample	IVT	35–500	6 h
	MessageAmp™ Premier	Sample	IVT	20–500	6 h
Clontech	Super SMART™	Sample	PCR	2–150	3 h
Enzo	BioArray Single-Round v2	Sample	IVT	500–5000	1 d
Epicenter	TargetAmp™ 1-Round Biotin 105	Sample	IVT	25–500	6 h
	TargetAmp™ 2 Round Biotin v3	Sample	IVT	0.05–0.1	2 d
	TargetAmp™ aRNA 2.0	Sample	IVT	0.01–0.5	2 d
Genisphere	3DNA™ FlashTag™	Signal	Dendrimer	500–3000	1 h
	RampUp™	Sample	Tandem IVT	1	2 d
	SenseAmp™	Sample	IVT	25–250	1 d
Illumina	TotalPrep™ RNA	Sample	IVT	50	1 d
Molecular Devices	RiboAmp® HS Plus	Sample	IVT	0.1–0.5	1 d
NuGen	Ovation® V2	Sample	RiboSPIA	5–100	4 h
	WT-Ovation®	Sample	RiboSPIA	5–50	4 h
	WT-Ovation® Pico	Sample	RiboSPIA	0.5–50	6 h
Sigma-Aldrich	Transplex™	Sample	PCR	5–50	4 h
System Biosciences	Full Spectrum™	Sample	PCR	20	3 h
PerkinElmer	MICROMAX™ TSA™	Signal	Tyramide deposition	500–1000	2 h
University Health Network	Iscove *et al.*, Global-RT-PCR	Sample	PCR	0.01–0.1	5 h

Sensitivity increases can be gained via one of two basic approaches: amplification of the amount of signal generated on the array or amplification of the amount of starting material (Table 5.2).

Signal amplification methodologies work by bringing additional detection moieties to the hybridization site, either through the use of branched dendrimeric molecules or through enzymatic deposition of fluorophores (Tyramide Signal Amplification (TSA); MICROMAX™ TSA™, PE) [13]). The advantages of these techniques lie in the fact that the original RNA sample does not need to be unduly manipulated, and as such have the potential for less bias due to enzymatic manipulation. In addition, signal amplification techniques tend to be fairly rapid, taking on the order of 1–3 h total. Despite these advantages, signal amplification technologies still are limited by the amount of RNA present in the initial sample (e.g. the dendrimers can only bind to the cDNA generated from the RNA, and so the fewer cDNA molecules there are, then so too the fewer dendrimers). As such, their ability to lower sample requirements tends to be limited to around one order of magnitude.

The greatest improvements in sample reduction are realized through sample (RNA) amplification techniques. RNA amplification methodologies (whether they produce RNA or DNA copies of the original mRNA) tend to fall into three categories: (i) linear amplification methodologies that use T7-mediated transcription (Figure 5.1), (ii) NuGen RiboSPIA™ amplification processes (Figure 5.2) and (iii) exponential amplification processes that incorporate polymerase chain reaction (PCR) methodologies (Figure 5.3) [14–16].

One of the first methodologies applied to the amplification of RNA for microarrays was the T7-mediated approaches (also known as IVT) first described by Eberwine and co-workers and others [7, 9, 17]. This technique has been used extensively in microarray research and is the *de facto* standard for most of the Affymetrix, Agilent and Illumina gene expression arrays. T7-based amplification has several inherent advantages, particularly in that it is has been shown to be reproducible [17–21], is theoretically linear in nature and is not affected by template sequence [16]. There have also been several modifications of the IVT protocol that aim to improve sensitivity, reliability or functionality of the technique [22, 23]. The major drawbacks of the IVT method are that it is time consuming (a single round of amplification takes 1.5–2 days) and there is a limit to its sensitivity. An additional concern is that T7/IVT produces RNA-based products for profiling which are inherently less stable.

Although the initial descriptions of the T7 amplification technique used single neurons to assay gene expression, this was not a global analysis [7]. Several groups have indicated that a single round of T7-based amplification will provide between 1000- and 2000-fold increases in RNA levels [15, 16, 24]. Multiple rounds of amplification can be performed, but usually each subsequent round leads to increased bias and noise [17, 24, 25]. Many groups have reported that, when working with less than 10 ng of total RNA, the results of microarray profiling tend to be less reliable [17, 26, 27]. As such, microarray profiling of single cells using T7-based amplification strategies has not been extensively conducted; however, one report showed that it was possible with four rounds of amplification [28].

While T7/IVT amplification strategies have undoubtedly been the most popular to date for microarray-based gene expression profiling, scientists looking to profile single-cell expression patterns have more often turned toward PCR-based amplification strategies. It has been demonstrated that global cDNA libraries and gene expression profiles could be obtained using PCR-based approaches [29–33]. It has also been shown that these methods can profile RNA samples down to the low picogram range [34]. Since these initial reports, several single-cell microarray profiles have been obtained using these or other PCR-based approaches [6, 35, 36]. PCR-based methodologies afford many advantages: they are quick,

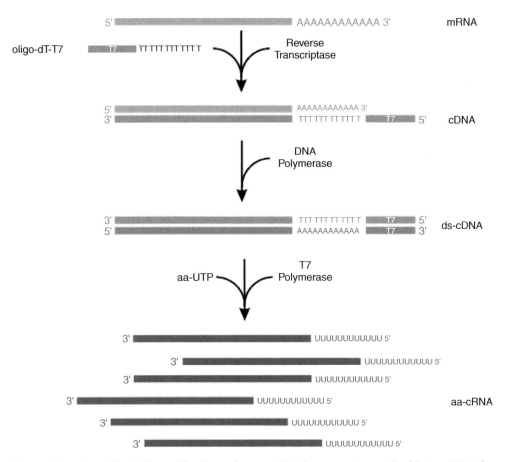

Figure 5.1 Generalized T7-amplification schema. mRNA is reverse transcribed into cDNA. The oligo-dT primer used for RT also has a T7-promoter sequence appended to it on the 5′ end. After second-strand cDNA synthesis is complete, an artificial gene with a T7-promoter is formed. Addition of T7-polymerase results in hundreds to thousands of copies of cRNA being generated. A modified nucleotide such as aminoallyl-UTP can be added during transcription to allow for subsequent chemical labeling with fluorophores.

inexpensive and highly sensitive. PCR amplification also results in DNA-based amplimers which are much more stable than their RNA counterparts. In some embodiments of the methodology, only a small portion of the amplified material is used for hybridization to the microarray and, as such, the technique is also archival. Despite these advantages, adoption of these approaches has been slower than for T7/IVT. Much of the skepticism surrounding PCR amplification for global gene expression stems from concerns that PCR shows GC-content biases, produces double-stranded products and is non-linear in nature [15]. Evolutions of the PCR-based approach have addressed many of these concerns. Some groups have opted to combine both PCR and T7/IVT approaches, which helps balance out some of the problems of both techniques [37–39].

Another approach to amplification is a proprietary technique call RiboSPIA™ from NuGen Technologies. This technique utilizes a chimeric RNA/DNA primer which in the presence of RNase H will displace the previous cDNA strand in order to create another

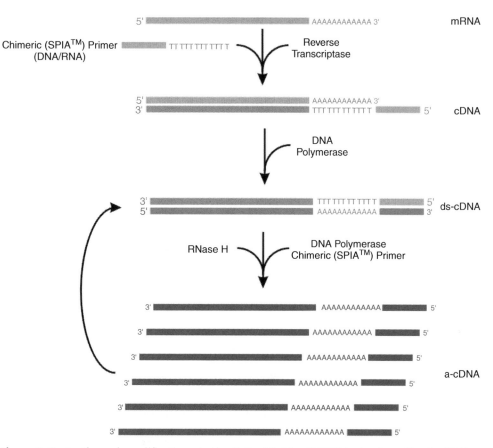

Figure 5.2 Isothermal amplification via the NuGen™ Ribo-SPIA™ technology (Ovation™ System). The NuGen™ Ovation™ system utilizes a special chimeric DNA/RNA primer (the SPIA™ primer) during RT. After second-strand synthesis is completed, RNase H is added, which degrades the RNA portion of the SPIA™ primer. Strand displacement by additional SPIA™ primers and DNA polymerase leads to several thousand-fold amplification creating DNA-based amplimers.

copy of the template (Figure 5.3). This technique is becoming increasingly popular, as it is relatively fast, sensitive to the mid to high picogram range and creates single-stranded DNA molecules which are highly stable [40].

5.1.2.1 Comparison of amplification methodologies

While there have been thousands of papers which use amplification methods for gene expression profiling by microarrays, there have been relatively few direct comparisons of the various methods available. Each technique has its proponents and usually it is the context of the downstream assay that will determine which technique outperforms the other. For example, in some cases it has been shown that IVT is preferable to PCR-based methodologies, particularly as IVT produces longer reads of the original template [21]. However, other reports indicate that PCR-based methods outperform IVT when picogram quantities of RNA are used as the starting template [24]. Thus, if one is using an array for which the arrayed probes target portions of the gene that are more 5′ or whole-gene biased, and

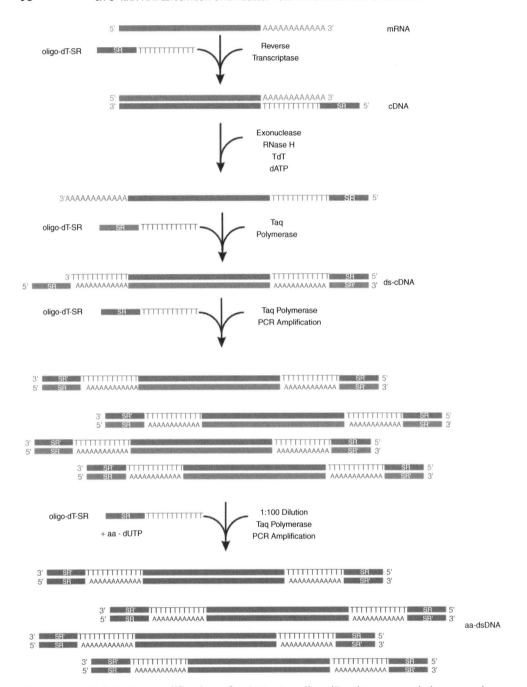

Figure 5.3   Global PCR amplification of mRNA. An oligo-dT primer appended to a unique sequence (SR) is used to reverse transcribe mRNA. Exonuclease treatment removes the excess primer and the cDNA products are tailed using terminal deoxynucleotidyl transferase. The same SR-oligo-dT primer is used to prime a second-strand synthesis and the resultant ds-DNA is then amplified via PCR using the oligo-dT-SR primer.

sufficient RNA is available, then IVT may be preferable. On the other hand, if one has a very small RNA sample, but is using a 3'-biased array, then PCR-based approaches may be more appropriate. Indeed, the techniques can often be altered as well to overcome some of the shortcomings, which are pointed out by some authors. With respect to shorter reads from the original mRNA template, it has been shown that random priming of reverse transcription (RT), rather than oligo-dT based priming, can also be used for PCR-based amplification strategies, thus affording better coverage of the gene [41]. What is important to note is that all amplification methodologies will introduce errors but that if such errors are reproducible then they can be negated, modeled and dealt with at the informatics stage [18, 24]. In our own investigations we have found that the various amplification methods are relatively reproducible, and often highly reflective of a non-amplified sample (Figure 5.4).

Figure 5.4 Reproducibility and reliability of amplification strategies. HeLa and Stratagene Universal Human Reference RNA were run on 19 000-element cDNA microarrays. 10 μg of each of the RNAs was used as a control condition (gray). RNA was then amplified by T7-amplification (blue), NuGen Ovation™ (orange) or Global-RT-PCR (yellow). (See Plate 5.4.)

5.2 Methods and approaches

While several methods for amplification exist, we have opted to focus on two different approaches: T7-based amplification using a commercial kit (MessageAmp™ from Ambion; Protocols 5.1–5.7) and our own in-house derived global-RT-PCR methodology [34]; Protocols 5.8–5.13).

5.2.1 T7 RNA polymerase-based *in vitro* transcription

While it is quite possible to put together a 'home-brew' version of the T7-amplification kits following the numerous publications available, we have chosen a commercial amplification kit that we have had good success with as an example for this chapter. Commercial kits have the advantage of overall quality control of the components as well as technical support for troubleshooting. As such, many researchers prefer to use a commercial kit for complex protocols such as this. The Ambion MessageAmp™ kit has been extensively used in the literature. Here we present only a single round amplification procedure. It has been our experience that if lower amounts of RNA need to be used, the global-RT-PCR procedure is more reliable than a two-round T7-amplification.

PROTOCOL 5.1   First-Strand cDNA Synthesis Using MessageAmp™ Kit (Ambion)

Equipment and reagents

- T7 Oligo(dT) primer (MessageAmp™ kit component, Ambion)
- Reverse transcriptase (MessageAmp™ kit component, Ambion)
- RNase inhibitor (MessageAmp™ kit component, Ambion)
- 10× first-strand buffer (MessageAmp™ kit component, Ambion)
- dNTP mix (MessageAmp™ kit component, Ambion)
- Nuclease-free water (Sigma)
- Thermal cycler set at 70 °C
- Hybridization oven or air incubator set at 42 °C.

Method

1 In a 0.2 ml microcentrifuge tube, combine the following reagents:

   - 100–1000 ng of total RNA^a
   - 1 µl of T7 Oligo(dT) primer
   - nuclease-free water to a final volume of 12 µl.

2 Incubate the reaction mixture at 70 °C for 10 min in a thermal cycler. Centrifuge briefly to collect the sample at the bottom of the tube and place on ice.

---

3 Meanwhile, prepare the RT master mix as follows (volumes shown are for one 20 μl reaction):

- 2 μl of 10× first-strand buffer

- 1 μl of RNase inhibitor

- 4 μl of dNTP mix

- 1 μl of reverse transcriptase.

Mix, centrifuge briefly to collect the reagents at the bottom of the tube and place on ice.

4 Add 8 μl of the RT master mix to each sample from step 2. Mix thoroughly, but gently,[b] and place the tubes at 42 °C for 2 h to allow synthesis of cDNA molecules from the mRNA template.

5 Centrifuge briefly to collect sample to bottom of tube and place tubes on ice.

6 Proceed with second-strand cDNA synthesis according to Protocol 5.2.

Notes

[a]It has been our experience that, when working with this kit, using a minimum of 200 ng of total RNA will improve reliability.

[b]Do not vortex.

PROTOCOL 5.2   Second-Strand cDNA Synthesis Using MessageAmp™ Kit (Ambion)

Equipment and reagents

- 10× second-strand buffer (MessageAmp™ kit component, Ambion)

- dNTP mix (MessageAmp™ kit component, Ambion)

- DNA polymerase (MessageAmp™ kit component, Ambion)

- RNaseH (MessageAmp™ kit component, Ambion)

- Nuclease-free water (Sigma)

- Thermal cycler set at 16 °C.

Method

1 Prepare the second-strand master mix in a nuclease-free tube in the order listed (volumes shown are for a single 100 μl reaction):

- 63 μl of nuclease-free water

- 10 μl of 10× second-strand buffer

- 4 μl of dNTP mix

- 2 μl of DNA polymerase

- 1 μl of RNaseH.[c]

2 Add 80 μl of second-strand master mix to each sample. Mix thoroughly, but gently,[d] and place the tubes at 16 °C for 2 h in a thermal cycler.

3 Proceed with cDNA purification according to Protocol 5.3.

Notes

[c]In order to prevent contamination, it is recommended that reactions featuring RNaseH steps are conducted on a separate bench away from where the first-strand cDNA synthesis and subsequent T7 reactions are carried out.

[d]Do not vortex.

PROTOCOL 5.3   cDNA Purification Using MessageAmp™ Kit (Ambion)

Equipment and reagents

- Nuclease-free water (Sigma) preheated to 50 °C

- cDNA filter cartridge (MessageAmp™ kit component, Ambion)

- Wash tube (MessageAmp™ kit component, Ambion)

- Binding buffer[e] (MessageAmp™ kit component, Ambion)

- American Chemical Society (ACS)-grade 100% (v/v) ethanol

- cDNA wash buffer[f] (MessageAmp™ kit component, Ambion)

- elution tube (MessageAmp™ kit component, Ambion).

Method

1 Equilibrate one cDNA cartridge for each sample by adding 50 μl of cDNA binding buffer to each filter. Incubate filters at room temperature for 5 min.

2 Add 250 μl of cDNA binding buffer to each sample from step 2 in Protocol 5.2. Mix thoroughly by vortexing gently.

3 Transfer the cDNA product and binding buffer (approximately 350 μl) onto the equilibrated cDNA filter cartridge (from step 1).

4 Centrifuge at 10 000g for 1 min.[g] Discard the flow-through and return the cDNA filter cartridge to the wash tube.

5 To each column, add 500 μl of cDNA wash buffer, centrifuge at 10 000g for 1 min and discard the flow-through.

6 Centrifuge the column for an additional 1 min at 10 000*g* to remove any residual ethanol.*[h]*

7 Transfer the column to a new cDNA elution tube and add 9 µl of 50 °C (preheated) nuclease-free water to the center of the membrane.

8 Incubate at room temperature for 2 min and then centrifuge for approximately 1.5 min at 10 000*g*, or until the nuclease-free water is through the filter.

9 Repeat the elution (steps 8 and 9) with an additional 9 µl of nuclease-free water. The double-stranded DNA is now in the eluate. Discard the cDNA filter cartridge.

10 Ensure that the volume of each sample is a minimum of 14 µl.

11 If necessary, add nuclease-free water to increase the volume of each sample to 14 µl. Place on ice.*[i]*

Notes

*[e]*If a precipitate has formed in the cDNA binding buffer, redissolve by heating the solution to 37 °C for up to 10 min and vortexing vigorously. Return the buffer to room temperature before proceeding.

*[f]*Prior to using the cDNA wash buffer for the first time, add 11.2 ml of 100% (v/v) ethanol to the bottle. Mix well and indicate that ethanol was added by marking the bottle label.

*[g]*If all the cDNA and wash buffer solution has not passed through the filter cartridge, perform additional centrifugations until all the mixture has passed through the column.

*[h]*Trace amounts of ethanol left on the column may interfere with downstream reactions.

*[i]*Purified double-stranded DNA samples are stable for several days at −20 °C, providing a convenient stopping place for users wishing to pause and resume subsequent steps at a later time.

PROTOCOL 5.4   *In Vitro* Transcription (IVT) Synthesis of Aminoallyl-Containing aRNA Using MessageAmp™ Kit (Ambion)

Equipment and reagents

- Air incubator*[j]* set at 37 °C

- ATP, CTP, GTP mix (25 mM each ; MessageAmp™ kit component, Ambion)

- Aminoallyl-UTP (aa-UTP, 50 mM ; MessageAmp™ kit component, Ambion)

- UTP solution (50 mM; MessageAmp™ kit component, Ambion)

- T7 10× reaction buffer (MessageAmp™ kit component, Ambion)

- T7 enzyme mix (MessageAmp™ kit component, Ambion)

- DNase I (MessageAmp™ kit component, Ambion).

Method

1 Prepare an IVT master mix (for each cDNA sample) by combining (at room temperature) the following reagents in this order:

   • 3 μl of aa-UTP (50 mM)

   • 12 μl of ATP, CTP, GTP Mix (25 mM)

   • 3 μl of UTP solution (50 mM)

   • 4 μl of T7 10× reaction buffer

   • 4 μl of T7 enzyme mix.

2 Mix gently by pipetting and centrifuge briefly to collect the reaction mixture at the bottom of the tube.

3 Transfer 26 μl of the IVT master mix to a 14 μl cDNA sample (prepared in Protocol 5.3). Mix thoroughly by flicking the tube and then centrifuge briefly to collect the reaction mixture at the bottom of the tube.

4 Incubate the reaction at 37 °C for 14 h in an air incubator.[j]

5 Add 2 μl of DNase I to each reaction. Gently mix and centrifuge briefly to collect the reaction mixture at the bottom of each tube.[k]

6 Incubate at 37 °C for 30 min.

7 Add 58 μl of nuclease-free water to each aRNA sample to adjust the final volume to 100 μl, and mix each reaction mixture thoroughly, but gently.[l]

Notes

[j]It is important to use an air incubator to prevent condensation formation in the tube cap. Formation of condensation would alter the concentration of reactants and could result in a reduced yield.

[k]Although the Ambion protocol suggests this is an optional step, we recommend always performing it as it improves reproducibility.

[l]aRNA samples are stable for several days at −20 °C, providing a convenient stopping place for users wishing to pause and resume subsequent steps at a later time.

PROTOCOL 5.5   aRNA Purification Using MessageAmp™ Kit (Ambion)

Equipment and reagents

• aRNA binding buffer (MessageAmp™ kit component, Ambion)

• ACS-grade 100% (v/v) ethanol

• aRNA filter cartridge (MessageAmp™ kit component, Ambion)

- aRNA collection tube (MessageAmp™ kit component, Ambion)
- aRNA wash buffer with ethanol added according to manufacturer's directions (MessageAmp™ kit component, Ambion)
- Nuclease-free water (Sigma) preheated to 50 °C
- Vacuum evaporator (SpeedVac; ThermoElectron)
- UV–V is spectrophotometer.

Method

1 Add 350 µl of aRNA binding buffer to an aRNA sample. Mix thoroughly but gently.

2 Add 250 µl of 100% (v/v) ethanol.[m]

3 Place an aRNA filter cartridge in an aRNA collection tube and transfer the sample mixture from step 2 onto the center of a filter.

4 Centrifuge at $10\,000g$ for 1 min.[n] Discard the flow-through and replace the aRNA filter cartridge in the aRNA collection tube.

5 Add 650 µl of aRNA wash buffer to the aRNA filter cartridge.

6 Centrifuge at $10\,000g$ for 1 min.[n] Discard the flow-through and replace the aRNA filter cartridge in the aRNA collection tube. Centrifuge at $10\,000g$ for an additional 1 min to ensure that all trace amounts of ethanol are removed.

7 Transfer the filter cartridge to a fresh aRNA collection tube.

8 Add 50 µl of 50 °C (pre-warmed) nuclease-free water[o] to the center of the aRNA filter cartridge membrane.

9 Incubate at room temperature for 2 min and then centrifuge at $10,000g$ for 1.5 min.

10 Repeat steps 8 and 9 with an additional 50 µl of nuclease-free water.

11 Determine the aRNA concentration by measuring the absorbance at 260 nm using a UV–vis spectrophotometer.

12 Aliquot the volume that contains 5 µg of aRNA into a new tube, and evaporate the 5 µg sample to dryness using a vacuum centrifugation on low-heat setting, taking care not to dry the sample excessively.[p] Resuspend the purified sample in 7 µl of nuclease-free water.

Notes

[m] Proceed immediately to step 3 after the ethanol is mixed with the aRNA sample, as once this occurs the aRNA will enter a semi-precipitated state and any delay may result in some loss of the sample.

[n] Additional centrifugation may be required if all the sample does not pass through the filter after the initial centrifugation. Continue centrifugation until the entire sample has passed through the filter.

[o] Maintain the nuclease-free water at 50 °C for the second elution in step 10.

[p] The remainder of the sample can be stored at −20 °C.

PROTOCOL 5.6   Coupling of Monofunctional Reactive Dyes to Aminoallyl-Containing aRNAs

Equipment and reagents

- Dimethyl sulfoxide (DMSO)

- Cyanine 5- and cyanine 3-NHS ester dyes (Enzo Life Sciences) *or* Alexa 647 and Alexa 555 monofunctional reactive dyes (Invitrogen) (dyes are individually packaged in vials for single reactions)

- 4 M hydroxylamine (MessageAmp™ kit component, Ambion)

- Coupling buffer (MessageAmp™ kit component, Ambion).

Method

1 For each aRNA sample, resuspend a tube of dye in 3 µl of DMSO immediately prior to use. Vortex to completely redissolve the dye and centrifuge briefly to collect the dye at the bottom of the tube.

2 To each purified 7 µl aRNA sample (from step 12 of Protocol 5.5), add 3 µl of cyanine or Alexa dye. Add 9 µl of coupling buffer. Mix by pipetting up and down, and incubate in the dark for 30 min at room temperature to allow chemical coupling between dye molecules and aminoallyl groups on the aRNA.

3 After the 30 min coupling reaction, quench the reaction by adding 4.5 µl of 4 M hydroxylamine and incubating at room temperature for 15 min in the dark.

PROTOCOL 5.7   Purification of Fluorescently Labeled aRNA

Equipment and reagents

- aRNA binding buffer (MessageAmp™ kit component, Ambion)

- ACS-grade 100% (v/v) ethanol

- aRNA filter cartridge (MessageAmp™ kit component, Ambion)

- aRNA collection tube (MessageAmp™ kit component, Ambion)

- aRNA wash buffer with ethanol added according to manufacturer's directions (MessageAmp™ kit component, Ambion)

- Nuclease-free water (Sigma) prewarm to 50 °C

- Vacuum evaporator (SpeedVac; ThermoElectron)

- UV–vis spectrophotometer.

Method

1 Add 73.5 µl of nuclease-free water to a dye-labeled aRNA sample (from step 3 of Protocol 5.6) so that each sample has a total volume of 100 µl.

2 For each aRNA sample, add 350 µl of aRNA binding buffer. Mix thoroughly, but gently.

3 Add 250 µl of 100% (v/v) ethanol.[q] Mix thoroughly, but gently.

4 Place an aRNA filter cartridge in an aRNA collection tube and transfer the sample mixture from step 2 onto the filter cartridge.

5 Centrifuge at 10 000g for 1 min.[r] Discard the flow-through and replace the aRNA filter cartridge in the aRNA collection tube.

6 Add 650 µl of aRNA wash buffer to the aRNA filter cartridge.

7 Centrifuge at 10 000g for 1 min.[r] Discard the flow-through and replace the aRNA filter cartridge in the aRNA collection tube. Centrifuge at 10 000g for an additional 1 min to ensure that all trace amounts of ethanol are removed.

8 Transfer the filter cartridge to a fresh aRNA collection tube.

9 Add 50 µl of 50 °C (pre-warmed) nuclease-free water[s] to the center of the aRNA filter cartridge membrane.

10 Incubate at room temperature for 2 min and then centrifuge at 10 000g for 1.5 min.

11 Repeat steps 8 and 9 with an additional 50 µl of nuclease-free water.[t]

12 Evaporate each aRNA sample to dryness using a vacuum centrifugation on low-heat setting, taking care not to dry the sample excessively.[u]

Notes

[q]Proceed immediately to step 3 after the ethanol is mixed with the aRNA sample, as once this occurs the aRNA will enter a semi-precipitated state and any delay may result in some loss of the sample.

[r]Additional centrifugation may be required if all the sample does not pass through the filter after the initial centrifugation. Continue centrifugation until the entire sample has passed through the filter.

[s]Maintain the nuclease-free water at 50 °C for the second elution in step 10.

[t]Determine the aRNA concentration by measuring the absorbance at 260 nm using a UV–vis spectrophotometer (optional).

[u]The reason for drying down the sample at this point is to allow for you to resuspend in the appropriate volume required for your downstream application (hybridization). Following purification, samples that will be co-hybridized can be combined.

5.2.2 Global-RT-PCR

The global-RT-PCR procedure we present has several advantages. First and foremost, it is extremely sensitive. We have now used this protocol extensively for the profile of single cells and have found it to be highly reliable and robust (Figure 5.5). The protocol is also very quick. T7-based amplification procedures generally take upwards of 2 days to complete; and when working with very small samples, multiple rounds of IVT are required. By contrast, the global-RT-PCR procedure can be completed in less than 8 h. Global-RT-PCR results in DNA-based products, which are more stable, compared with the RNA products produced as a result of the T7-reaction. In fact, in most cases, the samples can be archived after

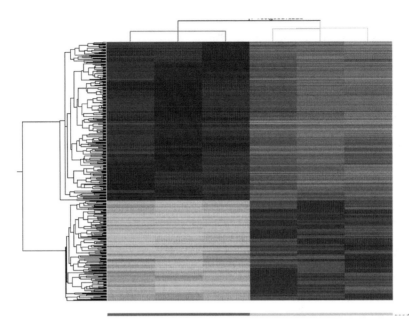

Figure 5.5 Single-cell profiling by Global-RT-PCR. Three individual cells from each of two groups were obtained and the RNA was amplified and profiled on Agilent 44k Whole Human Genome arrays. After a *t*-test was performed to identify a list of 358 genes which distinguished between the two groups, the gene panel was subjected to hierarchical clustering. Reproducible results from each of the cells used in each of the groups were obtained and provided a strong identifier panel. (See Plate 5.5.)

the first round of global-RT-PCR, providing an excellent resource for validation either by downstream quantitative PCR or use on another array type. A final advantage is that because the method is 'home-brewed' it is cost effective. The reaction components are inexpensive and the entire reaction can be run for under $50. One of the key developments that make the global-RT-PCR procedure possible is the ability to control the length of the cDNAs generated during the RT reaction. This is accomplished by running a very short (5 min) RT reaction compared with the 2 h used in the T7 amplification [31, 32, 34]. By keeping this step short, the average length of the products that result fall within a narrow range, on the order of only 300 bp in length. This ensures that, during the PCR step, each template is processed with similar efficiency, leading to improved reproducibility and preserving the abundance relationships of the original RNA pool.

PROTOCOL 5.8   Reverse Transcription

Equipment and reagents

- Lysis buffer (52 mM Tris-HCl, pH 8.3, 78 mM KCl, 3.1 mM MgCl$_2$, 0.52% (v/v) NP-40)
- SUPERase-In (Ambion)

- 100× acetylated bovine serum albumin (BSA) (dilute ultrapure acetylated BSA (Life Technologies) to 10 mg/ml)

- Reverse transcriptase and included reagents (200 U/μl SuperScript III, 0.1 M dithiothreitol (DTT); Life Technologies)

- SR-T24 primer (5'-GTT AAC TCG AGA ATT CTT TTT TTT TTT TTT TTT TTT TTTT-3', 90.4 mM)

- 4× 25 mM dNTPs (combine equal amounts of each 100 mM PCR-grade dNTP; Invitrogen)

- RNase H (Life Technologies)

- Exonuclease I and 10× exonuclease I buffer (Fermentas)

- 75 mM $MgCl_2$

- Terminal deoxynucleotidyl transferase (TdT) and included reagents (25 mM $CoCl_2$, 5× TdT buffer; Roche)

- Nuclease-free water (Sigma)

- 2× tailing buffer (80 μl of 5× TdT buffer, 3 μl of 100 mM dATP, 24 μl of 25 mM $CoCl_2$, 93 μl of nuclease-free water)

- Thermal cycler.

Method

1 Prepare first-strand buffer by combining:

- 94 μl of lysis buffer

- 5 μl of SUPERase-In

- 0.43 μl of 4× 25 mM dNTPs

- 1 μl of 100× acetylated BSA

- 1 μl of 0.1 M DTT

- 0.5 μl of 3.68 mM SR-T24.

2 In a 0.2 ml microcentrifuge tube, combine the following reagents:

- single cell[v]

- 4 μl of first-strand buffer

- 0.5 μl of SuperScript III.[w]

3 Run the following PCR program:

- 65 °C, 1 min 30 s

- 50 °C, 5 min

- 70 °C, 10 min.

4 Combine 1 μl of 10× exonuclease I buffer and 1 μl of exonuclease I with 8 μl of nuclease-free water. Add 1 μl to the RT reaction mixture. Incubate in the thermal cycler at 37 °C for 15 min, followed by 15 min at 80 °C.

5 Add 0.7 μl of 75 mM MgCl$_2$ and 0.5 μl of RNase H. Mix well and incubate at 37 °C for 15 min.

6 Add 6.5 μl of 2× tailing buffer and 0.7 μl of TdT. Mix well and incubate at 37 °C for 15 min. Heat inactivate the TdT by incubating at 65 °C for 10 min.

Notes

^{v}In addition to a single cell, this reaction can be performed on 10–1000 pg of total RNA in a volume of 0.5 μl. The final volume of this reaction should be 5 μl.

^{w}It is important to use SuperScript III, as this enzyme has been engineered to be stable at 50 °C. The increased temperature used during the RT helps to reduce secondary structure of the RNA, providing a more reliable representation of the entire mRNA pool.

PROTOCOL 5.9   Second-Strand cDNA Synthesis and PCR Amplification

Equipment and reagents

- Recombinant Taq DNA polymerase and included reagents (Taq DNA polymerase, 10× Taq buffer, 50 mM MgCl$_2$; Life Technologies)

- SR-T24 primer (5′-GTT AAC TCG AGA ATT CTT TTT TTT TTT TTT TTT TTT TTT T-3′; 425 mM)

- 4× 25 mM dNTPs (combine equal amounts of each 100 mM PCR-grade dNTP; Life Technologies)

- 75 mM MgCl$_2$

- Nuclease-free water (Sigma)

- Thermal cycler.

Method

1 Prepare a PCR master mix by combining (for each RT product):

- 2 μl of 10× Taq buffer

- 0.6 μl of 50 mM MgCl$_2$

- 0.7 μl of 4× 25 mM dNTPs

- 1 μl of SR-T24 primer (425 mM)

- 10.7 μl of nuclease-free water.

2 In a new 0.2 ml microcentrifuge tube, mix 15 µl of PCR master mix with 4 µl of RT product (from step 5 of Protocol 5.[x]) and 0.5 µl of Taq DNA polymerase.

3 To synthesize cDNA second strands, incubate in a thermal cycler as follows:

- 94 °C, 15 s

- 50 °C, 2 min

- 72 °C, 2 min.

4 To PCR amplify the double-stranded DNA, incubate for 35 cycles of[y]:

- 94 °C, 15 s

- 60 °C, 30 s

- 72 °C, 2 min.

5 In a new 0.2 ml microcentrifuge tube, combine 15 µl of fresh PCR master mix (prepared as described in step 1) with 3 µl of product from step 4. Add 1.2 µl of 75 mM $MgCl_2$ and 0.5 µl of Taq DNA polymerase. Complete a second round of five cycles of amplification, as follows:

- 94 °C, 15 s

- 60 °C, 30 s

- 72 °C, 2 min.

6 Check the size of the PCR product[z,aa] by 1.8% (w/v) agarose gel electrophoresis.

Notes

[x]If the RT is performed on a single cell, or less than 20 pg of total RNA, then 4 µl of the RT product are added to each of three tubes containing PCR master mix, so that nearly all of the RT reaction is carried forward to amplification. This is done to increase the chances that genes that are not highly expressed will be sufficiently amplified to maintain their representation when starting with a small pool of RNA.

[y]Create one PCR program to complete both the second-strand cDNA synthesis (step 3) and the PCR amplification of double-stranded DNA (step 4).

[z]The PCR product should appear as a smear between about 200 and 400 bp.

[aa]Store the double-stranded PCR product at −20 °C.

PROTOCOL 5.10   Aminoallyl Incorporation

Equipment and reagents

- Recombinant Taq DNA polymerase and included reagents (Taq DNA polymerase, 10× Taq buffer, 50 mM $MgCl_2$; Life Technologies)

- SR-T24 primer (5'-GTT AAC TCG AGA ATT CTT TTT TTT TTT TTT TTT TTT TTT T-3'; 7 µM)

- 2 mM dNTP mix (1 μl of 100 mM dGTP, 1 μl of 100 mM dCTP, 1 μl of 100 mM dATP and 47 μl of nuclease-free water) (100 mM PCR-grade dNTPs; Life Technologies)

- 1 mM dTTP (dilute from 100 mM PCR-grade dNTP; Life Technologies)

- 2 mM aminoallyl-dUTP (dilute from 50 mM stock; Enzo Life Sciences)

- Nuclease-free water (Sigma)

- Thermal cycler.

Method

1 Prepare a PCR master mix by combining (for each PCR product):

- 10 μl of 10× Taq buffer

- 10 μl of 2 mM dNTP mix

- 10 μl of 1 mM dTTP

- 10 μl of 2 mM aminoallyl-dUTP

- 1 μl of SR-T24 primer (7 μM)

- 3 μl of 50 mM MgCl$_2$

- 0.5 μl of Taq DNA polymerase

- 54.5 μl of nuclease-free water.

2 Dilute 1 μl of amplified double-stranded DNA PCR product (from step 5 of Protocol 5.11) 1 : 100 in nuclease-free water.

3 In a new 0.2 ml microcentrifuge tube, mix 1 μl of the diluted double-stranded DNA PCR product with 99 μl of PCR master mix. Incubate for 25 cycles of:

- 94 °C, 15 s[bb]

- 60 °C, 30 s

- 72 °C, 1 min.

Notes

[bb]Incubate reactions at 94 °C for 2 min before beginning PCR amplification if the double-stranded DNA PCR product has been stored at −20 °C prior to beginning aminoallyl incorporation.

PROTOCOL 5.11　Aminoallyl Double-Stranded DNA Purification

Equipment and reagents

- Illustra™ CyScribe™ GFX™ purification kit (GE Healthcare Life Sciences)

- 80% (v/v) ACS-grade ethanol

- 17 mM sodium bicarbonate, pH 9.0
- Vacuum evaporator (SpeedVac, ThermoElectron).

Method

1 Add 500 µl of capture buffer (GFX™ purification kit) to a GFX™ column.

2 Transfer a labeled double-stranded DNA PCR product (from step 3 of Protocol 5.12) to the column and pipette up and down several times to mix the DNA with the capture buffer.

3 Centrifuge the column at 13 800g for 30 s and discard the flow-through.

4 Add 600 µl of 80% (v/v) ethanol and centrifuge at 13 800g for 30 s, discarding the flow-through.

5 Repeat step 4 twice, for a total of three times.

6 Centrifuge the column at 13 800g for an additional 30 s to ensure that all the ethanol is removed.

7 Transfer the GFX column to a fresh tube and add 60 µl of 17 mM sodium bicarbonate, pH 9.[cc]

8 Incubate the GFX column at room temperature for 1 min.

9 Centrifuge at 13 800g for 1 min to elute purified, labeled double-stranded DNA.[dd]

10 Evaporate the double-stranded DNA sample to complete dryness using a vacuum centrifugation on high-heat setting, taking care not to over dry the sample. Resuspend the purified sample in 7 µl of nuclease-free water.

Notes

[cc]It is crucial that the elution buffer covers the membrane so that when the purified DNA is dried down, resuspended in 7 µl water, and added to 3 µl dye/DMSO, the final sodium bicarbonate concentration is 0.1 M in the 10 µl dye conjugation reaction.

[dd]If you wish to stop the protocol at any point and resume the next day you can do so after this purification step. Simply freeze the aminoallyl double-stranded DNA at −20 °C. The material is stable for several days.

PROTOCOL 5.12 Coupling of Monofunctional Reactive Dyes to Aminoallyl-Containing cDNA

Equipment and reagents

- DMSO

- Cyanine 5- and cyanine 3-NHS ester dyes (Enzo Life Sciences) *or* Alexa 647 and Alexa 555 monofunctional reactive dyes (Invitrogen) (dyes are individually packaged in vials for single reactions)

- 4 M hydroxylamine.

Method

1 For each cDNA sample, resuspend a tube of dye in 3 µl of DMSO immediately prior to use. Vortex to completely redissolve the dye and centrifuge briefly to collect the dye at the bottom of the tube.

2 To each purified 7 µl cDNA sample (from step 10 of Protocol 5.11), add 3 µl of cyanine, or Alexa, dye. Mix by pipetting up and down, and incubate in the dark for 60 min at room temperature to allow chemical coupling between dye molecules and aminoallyl groups on the cDNA.

3 After the 60 min coupling reaction, quench the reaction by adding 10 µl of 4 M hydroxylamine and incubating at room temperature for 15 min in the dark.

PROTOCOL 5.13   Purification of Fluorescently Labeled Double-Stranded DNA

Equipment and reagents

- Illustra™ CyScribe™ GFX™ purification kit (GE Healthcare Life Sciences)

- 80% (v/v) ACS-grade ethanol

- Nuclease-free water (Sigma)

- Vacuum evaporator (SpeedVac; ThermoElectron).

Method

1 Add 80 µl of nuclease-free water to each labeled double-stranded cDNA (from step 6 of Protocol 5.12) to adjust the volume of each cDNA sample to about 100 µl.

2 Add 500 µl of capture buffer (from GFX™ purification kit) to each GFX™ column.

3 Transfer the labeled double-stranded cDNAs (100 µl) to the column, pipetting up and down several times to mix.

4 Centrifuge each column at 13 800*g* for 30 s and discard the flow-through.

5 To each column, add 600 µl of 80% (v/v) ethanol. Centrifuge at 13 800*g* for 30 s and discard the flow-through.

6 Repeat step 5 twice more, for a total of three washes.

7 Centrifuge each column at 13 800*g* for an additional 30 s to remove any residual ethanol.

8 Transfer each GFX™ column to a fresh tube and add 60 µl of elution buffer (GFX™ purification kit).

9 Incubate each GFX™ column at room temperature for 1 min to render the cDNA soluble.

10 Centrifuge each column at 13 800*g* for 1 min to elute the purified labeled cDNA.*ee*

11 Evaporate each cDNA sample to dryness using a vacuum centrifugation on high-heat setting, taking care not to dry the sample excessively.*ff*

Notes

*ee*If you wish to stop the protocol at any point and resume the next day you can do so after this purification step. Simply freeze the fluor-labeled double-stranded cDNA at −20 °C. The material is stable for several days.

*ff*This allows you to bring the sample up in whatever volume is appropriate for your hybridization step based on the platform you are using.

5.3 Troubleshooting

- **Underrepresentation of RNA species by T7-amplification method**: Owing to the higher order secondary and tertiary structure of many mRNA species, it is not uncommon for some RNAs to be underrepresented in the T7-amplification method. This is not necessarily a problem that is unique to amplification; however, the effect tends to be magnified under such conditions. The effect is most prominent in the initial cDNA production step (during RT [42]). It is possible to mitigate the effects of secondary structures in RNA by using SuperScript III rather than SuperScript II and performing the RT reaction at 50 °C for more robust cDNA production.

- **Underrepresentation of RNA species by global-RT-PCR method**: The global-RT-PCR method uses SuperScript III at 50 °C for the RT step and, as such, the underrepresentation of certain RNAs is generally not attributable to higher order secondary and tertiary structures. In the global-RT-PCR reaction, the ability to control the amplimer length leads to a highly robust and reproducible technique; however, this does lead to an extreme 3′ bias which may not be compatible with all array types (probes which are much beyond 300 bp from the polyA tail will be less reliable). It is possible to substitute random primers rather than oligo-dT to provide better coverage of the entire gene, but we have not yet optimized such a procedure. Another option is to use arrays with extreme 3′-bias, such as the X3P arrays from Affymetrix (this would require altering the protocol to incorporate biotin rather than fluorescent moieties). This protocol also works well with cDNA arrays, which are less common now, but are by nature almost always inclusive of the 3′ end of the gene.

- **Low yield from T7-amplification method**: The RNA being produced during the IVT reaction may not be very stable at the optimal temperature for the T7 polymerase. It has been shown that significant RNA degradation can be detected after 4 h of IVT [43]. Interestingly, the impact of this finding has not been fully realized, as most of the commercial kits (including those for the Affymetrix and Agilent platforms) still recommend relatively long (6 h to overnight) incubations during IVT. If, however, it appears that RNA quality is a problem or that longer reactions are actually providing a diminished yield of aRNA, then it may be beneficial to try shorter IVT reactions of less than 4 h. The result will be decreased sensitivity, which may require additional rounds of T7-based amplification.

- **Data does not seem comparable to non-amplified data of the same cell type**: One issue that is critical to the success of any amplification procedure is that all samples must be treated the same [20, 44]. In order to ensure success, researchers should look to determine which samples that will be included in the study are the smallest. The appropriate amplification strategy for that sample should be determined and that method should be used to treat all samples regardless of whether some samples could be used without amplification or with fewer rounds of amplification. Sample input amounts should also be normalized. Most of the amplification procedures have an optimal range of input material, and will change slightly with the amount of RNA input into the reactions. When using a two-color array system, it is critical that both samples be treated the same as well. Thus, if using a reference RNA (such as those available from Stratagene, Clontech, ArrayIt), this reference RNA must also be amplified. This is a critical step. Each amplification procedure introduces some inherent bias; however, for good methods this bias is reproducible. As such, by amplifying both the reference and the experimental sample, the bias is equally applied to each sample, and when looking at ratiometric data, it is largely canceled out, greatly improving the reliability and usefulness of the technique.

References

1. Schena, M., Shalon, D., Davis, R.W. and Brown, P.O. (1995) Quantitative monitoring of gene expression patterns with a complementary DNA microarray. *Science*, **270**, 467–470.

2. Schena, M., Shalon, D., Heller, R. *et al.* (1996) Parallel human genome analysis: microarray-based expression monitoring of 1000 genes. *Proceedings of the National Academy of Sciences of the United States of America*, **93**, 10614–10619.

3. Szaniszlo, P., Wang, N., Sinha, M. *et al.* (2004) Getting the right cells to the array: gene expression microarray analysis of cell mixtures and sorted cells. *Cytometry A*, **59**, 191–202.

4. Symmans, W.F., Ayers, M., Clark, E.A. *et al.* (2003) Total RNA yield and microarray gene expression profiles from fine-needle aspiration biopsy and core-needle biopsy samples of breast carcinoma. *Cancer*, **97**, 2960–2971.

5. Assersohn, L., Gangi, L., Zhao, Y. *et al.* (2002) The feasibility of using fine needle aspiration from primary breast cancers for cDNA microarray analyses. *Clinical Cancer Research*, **8**, 794–801.

6. Chiang, M.K. and Melton, D.A. (2003) Single-cell transcript analysis of pancreas development. *Developmental Cell*, **4**, 383–393.

7. Eberwine, J., Yeh, H., Miyashiro, K. *et al.* (1992) Analysis of gene expression in single live neurons. *Proceedings of the National Academy of Sciences of the United States of America*, **89**, 3010–3014. This is the first time that the ability to measure transcript levels from a single cell was presented. The T7-amplification method is often referred to as Eberwine amplification because of this paper.

8. Kamme, F., Salunga, R., Yu, J. *et al.* (2003) Single-cell microarray analysis in hippocampus CA1: demonstration and validation of cellular heterogeneity. *Journal of Neuroscience*, **23**, 3607–3615.

9. Van Gelder, R.N., von Zastrow, M.E., Yool, A.A. *et al.* (1990) Amplified RNA synthesized from limited quantities of heterogeneous cDNA. *Proceedings of the National Academy of Sciences of the United States of America*, **87**, 1663–1667.

10. Siminovitch, L., Mcculloch, E.A. and Till, J.E. (1963) The distribution of colony-forming cells among spleen colonies. *Journal of Cell Physiology*, **62**, 327–336.

11. Schneider, T.E., Barland, C., Alex, A.M. *et al.* (2003) Measuring stem cell frequency in epidermis: a quantitative in vivo functional assay for long-term repopulating cells. *Proceedings of the National Academy of Sciences of the United States of America*, **100**, 11412–11417.

12. Wang, J.C. and Dick, J. (2005) Cancer stem cells: lessons from leukemia. *Trends in Cell Biology*, **15**, 494–501.

13. Badiee, A., Eiken, H.G., Steen, V.M. and Løvlie, R. (2003) Evaluation of five different cDNA labeling methods for microarrays using spike controls. *BMC Biotechnology*, **3**, 23.

14. Nygaard, V. and Hovig, E. (2006) Options available for profiling small samples: a review of sample amplification technology when combined with microarray profiling. *Nucleic Acids Research*, **34**, 996–1014. A helpful review of different amplification methods available for microarray analysis.

15. Glanzer, J.G. and Eberwine, J.H. (2004) Expression profiling of small cellular samples in cancer: less is more. *British Journal of Cancer*, **90**, 1111–1114. An excellent review on the need for and methods of amplification in gene expression studies.

16. Kawasaki, E. (2004) Microarrays and the gene expression profile of a single cell. *Annals of the New York Academy of Sciences*, **1020**, 92–100.

17. Wang, E., Miller, L.D., Ohnmacht, G.A. *et al.* (2000) High-fidelity mRNA amplification for gene profiling. *Nature Biotechnology*, **18**, 457–459. One of the first examples showing the utility of T7-amplification methods for spotted DNA arrays.

18. Puskás, L.G., Zvara, A., Hackler, L. and Van Hummelen, P.P (2002) RNA amplification results in reproducible microarray data with slight ratio bias. *BioTechniques*, **32**, 1330–1334, 1336, 1338, 1340.

19. Saghizadeh, M., Brown, D.J., Tajbakhsh, J. *et al.* (2003) Evaluation of techniques using amplified nucleic acid probes for gene expression profiling. *Biomolecular Engineering*, **20**, 97–106.

20. Schneider, J., Buness, A., Huber, W. *et al.* (2004) Systematic analysis of T7 RNA polymerase based *in vitro* linear RNA amplification for use in microarray experiments. *BMC Genomics*, **5**, 29.

21. Wadenbäck, J., Clapham, D., Craig, D. *et al.* (2005) Comparison of standard exponential and linear techniques to amplify small cDNA samples for microarrays. *BMC Genomics*, **6**, 61.

22. Schlingemann, J., Thuerigen, O., Ittrich, C. *et al.* (2005) Effective transcriptome amplification for expression profiling on sense-oriented oligonucleotide microarrays. *Nucleic Acids Research*, **33**, e29.

23. Shearstone, J., Allaire, N.E., Campos-Rivera, J. *et al.* (2006) Accurate and precise transcriptional profiles from 50 pg of total RNA or 100 flow-sorted primary lymphocytes. *Genomics*, **88**, 111–121.

24. Subkhankulova, T. and Livesey, F.J. (2006) Comparative evaluation of linear and exponential amplification techniques for expression profiling at the single-cell level. *Genome Biology*, **7**, R18.

25. Wilson, C.L., Pepper, S.D., Hey, Y. and Miller, C.J. (2004) Amplification protocols introduce systematic but reproducible errors into gene expression studies. *BioTechniques*, **36**, 498–506.

26. Baugh, L.R., Hill, A.A., Brown, E.L. and Hunter, C.P. (2001) Quantitative analysis of mRNA amplification by in vitro transcription. *Nucleic Acids Research*, **29**, E29.

27. Choesmel, V., Foucault, F., Thiery, J.P. and Blin, N. (2004) Design of a real time quantitative PCR assay to assess global mRNA amplification of small size specimens for microarray hybridisation. *Journal of Clinical Pathology*, **57**, 1278–1287.

28. Seshi, B., Kumar, S. and King, D. (2003) Multilineage gene expression in human bone marrow stromal cells as evidenced by single-cell microarray analysis. *Blood Cells, Molecules and Diseases*, **31**, 268–285.

29. Billia, F., Barbara, M., McEwen, J. *et al.* (2001) Resolution of pluripotential intermediates in murine hematopoietic differentiation by global complementary DNA amplification from single cells: confirmation of assignments by expression profiling of cytokine receptor transcripts. *Blood*, **97**, 2257–2268.

30. Brady, G., Barbara, M. and Iscove, N.N. (1990) Representative *in vitro* cDNA amplification from individual hematopoietic cells and colonies. *Methods in Molecular and Cellular Biology*, **2**, 17–25. The first demonstration of our global-RT-PCR approach to the study of transcripts from single cells.

31. Brady, G., Billia, F., Knox, J. *et al.* (1995) Analysis of gene expression in a complex differentiation hierarchy by global amplification of cDNA from single cells. *Current Biology*, **5**, 909–922. Application of the global-RT-PCR method to the study of heterogeneous populations of single cells.

32. Brady, G. and Iscove, N.N. (1993) Construction of cDNA libraries from single cells. *Methods in Enzymology*, **225**, 611–623. A detailed account of the utility of the global-RT-PCR method.

33. Dulac, C. and Axel, R. (1995) A novel family of genes encoding putative pheromone receptors in mammals. *Cell*, **83**, 195–206. Generation of single cell source cDNA libraries and downstream methods.

34. Iscove, N.N., Barbara, M., Gu, M. *et al.* (2002) Representation is faithfully preserved in global cDNA amplified exponentially from sub-picogram quantities of mRNA. *Nature Biotechnology*, **20**, 940–943. The first application of our global-RT-PCR methodology for microarray-based gene expression analysis.

35. Tietjen, I., Rihel, J.M., Cao, Y. *et al.* (2003) Single-cell transcriptional analysis of neuronal progenitors. *Neuron*, **38**, 161–175. Application of a global-RT-PCR strategy to gene expression analysis using Affymetrix GeneChips™.

36. Jensen, K. and Watt, F. (2006) Single-cell expression profiling of human epidermal stem and transit-amplifying cells: Lrig1 is a regulator of stem cell quiescence. *Proceedings of the National Academy of Sciences of the United States of America*, **103**, 11958–11963.

37. Kurimoto, K., Yabuta, Y., Ohinata, Y. *et al.* (2006) An improved single-cell cDNA amplification method for efficient high-density oligonucleotide microarray analysis. *Nucleic Acids Research*, **34**, e42. An extension of the global-RT-PCR technique to increase robustness.

38. Kurimoto, K., Yabuta, Y., Ohinata, Y. and Saitou, M. (2007) Global single-cell cDNA amplification to provide a template for representative high-density oligonucleotide microarray analysis. *Nature s*, **2**, 739–752. An extension of the global-RT-PCR technique to increase robustness.

39. Ohtsuka, S., Iwase, K., Kato, M. *et al.* (2004) An mRNA amplification procedure with directional cDNA cloning and strand-specific cRNA synthesis for comprehensive gene expression analysis. *Genomics*, **84**, 715–729.

40. Singh, R., Maganti, R.J., Jabba, S.V. *et al.* (2005) Microarray-based comparison of three amplification methods for nanogram amounts of total RNA. *American Journal of Physiology: Cell Physiology*, **288**, C1179–C1189.

41. Klur, S., Toy, K., Williams, M.P. and Certa, U. (2004) Evaluation of procedures for amplification of small-size samples for hybridization on microarrays. *Genomics*, **83**, 508–517.

42. Malboeuf, C.M., Isaacs, S.J., Tran, N.H. and Kim, B. (2001) Thermal effects on reverse transcription: improvement of accuracy and processivity in cDNA synthesis. *BioTechniques*, **30**, 1074–1078, 1080, 1082, passim.

43. Spiess, A.N., Mueller, N. and Ivell, R. (2003) Amplified RNA degradation in T7-amplification methods results in biased microarray hybridizations. *BMC Genomics*, **4**, 44.

44. Li, Y., Li, T., Liu, S. *et al.* (2004) Systematic comparison of the fidelity of aRNA, mRNA and T-RNA on gene expression profiling using cDNA microarray. *Journal of Biotechnology*, **107**, 19–28.

6

Real-Time Quantitative RT-PCR for mRNA Profiling

Stephen A. Bustin[1] and Tania Nolan[2]
[1]*Institute of Cell and Molecular Science, Barts and The London, Queen Mary's School of Medicine and Dentistry, University of London, Whitechapel, London, UK*
[2]*Sigma-Aldrich House, Haverhill, Suffolk, UK*

6.1 Introduction

The real-time reverse transcription quantitative polymerase chain reaction (RT-qPCR) [1] is characterized by its specificity, sensitivity, simplicity and speed. These attributes have made it the method of choice for the detection and quantification of RNA [2, 3], and have effected its extensive use in biotechnology [4], microbiology [5], virology [6] and molecular medicine [7] applications. The assay involves a conventional reverse transcription (RT) procedure, followed by the qPCR step, which makes use of fluorescent reporter molecules to combine the amplification and detection components of the PCR in a single tube format [8, 9]. An increase in fluorescent signal is proportional to the amount of DNA produced during each PCR cycle and produces a characteristic threshold cycle (C_t) or crossing point (C_p) for every reaction. The C_t or C_p is defined as the PCR cycle at which the signal first rises above background fluorescence. The more target there is in the starting material, the earlier the instrument can detect the fluorescence and the lower the C_t. This correlation between fluorescence and amount of amplified product permits accurate quantification of target molecules over a wide dynamic range and the homogeneous format significantly reduces hands-on time and the risk of contamination [10].

The consistency and reliability of RT-qPCR assays depend on the proper execution of a number of steps, in particular those relating to sample selection, template quality, assay design and data analysis [11], as well as the correct application of statistical models and methods for data analysis [12]. If correct procedure is adhered to, then results are generally robust, reproducible and accurate. Specifically, the reliability of the RT-qPCR assay is

Genomics: Essential Methods   Edited by Mike Starkey and Ramnath Elaswarapu
© 2011 John Wiley & Sons, Ltd.

determined by careful consideration of the quality of the RNA template [13–15], the choice of cDNA priming strategy [16] and reverse transcriptase enzyme [17, 18], the character- istics of the PCR primers [19, 20] and the validity of the normalization method [21–29]. Not surprisingly, the pervasive penetration of this technology has led to the development of numerous, distinct and frequently divergent experimental protocols that often generate dis- cordant results [30–32]. Fortunately, the resulting uncertainty has increased the awareness of a need for common guidelines, in particular those relating to quality assessment of every component of the RT-qPCR assay and appropriate data analysis [11]. This clear requirement for improved consistency of gene expression measurements is particularly relevant in rela- tion to human clinical diagnostic assays [7, 33]. Certainly, it is apparent that while qPCR is driving PCR-based innovations, effective guidelines regulating its use are essential for the future of molecular diagnostics within biomedical sciences [34, 35].

Detailed discussions of the many considerations that are essential for obtaining biologi- cally relevant data can be found elsewhere [36] and online (http://www.gene-quantification. info/).

6.2 Methods and approaches

A major reason for the popularity of RT-qPCR is its capacity to characterize and quantitate RNA templates that are present in very low copy numbers in minute amounts of sample. Although there are alternative technologies, both PCR-based (e.g. competitive methods such as StaRT-PCR [37]) and non-PCR-based methods [38], the ostensible simplicity and sheer ubiquity of real-time assays has ensconced it firmly as the method of choice in most areas of life, medical and agricultural sciences. Although undoubtedly a straightforward technology, many of the problems generally associated with quantification using conventional RT-PCR [39] remain, and a successful RT-qPCR assay is characterized by a sequential series of steps that must be executed carefully in order to complete a meaningful quantification experiment (see Figure 6.1).

6.2.1 Sample selection

Optimal sample quality is a prerequisite for generating valid quantitative data. Hence, sam- ple selection and collection, as well as RNA quality control, are critical parameters in test performance and must be optimized [40, 41]. In general, extraction of RNA from tissue culture, blood and serum is relatively straightforward, while there are significant problems associated with the extraction of RNA from solid tissue, feces, semen, plants and soil sam- ples. A critical consideration is the quantification of RNA from complex tissue samples, as these usually contain several different cell types that may express the target RNA(s) at different levels of abundance. This inevitably results in the averaging of the expression of the transcript from different cell types, and the expression profile of a specific cell type may be masked, lost or ascribed to and dismissed as illegitimate transcription. This is particularly relevant when comparing gene expression profiles between normal and cancer tissues, since normal cells adjacent to a tumor may be phenotypically normal, but genotypically abnormal or exhibit altered gene expression profiles due to their proximity to the tumor. This problem can be addressed by microdissection, in particular laser capture microdissection (LCM) [42], and significant differences have been detected in the gene expression profiles of microdis- sected and bulk tissue samples [43, 44]. A further refinement combines RT-qPCR with *in situ* hybridization on adjacent tissue sections and so allows simultaneous spatiotemporal

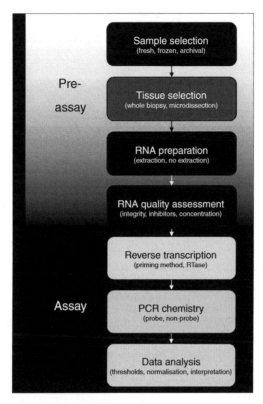

Figure 6.1 RT-qPCR experimental workflow. The assay divides into two main segments, the pre-assay and the assay, which can be subdivided into sample selection, tissue selection, RNA preparation, RNA quality assessment and reverse transcription (RT), PCR chemistry, data analysis respectively. These are highlighted in different colours.

monitoring and quantification of gene expression changes [45]. However, it is essential to realize that quantification from very few cells also poses problems, as different cells of the same type may not express the same set of mRNAs [46]. In addition, it could be expected that identical cell samples would contain identical copy numbers of a target mRNA, or more likely a normal distribution; however, it has been demonstrated that mRNA distribution is often in a lognormal fashion [47]. Microdissection can be carried out on either fresh/frozen or archival samples.

6.2.2 RNA extraction

RNA extraction is easiest from fresh or frozen material, with the main concern relating to the maintenance of RNA integrity. RNA is easily degraded and it is easy to co-purify inhibitors of the RT, or PCR steps, which will generate inaccurate results [48]. When dealing with very small amounts of sample (e.g. single cells or minute amounts of laser capture microdissected fresh/frozen sample) it is best to carry out the RT-qPCR directly from lysed tissue without going through an extraction process [49]. In our experience this results in an RNA quality indistinguishable from that obtained using conventional purification methods (see Figure 6.2). RNA integrity is best determined using a $3' : 5'$ assay, either target specific or using glyceraldehyde-3-phosphate dehydrogenase (GAPDH) as the target sequence [11]

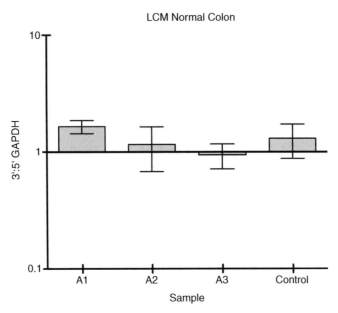

Figure 6.2 RNA integrity assessment from RNA obtained without extraction. Frozen samples of four colonic biopsies were catapulted into the caps of microfuge tubes, following LCM using a PALM LCM system. Three (A1–A3) were processed using Invitrogen's CellsDirect two-step system, with the RNA subjected to RT-qPCR without a separate extraction step. RNA from the fourth sample (control) was extracted and purified using Qiagen's RNeasy Mini Kit. RT-qPCR assays were carried out as described, with separate primers and probes (see Protocol 6.1) quantitating the 5′ and 3′ ends of the GAPDH mRNA [11]. It is apparent that there are no significant differences in the 3′ : 5′ ratios between the four samples. This suggests that a separate RNA extraction step is not required for preserving maximum RNA quality when processing frozen LCM samples.

(see Protocol 6.1). The data obtained are independent of ribosomal RNA (rRNA) integrity, provide a quantifiable measure of the degradation of the transcripts of interest and are modeled on the standard approach adopted by microarray users and long-accepted conventional techniques applied to end-point PCR assays [50]. This is in contrast to RNA quality assays performed using an RNA chip on the Agilent 2100 Bioanalyser or BioRad Experion platforms that provide a measure of rRNA integrity, which may or may not reflect in the quality of the (largely unmeasured) mRNA. The 3′ : 5′ assay described in Protocol 6.1 measures the integrity of the ubiquitously expressed mRNA specified by the GAPDH gene, which is taken as representative of the integrity of all mRNAs in a given RNA sample. However, since different mRNAs degrade at different rates, this may not always be the case and it may be necessary to design similar assays for specific targets. The RT reaction of the 1.3 kb GAPDH mRNA is primed using oligo-dT, and a separate multiplex PCR assay is used to quantitate the levels of three target amplicons. These are spatially separated with one towards the 5′ end, the second towards the center and the third towards the 3′ end of the mRNA sequence. The ratio of amplicons reflects the relative success of the oligo-dT primed RT to proceed along the entire length of the transcript. Clearly, the progress of the RT enzyme past the 5′ amplicon is dependent on the intactness of the mRNA, with the enzyme unable to reach it if the mRNA is degraded. Consequently, all things being equal, a 3′ : 5′ ratio of around unity indicates high integrity, whereas anything greater than five suggests degradation. The assay is designed as a triplex assay using a labeled hydrolysis probe for detection of each assay such that each amplicon is detected by a target-specific, differentially labeled probe.

PROTOCOL 6.1   Analysis of mRNA Integrity

Equipment and reagents

- Reverse transcription

  — reverse transcriptase enzyme (RT) 50 U/μl

  — 10× RT buffer (supplied with RT)

  — oligo-dT primer (500 ng/μl)

  — 100 mM dNTP mix (25 mM each of dATP, dCTP, dGTP and dTTP)

- qPCR

  — six GAPDH primers, two each for 5′, center and 3′ amplicons (10 μM each)

  — three GAPDH probes, one each for 5′, center and 3′ amplicons (5 μM each)

  — 2× commercial qPCR master mix buffer (containing dNTPs and thermostable^a polymerase) without $MgCl_2$

  — 25 mM $MgCl_2$

  — yeast tRNA (Invitrogen)

  — universal RNA (Stratagene human, mouse, rat)

  — real-time thermocycler with the capacity to detect multiplex reactions (e.g. Corbett 6000 or Stratagene MX3005p).

Oligonucleotides (5′ – 3′)

5′-GAPDH:

  P: **(FAM)**-CCTCAAGATCATCAGCAATGCCTCCTG-**(BHQ1)**

  F: GTGAACCATGAGAAGTATGACAAC

  R: CATGAGTCCTTCCACGATACC

Center GAPDH:

  P: **(HEX)**-CCTGGTATGACAACGAATTTGGCTACAGC-**(BHQ1)**

  F: TCAACGACCACTTTGTCAAGC

  R: CCAGGGGTCTTACTCCTTGG

3′-GAPDH:

  P: **(CY5)**-CCCACCACACTGAATCTCCCCTCCT-**(BHQ3)**

  F: AGTCCCTGCCACACTCAG

  R: TACTTTATTGATGGTACATGACAAGG

Method

Reverse transcription

1 Prepare an RT master mix of volume sufficient for the number of RNA samples being analysed. For each RNA sample, prepare 18.75 µl of RT master mix by combining the following:

- 12.75 µl of water

- 2.5 µl of 10× RT buffer

- 2.5 µl of 50 ng/µl oligo-dT

- 1 µl of 4 mM dNTP mix.

2 Make up two 'minus RT' controls for each RNA sample by combining 18.75 µl of master mix with 6.25 µl of water.

3 Add 1.25 µl of RT to every 18.75 µl of RT master mix to give a final concentration of 2.5 U/µl.[b]

4 Add 20 µl of master mix to 5 µl of each RNA sample (preferably testing each sample in duplicate), aiming for a target concentration of 50–500 ng/µl (ensuring that each reaction contains the same RNA concentration). Ensure that the final reaction volume is 25 µl.

5 Incubate at 20 °C for 10 min followed by 50 °C for 60 min.

6 Terminate the reactions by heating to 85 °C for 5 min and place them on ice for 2 min, or until required. Collect by brief centrifugation prior to continuing with step 10.

qPCR

7 For each single-stranded cDNA sample, prepare 20 µl of qPCR master mix by combining the following:

- 12.5 µl of 2× qPCR master mix buffer

- 4.5 µl of 25 mM $MgCl_2$ (4.5 mM final concentration)

- 0.25 µl of each 10 µM primer (six primers)

- 0.5 µl of each 5 µM probe (three probes).

8 Prepare a standard curve template as follows. Prepare seven 10-fold serial dilutions of target-specific amplicon in yeast tRNA (100 ng/µl) using a range of approximately 10^7 to 10^1 copies. Alternatively, use six fivefold dilutions of cDNA (where the highest concentration is 1 : 2 dilution of cDNA synthesis) prepared from universal RNA, or sample RNA. In each case, carry out the reactions in duplicate.

9 Add 5 µl of standard curve template to the appropriate qPCR reaction tubes, or microtiter plate wells.

10 Prepare 1 : 10 dilutions (in water) of the cDNA from step 6 and add 5 µl of each diluted sample to two qPCR reaction tubes.

11 Add 20 µl of qPCR master mix to each reaction tube and mix gently by repeatedly pipetting up and down, avoiding bubbles. Briefly centrifuge to collect the reactions at the base of each tube.

12 Perform qPCR using the following three-step protocol:

1 cycle:	Taq activation	95 °C, 10 min[c]
45 cycles:	Denaturation	95 °C, 30 s
	Annealing	56 °C, 30 s
	Extension	62 °C, 30 s (collect data)

13 Analyse the data from the standard curve samples.[d]

14 Analyse each assay individually (i.e. 5′, center or 3′) by referring to the standard curve[e] generated in step 13 to quantify the number of copies of each target.

15 Calculate the ratio of each target.[f]

Notes

[a]Chemically or antibody-inactivated hot-start polymerases are available. Note differences in the activation conditions.

[b]Do not add RT enzyme to the minus RT controls.

[c]Shorter activation times are required for antibody hot start enzymes.

[d]In order to ensure that the baseline settings define the appropriate start and end cycles, examine the amplification plots without software analysis (raw data view) and check the range of the linear, horizontal baseline region. Check that the software settings reflect this linear region, such that noise in early cycles is ignored and amplification occurs after the baseline end cycle (refer to instrument handbook for further details). View amplification plots on baseline corrected, normalized view and set threshold in the logarithmic phase of amplification. The log phase is viewed most easily by examination of amplification plots on a plot of log fluorescent signal versus cycle and selecting the linear phase of amplification.

[e]Using a standard curve controls for differences in reaction efficiencies and ensures more accurate quantification.

[f]If the 5′ amplicon has a calculated copy number of 1.0×10^6 and the 3′ amplicon has a calculated copy number of 1.5×10^6, then the 3′ : 5′ ratio of that sample is $1.5 \times 10^6/1.0 \times 10^6$, or 1.5, and is indicative of good quality RNA. Conversely, if the relative copy numbers are 1.0×10^5 (5′) and 1.5×10^7 (3′), then the resulting ratio of 150 suggests that the mRNA is degraded. Whilst it may still be acceptable to use such RNA in an RT-qPCR assay, especially when using target specific primers, it is preferable not to directly compare the results obtained from such RNA with the results obtained from high quality RNA. As a general guideline, consider RNA with a 3′ : 5′ ratio of <5 to be of high quality and suitable for any downstream application.

6.2.3 Clinical and environmental samples

Formalin-fixed paraffin-embedded tissue (FFPE tissue) is the most widely available material for retrospective clinical studies and, together with clinical data, represents an important resource for the elucidation of disease mechanisms and validation of differentially expressed genes as novel therapeutic targets or prognostic indicators.

RNA extracted from FFPE material is a little more difficult to analyze [51], as extensive degradation can occur before [52] or during [53] the formalin fixation process. Furthermore, formalin fixation generates cross-links between nucleic acids and proteins and covalently modifies RNA, making subsequent RNA extraction, RT and quantification analysis problematic [54]. Not surprisingly, fixatives are important [55] and different tissue preparation methodologies will invariably lead to different results from different laboratories. However, since real-time RT-PCR amplification generates amplicons that are as small as 60 bp, this technique is suitable for estimating mRNA levels from such tissue samples with best results when a specific gene reverse transcription primer is used [56–60].

Various methods can be used to assess the presence of inhibitors within biological samples. The PCR efficiency in a test sample can be assessed by serial dilution of the sample [61], although this is impossible when using very small amounts of RNA extracted, for example, from single cells or from laser capture microdissected sections. Alternatively, there are mathematical algorithms that provide a measure of PCR efficiency from analysis of the amplification response curves [62–64]. Internal amplification controls (IACs) that co-purify and co-amplify with the target nucleic acid can detect inhibitors as well as indicate template loss during processing [65]. Another approach utilizes a whole bacterial genome to detect inhibition from clinical samples [66]. A recently described qPCR reference assay identifies inhibitors of the RT or PCR steps by recording the C_t values characteristic of a defined number of copies of an artificial sense-strand amplicon [15] (see Protocol 6.2). Inhibitors in the RNA sample will result in an increased C_t value when the amplicon is run in the presence or absence of RNA samples.

PROTOCOL 6.2   Detection of RT-qPCR Inhibitors

Equipment and reagents

- RNA template (50–500 ng/ul) or cDNA (1 : 10 dilution of cDNA synthesis)[g]

- 2× commercial qPCR master mix buffer (containing dNTPs and chemically or antibody inactivated thermostable polymerase) without $MgCl_2$

- 25 mM $MgCl_2$

- SPUD synthetic DNA amplicon (5′–3′):

  AACTTGGCTTTAATGGACCTCCAATTTTGAGTGTGCACAAGCTA

  TGGAACACCACGTAAGACATAAAACGGCCACATATGGTGCCATGTAAGGATGAATGT

  (Sigma Genosys; prepare a stock 20 µM solution which contains 1.2×10^{13} molecules/µl; dilute to an approximately 20 000 copies/µl working solution)

- SPUD F and R primers (10 µM each) (5′–3′):

  forward primer: AACTTGGCTTTAATGGACCTCCA

  reverse primer: ACATTCATCCTTACATGGCACCA

- SPUD probe (5 μM) (5'–3'):

    FAM-TGCACAAGCTATGGAACACCACGT-(**BHQ1**)

- Real-time thermocycler (e.g. Corbett 6000 or Stratagene MX3005p).

Method

1 Prepare a qPCR master mix sufficient for all RNA samples to be tested in duplicate. Include a minimum of two 'no RNA sample' controls where the only template is the SPUD amplicon (SPUD-A). For each RNA sample, prepare 20 μl of master mix by combining the following:

   - 0.5 μl of water

   - 12.5 μl of 2× master mix buffer

   - 5.0 μl of 25 mM $MgCl_2$

   - 1.0 μl of 20 000 copies/μl SPUD-A

   - 1.0 μl of 5 μM SPUD P

   - 0.5 μl of each of 10 μM SPUD F and SPUD R.

2 Add 5 μl of each RNA or cDNA template to be tested to each of two qPCR tubes.

3 Add 20 μl of master mix to each RNA sample.

4 Run in the qPCR instrument using a two-step protocol:

1 cycle:	Activation	95 °C, 10 min
40 cycles:	Denaturation	95 °C, 30 s
	Annealing/extension	60 °C, 60 s (collect data)

5 Determine the C_t value for the control reactions containing SPUD-A only (i.e. no sample RNA).

6 Determine the C_t for reactions containing test samples and compare with the no sample control.[h]

Notes

[g]When a two step RT-PCR procedure is adopted it may be preferable to identify only the factors which inhibit the qPCR and include cDNA (at the concentration desired for subsequent qPCR reactions) as the test sample.

[h]The presence of inhibitors is indicated by a higher C_t being recorded for test samples than for the control containing SPUD-A alone. The distribution of the C_t values from the duplicate samples containing SPUD-A alone imply the assay coefficient of variance (CV) and this defines the acceptable range of ^bC_t values for samples which do not contain inhibitors. Samples with a C_t shift greater beyond the CV for the control samples are considered to contain inhibitors, and should be purified, or a fresh RNA sample extracted.

6.2.4 Reverse transcription

Primer selection, especially of the reverse primer used in the RT step, is critical since it affects the sensitivity of the RT-PCR assay [67]. It has also been shown that tissue-specific factors such as polysaccharides or proteins can influence amplification kinetics in a sequence-specific manner and that this effect can be mitigated, in part, by appropriate primer selection [68]. The structure of the RNA target at the primer-binding site must be taken into account, as this affects the accessibility of the target to the primers. Even so-called single-stranded RNA is typically extensively folded back onto itself, and selection of a primer binding site in a double-stranded target site that is folded will result in a very inefficient assay. For RNA viruses there is the additional problem of different viral serotypes resulting in sequence variability, and it may be necessary to use a nested RT-PCR assay with universal primers that bind to target sequences which are shared by all the serotypes followed by a serotype-specific primer pair [69].

The most appropriate method for cDNA priming remains a contentious issue. Direct comparison of the different methods available reveals that there is no one universally best method, and that results are target- and enzyme-dependent [17, 18]. Random priming or oligo-dT priming both allow a representative pool of cDNA to be produced during a single reaction (see Protocols 6.3 and 6.4 respectively). However, it has been shown that priming using random hexamer primers does not result in equal efficiencies of RT for all targets in the sample and that there is not a linear correlation between input target amount and cDNA yield when specific targets are measured [16, 70]. A recent comparison of the efficiency of RT priming by random primers of varying lengths showed that 15-nucleotide-long random oligonucleotides consistently yielded at least twice the amount of cDNA as random hexamers [71]. Oligo-dT primers should only be used with intact RNA. Even when using high-quality RNA, the cDNA molecules may be truncated, since the RT enzyme cannot proceed efficiently through highly structured regions. Qiagen claim that their Omniscript (catalogue number 205110) and Sensiscript (catalogue number 205211) reverse transcriptases can open up and read through regions of secondary RNA structure. Nonetheless, to be on the safe side, oligo-dT-primed assays should be targeted towards the $3'$ end of the transcript. This is an unsuitable choice for experiments that require examination of splice variants, sequences with long $3'$ untranslated regions, or those without polyA sequences. Furthermore, oligo-dT priming is not recommended when using RNA extracted from paraffin tissue sections, since formalin fixation results in the loss of the polyA tails on mRNA [72]. Target-specific primers are the most specific and, in general, the most sensitive method for converting mRNA into cDNA [16, 73] (see Protocol 6.5). An important argument against the use of specific priming is that it requires a lot of target RNA and, hence, is unsuitable for the detection of numerous RNA targets from limited amounts of RNA. However, a recent report demonstrates the use of specific primers for the reliable and specific amplification of 72 genes from limiting amounts of RNA using a multiplexed tandem (mt-) PCR approach [74]. Nevertheless, as with random priming, there may be differences in the efficiencies with which individual RT reactions occur.

PROTOCOL 6.3 Reverse Transcription Using Random Primers

Equipment and reagents

- RNA (10–500 ng)[i]
- Random primers (6-mer, 9-mer or 15-mer; 50 ng/μl)
- Reverse transcriptase (RT) 200 U/μl
- 10× RT buffer (supplied with RT)
- 25 mM MgCl$_2$
- 100 mM dithiothreitol (DTT).

Method

1 Briefly centrifuge the RNA and primers and combine to prepare the following pre-reaction mixture:

   - 1.0–9.0 μl of RNA
   - 1.0 μl of 50 ng/μl random primers
   - 0–9.0 μl of water (to adjust the total volume to 10 μl).

2 Incubate at 65 °C for 10 min and snap cool on ice for 5 min.

3 Prepare an RT master mix combining, for each RNA sample:

   - 2.5 μl of 10× RT buffer
   - 5.0 μl of 25 mM MgCl$_2$
   - 2.5 μl of 100 mM DTT
   - 1.0 μl of 200 U/μl RT[j]
   - 4.0 μl of water.

4 Add 15 μl of theRT master mix to the RNA/primer mix to make a total volume of 25 μl.[k] Gently mix the tube contents and briefly centrifuge.

5 Incubate at 20 °C for 10 min followed by 50 °C for 60 min.

6 Terminate the reactions by incubating at 85 °C for 5 min, and then place on ice for 5 min. Collect the reaction mixtures by brief centrifugation.[l]

Notes

[i]Each reaction must contain the same final concentration of RNA (maximum 20 ng/μl). Hence the most dilute sample will determine the total concentration used for all reactions.

[j]Ensure that manufacturers recommendations for specific enzymes are consulted.

[k]Prepare no-RT controls in duplicate by adding 14 μl of master mix and 1 μl of water prior to adding the RT enzyme to the master mix. Add this 'no enzyme' mixture to the primed RNA from step 2.

[l]First strand cDNAs can be stored at −20 °C for at least six months.

PROTOCOL 6.4   Reverse Transcription Using Oligo-dT Priming

Equipment and reagents

- RNA (10–500 ng)[m]
- Oligo-dT (500 ng/μl)
- Reverse transcriptase (RT) 200 U/μl[n]
- 10× RT buffer (supplied with RT)
- 25 mM $MgCl_2$
- 100 mM DTT.

Method

1 Briefly centrifuge the RNA and primers and combine to prepare the following pre-reaction mixture:

  - 1.0–9.0 μl of RNA

  - 1.0 μl of 500 ng/μl oligo-dT

  - 0–9.0 μl of water (to adjust the total volume to 10 μl).

2 Incubate at 65 °C for 10 min, and snap cool on ice for 5 min.

3 Prepare an RT master mix combining, for each RNA sample:

  - 2.5 μl of 10× RT buffer

  - 5.0 μl of 25 mM $MgCl_2$

  - 2.5 μl of 100 mM DTT

  - 1.0 μl of 200 U/μl RT[o]

  - 4.0 μl of water.

4 Add 15 μl of the RT master mix to the RNA/primer mix to make a total volume of 25 μl.[n] Gently mix the tube contents and briefly centrifuge.

5 Incubate at 20 °C for 10 min followed by 50 °C for 60 min.

6 Terminate the reaction by incubating at 85 °C for 5 min, and then place on ice for 5 min. Collect the reaction mixtures by brief centrifugation.[p]

Notes

[m]Each reaction must contain the same final concentration of RNA (maximum 20 ng/μl). Hence the most dilute sample will determine the total concentration used for all reactions.

[n]Each reaction must contain the same final concentration of RNA (maximum 20 ng/μl). Hence the most dilute sample will determine the total concentration used for all reactions.

[o]Ensure that manufacturers recommendations for specific enzymes are consulted.

[p]First strand cDNAs can be stored at −20 °C for at least six months.

PROTOCOL 6.5 Reverse Transcription Using Target-Specific Primers

Equipment and reagents

- RNA (1–200 ng)[q]
- Target-specific (antisense) primer (2 μM)
- Reverse transcriptase (RT) 200 U/μl
- 10× RT buffer (supplied with RT)
- 25 mM MgCl$_2$
- 100 mM DTT.

Method

1 Briefly centrifuge the RNA and primers and combine to prepare the following pre-reaction mixture:

- 1.0–9.0 μl of RNA
- 1.0 μl of 2 μM target-specific (anti-sense) primer
- 0–9.0 μl of water (to adjust the total volume to 10 μl).

2 Incubate at 65 °C for 10 min, and then snap cool on ice for 5 min.

3 Prepare an RT master mix combining, for each RNA sample:

- 2.5 μl of 10× RT buffer
- 5.0 μl of 25 mM MgCl$_2$
- 2.5 μl of 100 mM DTT
- 1.0 μl of 200 U/μl RT[r]
- 4.0 μl of water.

4 Add 15 μl of the RT master mix to the RNA/primer mix to make a total volume of 25 μl.[s] Gently mix the tube contents and briefly centrifuge.

5 Incubate at 50–65 °C for 5–15 min.

6 Terminate the reaction by incubating at 85 °C for 5 min, and then place on ice for 5 min. Collect the reaction mixtures by brief centrifugation.[t]

Notes

[q]It is preferable (but not essential) for each reaction to contain the same final concentration of RNA (maximum 20 ng/μl). Hence the most dilute sample may determine the total concentration used for all reactions.

[r]Ensure that manufacturers recommendations for specific enzymes are consulted.

[s]Prepare no-RT controls in duplicate by adding 14 μl of master mix and 1 μl of water prior to adding the RT enzyme to the master mix. Add this 'no enzyme' mixture to the primed RNA from step 2.

[t]First strand cDNAs can be stored at −20 °C for at least six months.

6.2.5 qPCR using SYBR green I dye detection

There are numerous non-probe- and probe-based chemistries, described in detail elsewhere [36]. The most suitable detection chemistry depends on the application; in general, the use of a double-stranded DNA binding dye, typically SYBR Green I, is the most cost-effective chemistry for primer optimization experiments and for most routine investigations [75] (see Protocol 6.6). An example of a typical qPCR result obtained using SYBR Green I dye as the reporter system is shown in Figure 6.3. Although inadvisable, legacy assays can

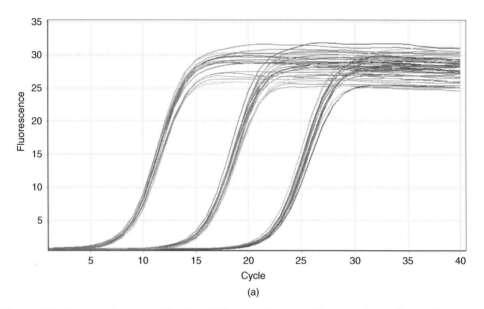

(a)

Figure 6.3 Comparative quantification. This procedure provides an alternative to the more usual dilution curve analysis for determining the fold change of a gene of interest relative to a calibrator sample. (a) Three amplification plots derived from a 100-fold serial dilution series of an RNA template used to optimize a new primer set. Note that there is no threshold line, which would be used to determine the C_q using the usual dilution curve method. (b) How the relative concentration of a sample is derived from the amplification plots shown in (a). (1) The second derivatives of the three amplification plots are calculated. These produce peaks corresponding to the maximum rate of fluorescence increase in the reaction denoted by 1, 2 and 3. (2) The 'takeoff' points (labelled 4, 5 and 6) are determined for each curve. A takeoff point is defined as the cycle at which the second derivative is at 20% of the maximum level, and indicates the end of the noise and the transition into the exponential phase. (3) The average increase in signal four cycles following the takeoff point (denoted by three bars labeled a, b and c) is used to calculate a slope, which provides a measure of the amplification efficiency for each curve. A 100% efficient reaction should double the signal in the exponential phase. So, if the signal was 10 at cycle 15, then went to 11 at cycle 16, it should go to 13 fluorescence units at cycle 17. (4) All of the amplification values for each sample are averaged to give a mean efficiency of a group of cycling curves for each sample (three in this example). The more variation there is between the estimated amplification values of each sample, the larger the confidence interval will be. In this example, the average amplification is 1.68 ± 0.02 for the neat template and 1.76 ± 0.01 and1.76 ± 0.02 for the two dilutions. (5) The same procedure is carried out for the calibrator sample and a fold change can then be calculated according to the formula Fold change = Efficiency$^{(\text{Calibrator takeoff} - \text{Target sample takeoff})}$. (See Plate 6.3.)

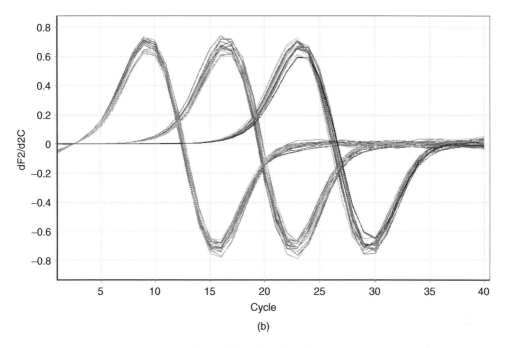

Figure 6.3 *(continued)*

be transferred to a SYBR Green I real-time format, but new assays are simple to design and this is recommended, especially if the legacy assay turns out to be suboptimal. Primer design for real-time PCR largely follows the same recommendations as for end-point PCR. The main difference is that the amplicon size is best kept below 250 bp, ideally between 75 and 150 bp. It is important to select primer pairs that do not show significant potential for self-dimerization, because primer dimers also contribute to the SYBR Green I quantification signal. Primers should also be specific for the desired target, particularly avoiding 3′ hybridization to reduce mispriming and linear amplification events. There are a variety of software options, with Primer3 (http://primer3.sourceforge.net/) being freely available, or more sophisticated options such as Beacon Designer (Premier Biosoft). In addition, many companies that provide oligos also offer assay design services (e.g. www.sial.com/designmyprobe from Sigma-Genosys). An alternative to designing a new assay is to identify a previously designed, validated set of primers and reaction conditions. The best sources of primers and probes specifically designed for RT-qPCR assays are public primer and probe databases such as RTPrimerDB (http://medgen.ugent.be/rtprimerdb/), PrimerBank (http://pga.mgh.harvard.edu/primerbank/index.html), or Real Time PCR Primer Sets (http://www.realtimeprimers.org/). RTPrimerDB lists validated qPCR assays submitted by researchers for the commonly used chemistries and includes all the information required to understand the purpose of an assay and to implement them in an experiment.

Although the primers alone determine the specificity of product detection, information regarding product size and population is easily determined from a melt curve analysis. Melt curves are a means of providing identification of amplified products and distinguishing them from primer dimers and other small amplification artifacts. The melting temperature (T_m) of DNA is defined as the temperature at which half of the DNA helical structure is lost.

The melting temperature of a DNA molecule depends on both its size and its nucleotide composition; hence, GC-rich amplicons have a higher T_m than those having an abundance of AT base pairs. During melt curve analysis, the real-time instrument continuously monitors the fluorescence of each sample as it is slowly heated from a user-defined temperature below the T_m of the products to a temperature above their melting point. Fluorescent dye is released upon melting (denaturation) of the double-stranded DNA, providing accurate T_m data for every single amplified product. Melting peaks are calculated by taking the differential (the first negative derivative, $-dF/dT$) of the melt curve. These peaks are analogous to the bands on an electrophoresis gel and allow for the qualitative monitoring of products at the end of a run. Short primer dimers will melt at lower temperature than longer, target amplicon products.

PROTOCOL 6.6   qPCR Assay Using SYBR Green I Dye as the Reporter System

Equipment and reagents

- 2× commercial qPCR master mix buffer, including dNTPs, thermostable DNA polymerase and SYBR Green I dye[u,v]

- 25 mM MgCl₂ (if required for optimization)

- 10 μM primers

- Real-time thermocycler with melt curve analysis option.

Method

1 Make a master mix combining, for each cDNA sample, the reagents in the order shown below:

   - 6.5 μl of water[w]

   - 0.5 μl of each 10 μM primer[x]

   - 12.5 μl of 2× qPCR master mix buffer.

   Mix gently by repeatedly pipetting up and down (making sure there are no air bubbles).

2 Add 20 μl of master mix to 5 μl of each template.[y]

3 Centrifuge briefly to ensure there are no bubbles, and place into a thermocycler.

4 Perform a three-step PCR reaction according to the following thermal profile:

1 cycle:	Activation	95 °C, 10 min	
40 cycles:	Denaturation	95 °C, 15 s	
	Annealing[z]	60 °C, 30 s	Collect data
	Extension	72 °C, 30 s	Collect data

5 Obtain a melting profile (as below, or according to the thermocycler manufacturer's instructions)[aa]

1 cycle:	95 °C, 1 min	
40 cycles:	55 °C, 30 s	Collect data
	Repeat and increase the temperature by 1 °C per cycle	Collect data

Notes

[u]Most commercially available blends also contain $MgCl_2$, but some reactions benefit from optimizing the $MgCl_2$ concentration.

[v]This protocol does not describe the use of a reference dye such as ROX, but such a reference dye may be included in the master mix. Many real-time thermocyclers require a constant reference dye in order to normalise optical differences in the detection of fluorescence from each well. The dye is included in the master mix and is detected by the instrument. If this is required (refer to instrument manufacturer) and is not contained in the 2 × master mix buffer, adjust the volume of water accordingly.

[w]Adjust the volume of water according to the volume of primers and template used. In this example, the primers are at a final concentration of 200 nM each, and 5 µl of the template is used.

[x]The volume of the primers required will depend upon the outcome of initial optimisation experiments; for a pilot test use a final concentration of 200 nM. Most assays are more efficient and sensitive when run under optimal conditions, and primer concentration greatly influences assay performance. The protocol described above can also be used to optimise primer concentration. In this case, cDNA is added to the reaction mix and varying primer concentrations are added to each reaction. It is advisable to select primer concentrations between 100 and 300 nM and test all combinations of forward and reverse primers. The primer concentration conditions selected are those resulting in the lowest C_t in the absence of primer dimers [11].

[y]When first strand cDNA is used, a target that is believed to be expressed at a medium to high level will be detected using 5 µl of a 1 : 10 dilution of the RT reaction.

[z]In most cases the annealing temperature for the primers is designed to be 60 °C. If this is not the case adjust the annealing temperature component of step 2.

[aa]The melt curve protocol is usually set automatically using the appropriate selection from the thermocycler software. Alternatively, a series of incubations are created such that the reaction is held for 30 seconds at increasing temperatures. The melt would begin by holding the PCR product at the start temperature, for example 55 °C. After 30 seconds the temperature would be increased to 56 °C and the products held for 30 seconds. This would be repeated until an incubation temperature of 95 °C is reached.

6.2.6 qPCR using labeled oligonucleotide probe detection

As an alternative to using DNA binding dyes, probe-based assays using reporter oligonucleotides may be used for amplicon detection (see Protocol 6.7). These probes are labeled oligonucleotides that are targeted towards the amplified target. At very low (<1000 copies) target concentration there is a greater probability of non-specific amplification and problems with primer dimer products becoming more pronounced than when using non-probe-based chemistries. In this case the use of a probe to detect amplicons may be preferable. The most popular format is the dual labelled fluorescent probe, popularly referred to as TaqMan, consists of a single-stranded oligonucleotide that is complementary to a sequence within the target template. It has a fluorescent dye at its 5' end and a quencher moiety at the 3' end. Following hybridization of the dual labelled probe to one of the template strands, it is digested by the exonuclease activity of the *Taq* DNA polymerase as it extends the amplification primer. This releases the fluorescent dye from the proximity of the quencher, resulting in an irreversible increase in the fluorescence signal. The addition of a DNA minor groove-binding (MGB) moiety at the 3' end of the dual labelled probe increases its stability and specificity and allows the probes to be shorter. A variation on this are MGB Eclipse probes, where the quencher and the MGB moieties are positioned at the 5' end of the probe, with the fluorescent reporter dye located on the 3' end [76]. When the probe is in solution during the denaturation phase, the quencher is in close proximity with the reporter dye and its fluorescence is quenched. However, when the probe anneals to a target sequence, the probe unfolds and the quencher becomes spatially separated from the reporter dye, resulting in fluorescence. An additional solution, designed to increase probe specificity by increasing the stability of the probe–target duplex, includes the incorporation of locked nucleic acids (LNAs). Probes containing LNA modifications are even shorter than MGB probes and offer greater flexibility in design [77].

There are other detection chemistries, such as Molecular Beacons, which incorporate a hairpin loop structure; these consist of a single-stranded loop complementary to the target template and a double-stranded stem, about six bases in length, with a fluorophore at one end (usually 5') and a quencher at the other end (usually 3') [78]. When the probe is in a hairpin configuration, the fluorophore and quencher are in close proximity. The probes are designed in such a way that they will bind to the amplicon at a specified temperature because the probe–target duplex is thermodynamically more stable than the hairpin structure. Upon binding, the stem comes apart and the fluorophore and quencher are separated, permitting fluorescence emission. Unlike with dual labelled probes, the fluorescence produced at each amplification cycle is reversible, as the probe is not destroyed, resulting in a lower overall background. The Scorpions probe system takes this principle a stage further and uses a combined primer and hairpin structured probe molecule. In this system the probe detects the DNA strand that is newly formed after PCR amplification. The advantage of the Scorpion system is that the probe detects the target *after* priming and, therefore, at lower temperature. This facilitates shorter probes with extreme sensitivity and specificity.

The design of primers for real-time assays was described above and is largely in common with primer design for conventional PCR assays. However, probe design requires a number of further considerations. On the whole these are addressed when using commercially available software or design services. In some cases, such as when targeting sequences that are AT or GC rich, or specific regions for genotyping or species-specific detection, the design software fails to identify suitable sequences for probe hybridization. In this situation it is critical to be aware of the factors that influence probe detection. Hydrolysis probes should be designed such that the 5' of the probe is 5–10 bp from the 3' of the forward primer, but ensuring that the 5' base is not G (G causing inherent quenching of the fluorophore). The probe should hybridize at 7–10 °C higher than the primers and so is usually 25–35 bases long. It is critical to check that the probe will not form secondary structure in solution by hybridization of tracts of self-complementarity (using a folding prediction algorithm such as mfold), or that the probe will hybridize to the primers (using BLASTn). Folding and mishybridization can be avoided by ensuring that runs of the same base are avoided (ensuring that there are fewer than four of the same base, especially G, consecutively), and aiming for a greater number of C bases than G (as G hybridization is more promiscuous and can cause secondary structure in the probe). Finally, it is important to ensure that an appropriate fluorophore and quencher combination have been requested when a probe is synthesized.

PROTOCOL 6.7 qPCR Assay Using Fluorescently Labeled Probes

Equipment and reagents

- 2× commercial qPCR master mix buffer (containing dNTPs and thermostable DNA polymerase) without MgCl₂[bb]
- 25 mM MgCl₂
- 10 μM primers
- 5 μM probe
- Real-time thermocycler.

Method

1 Make a master mix combining, for each cDNA sample, the reagents in the order shown below:

- water[cc]
- 2.5 μl of 50 mM MgCl₂[dd]
- 0.5μl of each 10 μM primer[ee]
- 1 μl of 5 μM probe (exact volume should be determined from optimization[dd])
- 12.5 μl of 2× qPCR master mix buffer.

Mix gently by repeatedly pipetting up and down (making sure there are no bubbles).

2 Add 20 μl of reaction mix to 5 μl of each template.[ff]

3 Centrifuge briefly to ensure there are no bubbles and place into a thermocycler.

4 Perform a two-step[gg] PCR reaction according to the following thermal profile:

1 cycle:	Activation	95 °C, 10 min	
40 cycles:	Denaturation	95 °C, 15 s	
	Annealing/extension	62 °C, 1 min	Collect data

Notes

[bb]This protocol does not include the use of the reference dye ROX; the requirement for a reference dye is primarily instrument dependent. Refer to instrument manufacturer's recommendations and if required, adjust the volume of water accordingly.

[cc]The volume required to adjust the total volume to 20 µl.

[dd]MgCl$_2$ concentration may need to be optimized for different chemistries and for different reactions. Most reactions using hydrolysis probes work well in a concentration between 3.5 and 6 mM MgCl$_2$, Molecular Beacons in 3.5 mM and Scorpions in 2.5 mM.

[ee]The volume of the primers required will depend upon the outcome of initial optimisation experiments; for a pilot test use a 200 nM final concentration. Most assays are more efficient and sensitive when run under optimal conditions and primer concentration greatly influences assay performance. The protocol described above can also be used to optimise primer concentration. In this case cDNA is added to the reaction mix and varying primer concentrations are added to each reaction. It is advisable to select primer concentrations between 100 and 300 nM and test all combinations of forward and reverse primer. The primer concentration conditions selected are those resulting in the lowest C$_t$ in the absence of primer dimers [11]. In addition, different probe concentrations may result in different signal intensity and assay sensitivity. After primer optimisation, identical assays are run including different probe concentrations from 50 to 300 nM [11].

[ff]A medium to highly expressed gene will be effectively detected when using the equivalent of 0.5 µl of the first strand cDNA (preferably 5 µl of a 1 : 10 dilution of the first strand cDNA).

[gg]Reactions containing hydrolysis probes are usually performed using a two-step PCR profile. During this reaction the double-stranded template is melted at 95 °C and the annealing and extension steps both occur during a single incubation at 62 °C. Although this is suboptimal for amplification by *Taq* polymerase, it is believed to encourage more efficient cleavage of the internal probe, resulting in maximum fluorescent signal per cycle. For optimal amplification and signal detection, use a three-step protocol for experiments including other probe-based chemistries.

6.2.7 Quantification methods

There is a wide-ranging debate about which methods are the most appropriate for calculating PCR efficiency in general and, more specifically, how to determine the efficiency of individual reactions (http://www.gene-quantification.info). The conventional approach is to construct a calibration curve by amplifying a target gene of interest in samples consisting of a serial dilution of template. The template may be cDNA, genomic DNA, PCR product or an artificial oligo. A plot of the logarithm of the template concentration (or relative concentration) against the $-Ct$ results in a linear graph of negative gradient. An assay with efficient doubling at each PCR cycle should yield a graph of slope -3.323 (it takes ~3.3 cycles to produce a 10-fold increase in PCR product). However, when using this method it is assumed that the efficiency of the reaction when using the samples as template is identical to when using the standard template. In order to address this uncertainty there has been a move to measure each individual reaction by analysis of the entire amplification plot and attempts to fit a curve which allows predictions about efficiency. Corbett Research has

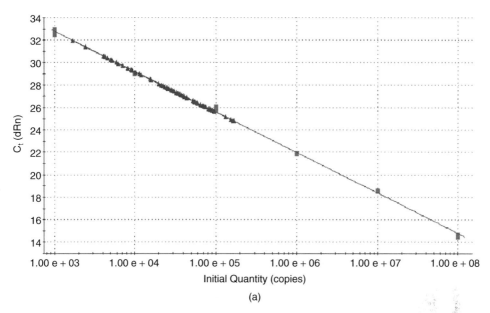

(a)

Figure 6.4 (a) Typical standard curve used to quantitate target mRNA from colonic biopsies. All the C_q quantification data from the test samples (blue triangles) in the upper picture are contained within the dynamic range of the standard curve, which is demarcated by the two outermost points of the standard derived from samples of a defined concentration and represented by red squares. This allows accurate quantification of the corresponding mRNAs. (b) Typical amplification plot obtained using a SYBR Green I assay. A single transcript has been quantified in a number of test samples and a serial dilution of standard material using SYBR Green I as the reporter. The two replicates for the three most concentrated standard samples (traces on left of the graph colored blue, red and green) illustrate a good standard of pipetting. The slopes of all the amplification plots are identical, indicating that the amplification efficiencies of every sample are the same. The high relative fluorescence (ΔR_n) value is typical of SYBR Green I assays. (See Plate 6.4.)

incorporated an algorithm into their data analysis software that allows the user to determine the efficiency of the PCR in each individual tube based on the fluorescence history of each reaction. The software uses a second derivative of the raw amplification data to determine the 'takeoff' point of a reaction, and the slope of the line from the takeoff point until exponential amplification ends is used to calculate the amplification efficiency. This value is then used when calculating the relative quantity of target in sample and calibrator reactions (see Figure 6.4). At present, this approach is most useful when using SYBR Green I dye as the reporter, because this gives a higher quantum yield per cycle, which is required to estimate the range of logarithmic amplification.

The common measure of all RT-qPCR assays, regardless of how the sample was obtained, reverse transcribed, or how the amplicon was detected is the C_q (C_t or C_p). The C_q is defined as the cycle when sample fluorescence exceeds a chosen threshold above calculated background fluorescence. Different systems may refer to the generic term C_q using alternative terminology, such as C_t or C_p. Background fluorescence is not a constant or absolute

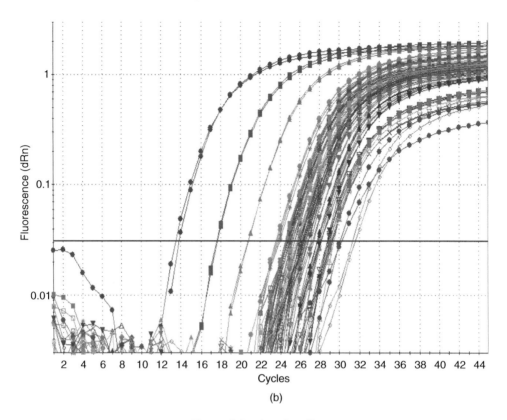

(b)

Figure 6.4 *(continued)*

value but is influenced by changing reaction conditions. Hence, if background fluorescence varies, the value of a C_q recorded for any particular sample is also going to be variable. Therefore, it is essential to understand that, on its own, a C_q value is meaningless; consequently, quoting a C_q is not sufficiently informative to allow a confident assessment of any conclusion drawn from the RT-qPCR experiment.

The C_q must be used to calculate a corresponding quantitative measurement, such as a copy number. This can be done in several ways, most commonly by using a standard curve obtained from a serially diluted standard solution of RNA or DNA to report copy numbers relative to that standard curve ('absolute quantification') [2] or by expressing the difference in the C_q values of a target RNA and a calibrator sample RNA and normalizing these as a ratio to one or (preferably) more internal reference RNA samples (relative quantification) [79]. Other methods for quantification of RNA levels exist, especially those based on reporting relative gene expression ratios based on calculating individual amplification efficiencies for each PCR assay [80, 81], but these are not in common use, as yet.

'Absolute' quantification is not really absolute, but is usually a measure relative to a standard curve; nevertheless, the term is in general use for this method of quantification. It is based on the use of an external standard dilution series with a known concentration of initial target copy number, which can be used to generate a standard curve of cycle threshold number (C_q) against initial target copy number [82]. To maximize accuracy, the dilutions are made over the range of copy numbers that include the amount of target mRNA expected in the experimental RNA samples. The C_q value is inversely proportional to the log of the initial copy number. Therefore, a standard curve is generated by plotting the C_q values against the logarithm of the initial copy numbers. The copy numbers of experimental RNAs can be calculated after real-time amplification from the linear regression of that standard curve, with the y-intercept giving a measure of the sensitivity and the slope a measure of the amplification efficiency. Standard curves can be constructed from PCR fragments, *in vitro* RNA polymerase-transcribed sense RNA transcripts, artificial, synthesized single-stranded sense-strand oligodeoxyribonucleotides, or from commercially available universal reference RNAs [83].

Absolute quantification is most obviously used for determination of RNA copy numbers as surrogates for quantifying tumor cells, or infectious particles like viruses or bacteria in body fluids, but it is also usefully applied to quantify changes in mRNA levels. The accuracy of absolute quantification depends entirely on the accuracy of the standards. In general, standard curves are highly reproducible and allow the generation of specific and reproducible results. Nevertheless, it is difficult to calibrate these standards so that they permit universal, absolute quantification, and results may not be comparable to those obtained using different probe/primer sets for the same markers, and will be different from results obtained using different techniques. Furthermore, external standards cannot detect or compensate for inhibitors that may be present in the samples. For this it is necessary to spike the sample with an internal control (e.g. a synthetic amplicon) or perform a prior screen to detect inhibitors using the SPUD assay (see Protocol 6.2).

Relative quantification is the most widely used quantification method, although its use is associated with a number of complications [75]. Perhaps the most obvious one concerns the choice of reference gene to use for expressing the relative quantity of any target gene(s). The 'gold standard' for relative quantification normalizes the C_q values from target RNAs to the geometric means of approximately three internal reference genes found to show the least variability [21] and results are expressed as relative fold over- or under-expression compared with the reference sample. This produces a corrected relative value for the target-specific RNA product that can be compared between samples and allows an estimate of the relative expression of target mRNA in those samples. It is crucial that the amplification efficiencies of target and reference are similar, since this directly affects the accuracy of any calculated expression result. Several models have been published that use different algorithms to correct for efficiency and claim to allow a more reliable estimation of the real expression ratio (see above). However, since the expression of most reference genes is regulated and their levels usually vary significantly with treatment or between individuals, relative quantification can be misleading [84]. Furthermore, if the relative levels of the reference and target genes vary by orders of magnitude, then the former may have entered its plateau phase by the time a C_q for the target becomes apparent. This is likely to interfere with the accurate quantification of the target mRNA.

6.2.8 RT-qPCR standardization

Any publication involving the quantification of mRNA using an RT-qPCR assay should include the following information [85]:

- RNA quality data (quantity, integrity, absence of inhibitors);
- sequence database accession number of the target gene;

- sequence and position of primers and probe (if appropriate);
- precise details of the RT reaction, especially the amount of RNA and priming strategies used;
- amount of RT reaction transferred to qPCR reaction;
- qPCR conditions;
- efficiency and lowest limit of linearity for the assay (molecules and C_q);
- error at that limit;
- normalization procedure and justification for reference genes selection;
- when reporting relative quantification data, the range of C_q values covered should be included.

6.3 Troubleshooting

6.3.1 No/Poor/Late amplification

No amplification is characterized by no significant increase in fluorescence above background. Poor amplification is typified by very low relative fluorescence values (ΔR_n; <0.05) with either the control templates or samples, or by a slope that varies significantly between samples amplifying the same target. A sample generating a $C_q > 35$ should be regarded with care, since this is very late and indicative of a theoretical, initial concentration of approximately 10 copies.

- Check the qPCR reaction conditions.
  - Check the amplification product on a gel. If no product is present, then repeat the assay, checking that all reagents are added and that the thermal cycling conditions are correct. If a product is present, then the instrument detection settings may be incorrect; for example, the wrong filter may be used to detect the light emitted from the reporter fluorophore. Collect all emitted data using all available data-collecting channels or filters. Check detection by substituting a probe with SYBR Green I dye to determine if the primers are amplifying.
  - Ensure that all required reagents (i.e. enzyme(s), reaction buffer, primers, probe and template) were added.
  - Check the annealing temperature (<60 °C), elongation times and cycle number. Ensure that the correct activation time for the DNA polymerase is used (antibody-inactivated enzymes usually require shorter initial incubation periods than chemical-inactivated DNA polymerases)
  - Check that the primers and other reactants have been diluted correctly.
  - Ensure that the pipettes are calibrated.
  - Ensure that the thermal cycler is programmed to detect fluorescence during the annealing and extension stages of the PCR.

—Consider if there is too little cDNA template in the qPCR. For a medium to highly expressed gene, use the equivalent of 0.5 μl of first-strand cDNA synthesis reaction). If a very late C_q is recorded, add up to 2.5 μl of cDNA synthesis reaction in a 25 μl qPCR.

—Consider if there is there too much cDNA template in the qPCR. If the final concentration of the cDNA synthesis reaction components in the PCR is too high this is inhibitory to PCR. Therefore, limit the volume of the cDNA synthesis reaction added to 10 % of the volume of the PCR.

—Check to see if the RNA is degraded. Use control cDNA to confirm that the assay works and an alternative qPCR assay to confirm that the samples support RT-qPCR.

—Consider the length of the amplicon. Ideally this should be <120 bp, especially for FFPE tissue samples. Longer amplicons amplify with lower efficiency and should be avoided.

—During RT, after denaturation of RNA, ensure that the RNA is kept on ice.

—If using homemade buffers, ensure that the salt and buffer concentrations are correct.

—Incomplete thawing of frozen 2× master mix buffers will change the salt concentration in the remaining buffer. Increases in $MgCl_2$ concentrations will affect the efficiency of primer binding and may cause the appearance of primer dimers and reduce the efficiency of the PCR. Ensure that master mixes and stock reaction components (oligos, $MgCl_2$) are completely defrosted and thoroughly mixed before use.

—If there have been changes in the reaction volumes or number of reactions for which master mix has been prepared, then errors may have been made during pipetting. Changing the volumes of the reaction mixtures can cause different amounts of error in the volumes being dispensed. This is partially due to differences in the tolerances between large- and small-volume pipettes. This error may not be propagated linearly during scale up. Highly sensitive methods such as PCR can significantly magnify these problems. It is best to scale up in stages if absolutely required. A pipetting robot can be a wonderful addition to the routine laboratory and these are particularly useful for high-throughput applications and to avoid variability in reaction mixes.

• Check the reagents and thermocycler.

—If the assay is probe based and has previously worked well, then the probe may have degraded. For example, it may have been photo bleached if it has been left in the light. In order to check a probe function, digest an aliquot (alongside a non-digested control) using DNAse I (use extreme care when handling DNAse I in close proximity to oligonucleotides). Examine the fluorescent yield and compare with a functioning assay. If the fluorescence is significantly lower then a new probe must be used. To prevent degradation, always store fluorophore-labeled oligonucleotides in aliquots in the dark at −20 °C. These can be used for at least 6 months after resuspension.

—If the background fluorescence level of the probe is very high, then it is possible that it may have become hydrolysed; for example, if subjected to repeated freeze–thaw cycles. See above for a diagnostic test.

— If the assay is a new assay, then it may well be that the probe manufacturer is to blame. Test the quality of the probe using a DNAse I digestion assay and compare with the fluorescence from a functioning probe (as above); following digestion the fluorescence should greatly increased due to reporter and quencher becoming separated. When the fluorescent yield is negligible, return the probe to the manufacturer and request a replacement.

— SYBR Green I dye loses fluorescent emission after storage. After SYBR Green I has been diluted, it goes off very quickly and can only be kept in the fridge for approximately 2 weeks. It must also be kept in the dark. For best results, dilute a fresh batch of SYBR Green I dye each day or use a commercial master mix, which usually contains components that stabilize the fluorescence.

— If using a new batch of thermostable DNA polymerase (or 2× master mix buffer containing DNA polymerase), be aware that different batches of DNA polymerase (quite apart from DNA polymerases from different suppliers) can have different polymerase and exonuclease activity, but still be within the manufacturer's specifications. Where possible, use the same batch of reagents and plan experimentation such that target quantification in all samples to be compared is carried out in as short a time period as possible. Use positive controls and standard curves to compare batches of master mix or enzymes.

— Different thermocyclers (particularly if they feature heating blocks) and thermocyclers from different manufacturers have different ramping kinetics and heating efficiencies and, hence, can affect the efficiency of the PCR. Variability in heating efficiency may also occur in different wells, and particularly so in 96- or 384-well instruments. When adapting a protocol, which has been optimized for a given instrument, ensure that conditions are optimized for the instrument to be used.

• Ensure the optimal design of RT primer, and qPCR primers (and probe), and optimal reaction conditions.

— Ensure that the RT primer is not binding to a region of the RNA molecule that could be double stranded. If this is the case, move the assay to an alternative region of the gene and use an alternative gene-specific primer.

— When using gene-specific RT primers, make sure they specify the anti-sense sequence.

— When using oligodT RT priming, ensure that qPCR assays are directed to approximately 1 kb from the polyA tail or 3′ end of the target sequence in order to ensure that the cDNA contains the representative sequence.

— Denature/anneal RNA and primers in the absence of salts and buffer.

— Ensure that RT was performed at the right temperature (check manufacturer's recommendations for each enzyme).

— Raise the RT reaction temperature to 65 °C if the RT primer T_m and the reverse transcriptase allows.

— When performing a two-step RT-qPCR protocol, ensure that the volume of RT reaction mixture added to the qPCR does not exceed 10 % of the volume of the qPCR.

—Reagents such as sodium dodecyl sulfate, EDTA, glycerol, sodium pyrophosphate, spermidine, formamide, guanidinium salts and dimethyl sulfoxide can inhibit reverse transcriptase (and *Taq* DNA polymerase). Dilute the RNA sample (1 : 10), as this may dilute out the inhibitor. If this fails, remove inhibitor by ethanol precipitation of the RNA. Include a 70 % (v/v) ethanol wash of the RNA pellet. Ensure all ethanol is removed, since this will also inhibit downstream reactions. Glycogen (0.2–0.4 µg/µl) can be included to aid recovery of small RNA samples.

—Optimize the PCR primer concentrations; this is essential for optimal assay efficiency [11].

—Confirm that the annealing temperature is appropriate for the qPCR primers. Most qPCR assays are designed for an annealing temperature of 60 °C, but some primers may perform better at different temperatures.

—Consider whether the annealing and/or extension times may be too short.

—Optimize the $MgCl_2$ concentration.

—Consider whether the probe is too long. Probes are usually designed to have an annealing temperature of 7–10 °C higher than primers. In A–T-rich regions, this may lead to very long probes with inefficient quenching and a tendency to form secondary structures. Probes are best kept to approximately 35 nucleotides and modifications such as LNA incorporated to reduce length while maintaining T_m.

—Consider whether the probe may have secondary structure by submitting the sequence to a folding analysis algorithm such as mfold.

—Consider redesigning the qPCR assay. When all possibilities have been explored and the assay still performs poorly on control targets, it may be that the most effective solution is to redesign the assay to an alternative region of the gene sequence

• Consider the possibility that the target mRNA is not expressed in the tissue under analysis.

—Always include the relevant controls to facilitate the troubleshooting process and to be assured that all data are valid. If the positive control sample is positive then the negative result for a test sample can be accepted with confidence. Similarly, a negative control ensures that a positive result in a test sample is more reliable. This is particularly critical when the positive result is of high C_q (i.e. low copy number).

6.3.2 No-template, negative control yields an amplification product

The master mix may be contaminated with DNA template or PCR product from previous reactions. Since the products from qPCR are rarely analyzed further, the latter is a relatively lower risk. However, if this is possible, or if PCR is used to generate products for the standard curve, it may be advantageous to use a dUTP/thermolabile uracil-N-glycosylase-utilizing protocol. In this case, all PCR reactions are carried out in the presence of dUTP (replacing dTTP). Prior to amplification of reactions containing native DNA or cDNA (containing T residues) the reactions are incubated in the presence of thermolabile uracil-N-glycosylase. This digests potential contaminants containing U residues.

- Use fresh aliquots of all reagents, including sterile water.
- Only use pipettes, tips, solutions (especially water) and racks dedicated to setting up qPCR. When possible, use a protective laminar flow hood. Do not use pipettes and accessories that have been exposed to amplicons.
- Decontaminate surfaces, pipettes and racks using UV exposure for 5–10 min prior to reaction set-up.
- Change gloves (after a maximum of 30 min wear) and tips frequently.
- Change the location of PCR set-up.
- When using SYBR Green I detection, check the melt curves. If the non-template control melt curve is different from the amplicon melt curve, you may still be able to use the template-derived data.
- An excess of probe can generate an artifactual positive result. This is often distinguished from a genuine positive by examination of raw data and comparison of background fluorescence. An artifact is apparent by a linear increase in background fluorescence that does not have a logarithmic phase. Alternatively, where a result is in question, repeat the assay on the sample using lower probe concentration (down to 50 nM).

6.3.3 No reverse transcriptase control yields an amplification product

A positive 'No RT control' indicates that the RNA sample contains contaminating DNA.

- To remove gDNA contamination, treat RNA samples with RNAse-free DNase I.
- If possible, design PCR primers to amplify across introns to avoid detection of genomic DNA template.

6.3.4 Primer dimers formed

Primer dimers are more likely to occur in no template controls or when there is very little target RNA. Even small amounts of target can suppress primer dimer formation.

- If dimers occur in the presence of normal amounts of template RNA, then redesign PCR primers. Ensure that complementary sequences are absent from the 3' ends of the primers (see Section 6.2.5 for details).

6.3.5 Multiple peaks in SYBR green I melt curve

- Improve the stringency of the qPCR to avoid non-specific hybridization by raising the annealing temperature or lowering the $MgCl_2$ concentration.
- Treat the RNA with RNase-free DNAse I.
- Primers may be amplifying multiple genuine products (e.g. splice variants, pseudogenes) and the assay may require primer redesign in order to detect the specific target. Analyze the reaction products by gel electrophoresis to check for specificity and consistent presence of additional bands. Perform BLASTn analysis to identify potential *in silico* products when target sequences are fully described.

- Shoulders in the melt curves do not necessarily mean that the assay is non-specific. The amplicon may contain AT-rich subdomains, and so when the products re-anneal they may not align correctly. Analyze the reaction products by gel electrophoresis to check for specificity.

6.3.6 Standard curve is unreliable (correlation coefficient <0.98 over at least 5 log dilution and with samples repeated in triplicate)

The low (or high) concentration point(s) of the dilution series can sometimes be removed to improve the correlation coefficient. However, all sample data must be encompassed by the standards. If the unknown samples fall in the low range beyond the limit of the standard curve, then quantification will be unreliable and the experiment may need to be repeated.

6.3.7 Erratic amplification plots/high well-to-well variation

- Use extreme care when pipetting solutions containing low amounts of target. To avoid working with low volumes, dilute target samples and pipette larger volumes. When possible, prepare samples in 5 µl volumes.

- Ensure that reaction master mixes are thoroughly mixed prior to adding to samples.

- Consider whether the baseline is set using wrong cycle range. When defining the baseline, examine raw data and instruct the instrument software to consider the range of cycles where there is no change in fluorescent signal.

- Investigate whether there is possibly sample evaporation due to loose lids or poor heat sealing.

- Consider if there may have been incomplete mixing of reagents.

- Consider if there could have been air bubbles at the bottom of the reaction tubes.

- Consider whether the frozen stocks of RNAs, primers or reaction buffer may not have been completely thawed or mixed when used.

- Check whether spikes in the signals could have been caused by problems with the instrument lamp, misaligned optics or other mechanical and/or electronic issues.

References

1. Gibson, U.E., Heid, C.A. and Williams, P.M. (1996) A novel method for real time quantitative RT-PCR. *Genome Research*, **6**, 995–1001. The first description of the RT-qPCR assay.

2. Bustin, S.A. (2000) Absolute quantification of mRNA using real-time reverse transcription polymerase chain reaction assays. *Journal of Molecular Endocrinology*, **25**, 169–193. Most cited, definitive review describing the principles and problems associated with RT-qPCR assays.

3. Ginzinger, D.G. (2002) Gene quantification using real-time quantitative PCR: an emerging technology hits the mainstream. *Experimental Hematology*, **30**, 503–512.

4. Kunert, R., Gach, J.S., Vorauer-Uhl, K. *et al.* (2006) Validated method for quantification of genetically modified organisms in samples of maize flour. *Journal of Agricultural and Food Chemistry*, **54**, 678–681.

5. Mackay, I.M. (2004) Real-time PCR in the microbiology laboratory. *Clinical Microbiology and Infection*, **10**, 190–212.

6. Mackay, I.M., Arden, K.E. and Nitsche, A. (2002) Real-time PCR in virology. *Nucleic Acids Research*, **30**, 1292–1305.

7. Bustin, S.A. and Mueller, R. (2005) Real-time reverse transcription PCR (qRT-PCR) and its potential use in clinical diagnosis. *Clinical Science (London)*, **109**, 365–379.

8. Higuchi, R., Dollinger, G., Walsh, P.S. and Griffith, R. (1992) Simultaneous amplification and detection of specific DNA sequences. *Biotechnology (N. Y.)*, **10**, 413–417. First description of a closed-tube PCR assay.

9. Higuchi, R., Fockler, C., Dollinger, G. and Watson, R. (1993) Kinetic PCR analysis: real-time monitoring of DNA amplification reactions. *Biotechnology (N. Y.)*, **11**, 1026–1030. First description of the monitoring of multiple polymerase chain reactions simultaneously over the course of thermocycling.

10. Wong, M.L. and Medrano, J.F. (2005) Real-time PCR for mRNA quantitation. *Biotechniques*, **39**, 75–85.

11. Nolan, T., Hands, R.E. and Bustin, S.A. (2006) Quantification of mRNA using real-time RT-PCR. *Nature Protocols*, **1**, 1559–1582.

12. Fu, W.J., Hu, J., Spencer, T. *et al.* (2006) Statistical models in assessing fold change of gene expression in real-time RT-PCR experiments. *Computational Biology and Chemistry*, **30**, 21–26.

13. Fleige, S. and Pfaffl, M.W. (2006) RNA integrity and the effect on the real-time qRT-PCR performance. *Molecular Aspects of Medicine*, **27**, 126–139.

14. Fleige, S., Walf, V., Huch, S. *et al.* (2006) Comparison of relative mRNA quantification models and the impact of RNA integrity in quantitative real-time RT-PCR. *Biotechnology Letters*, **28**, 1601–1613.

15. Nolan, T., Hands, R.E., Ogunkolade, B.W. and Bustin, S.A. (2006) SPUD: a quantitative PCR assay for the detection of inhibitors in nucleic acid preparations. *Analytical Biochemistry*, **351**, 308–310.

16. Bustin, S.A. and Nolan, T. (2004) Pitfalls of quantitative real-time reverse-transcription polymerase chain reaction. *Journal of Biomolecular Techniques*, **15**, 155–166.

17. Stahlberg, A., Hakansson, J., Xian, X. *et al.* (2004) Properties of the reverse transcription reaction in mRNA quantification. *Clinical Chemistry*, **50**, 509–515.

18. Stahlberg, A., Kubista, M. and Pfaffl, M. (2004) Comparison of reverse transcriptases in gene expression analysis. *Clinical Chemistry*, **50**, 1678–1680.

19. Pattyn, F., Robbrecht, P., De Paepe, A. *et al.* (2006) RTPrimerDB: the real-time PCR primer and probe database, major update 2006. *Nucleic Acids Research*, **34**, D684–D688.

20. Pattyn, F., Speleman, F., De Paepe, A. and Vandesompele, J. (2003) RTPrimerDB: the real-time PCR primer and probe database. *Nucleic Acids Research*, **31**, 122–123.

21. Vandesompele, J., De Preter, K., Pattyn, F. *et al.* (2002) Accurate normalization of real-time quantitative RT-PCR data by geometric averaging of multiple internal control genes. *Genome Biology*, **3**, 0034.0031–0034.0011. Seminal paper providing a practical method for normalization.

22. Pfaffl, M.W., Horgan, G.W. and Dempfle, L. (2002) Relative expression software tool (REST) for group-wise comparison and statistical analysis of relative expression results in real-time PCR. *Nucleic Acids Research*, **30**, e36.

23. Szabo, A., Boucher, K., Carroll, W.L. *et al.* (2002) Variable selection and pattern recognition with gene expression data generated by the microarray technology. *Mathematical Biosciences*, **176**, 71–98.

24. Pfaffl, M.W., Tichopad, A., Prgomet, C. and Neuvians, T.P. (2004) Determination of stable housekeeping genes, differentially regulated target genes and sample integrity: BestKeeper – Excel-based tool using pair-wise correlations. *Biotechnology Letters*, **26**, 509–515.

25. Akilesh, S., Shaffer, D.J. and Roopenian, D. (2003) Customized molecular phenotyping by quantitative gene expression and pattern recognition analysis. *Genome Research*, **13**, 1719–1727.

26. Andersen, C.L., Jensen, J.L. and Orntoft, T.F. (2004) Normalization of real-time quantitative reverse transcription-PCR data: a model-based variance estimation approach to identify genes suited for normalization, applied to bladder and colon cancer data sets. *Cancer Research*, **64**, 5245–5250.

27. Haller, F., Kulle, B., Schwager, S. *et al.* (2004) Equivalence test in quantitative reverse transcription polymerase chain reaction: confirmation of reference genes suitable for normalization. *Analytical Biochemistry*, **335**, 1–9.

28. Abruzzo, L.V., Wang, J., Kapoor, M. *et al.* (2005) Biological validation of differentially expressed genes in chronic lymphocytic leukemia identified by applying multiple statistical methods to oligonucleotide microarrays. *Journal of Molecular Diagnostics*, **7**, 337–345.

29. Kanno, J., Aisaki, K., Igarashi, K. *et al.* (2006) "Per cell" normalization method for mRNA measurement by quantitative PCR and microarrays. *BMC Genomics*, **7**, 64.

30. Bustin, S.A. (2002) Quantification of mRNA using real-time reverse transcription PCR (RT-PCR): trends and problems. *Journal of Molecular Endocrinology*, **29**, 23–39.

31. Bustin, S.A. (2005) Real-time, fluorescence-based quantitative PCR: a snapshot of current procedures and preferences. *Expert Review of Molecular Diagnostics*, **5**, 493–498.

32. Bustin, S.A., Benes, V., Nolan, T. and Pfaffl, M.W. (2005) Quantitative real-time RT-PCR – a perspective. *Journal of Molecular Endocrinology*, **34**, 597–601.

33. Bustin, S.A. and Mueller, R. (2006) Real-time reverse transcription PCR and the detection of occult disease in colorectal cancer. *Molecular Aspects of Medicine*, **27**, 192–223.

34. Valasek, M.A. and Repa, J.J. (2005) The power of real-time PCR. *Advances in Physiology Education*, **29**, 151–159.

35. Picard, C., Silvy, M. and Gabert, J. (2006) Overview of real-time RT-PCR strategies for quantification of gene rearrangements in the myeloid malignancies. *Methods in Molecular Biology*, **125**, 27–68.

36. Bustin, S.A. (2004) *A–Z of Quantitative PCR*, IUL Press, La Jolla, CA. Complete guide to all aspects of qPCR analysis.

37. Willey, J.C., Crawford, E.L., Jackson, C.M. *et al.* (1998) Expression measurement of many genes simultaneously by quantitative RT-PCR using standardized mixtures of competitive templates. *American Journal of Respiratory Cell and Molecular Biology*, **19**, 6–17.

38. Flagella, M., Bui, S., Zheng, Z. *et al.* (2006) A multiplex branched DNA assay for parallel quantitative gene expression profiling. *Analytical Biochemistry*, **352**, 50–60.

39. Freeman, W.M., Walker, S.J. and Vrana, K.E. (1999) Quantitative RT-PCR: pitfalls and potential. *Biotechniques*, **26**, 112–115.

40. Müller, M.C., Hördt, T., Paschka, P. *et al.* (2004) Standardization of preanalytical factors for minimal residual disease analysis in chronic myelogenous leukemia. *Acta Haematologica*, **112**, 30–33.

41. Benoy, I.H., Elst, H., Van Dam, P. *et al.* (2006) Detection of circulating tumour cells in blood by quantitative real-time RT-PCR: effect of pre-analytical time. *Clinical Chemistry and Laboratory Medicine*, **44**, 1082–1087.

42. Emmert-Buck, M.R., Bonner, R.F., Smith, P.D. *et al.* (1996) Laser capture microdissection. *Science*, **274**, 998–1001.

43. Fink, L., Kohlhoff, S., Stein, M.M. *et al.* (2002) cDNA array hybridization after laser-assisted microdissection from nonneoplastic tissue. *American Journal of Pathology*, **160**, 81–90.

44. Oji, Y., Yamamoto, H., Nomura, M. *et al.* (2003) Overexpression of the Wilms' tumor gene WT1 in colorectal adenocarcinoma. *Cancer Science*, **94**, 712–717.

45. Haupt, C., Tolner, E.A., Heinemann, U. *et al.* (2006) The combined use of non-radioactive *in situ* hybridization and real-time RT-PCR to assess gene expression in cryosections. *Brain Research*, **1118**, 232–238.

46. Peixoto, A., Monteiro, M., Rocha, B. and Veiga-Fernandes, H. (2004) Quantification of multiple gene expression in individual cells. *Genome Research*, **14**, 1938–1947.

47. Bengtsson, M., Ståhlberg, A., Rorsman, P. and Kubista, M. (2005) Gene expression profiling in single cells from the pancreatic islets of Langerhans reveals lognormal distribution of mRNA levels. *Genome Research*, **15**, 1388–1392.

48. Hoadley, M.E. and Hopkins, S.J. (2006) Comparison of 'real-time' and immunometric RT-PCR: RNA interference of reverse transcriptase-PCR. *Journal of Immunological Methods*, **312**, 40–44.

49. Hartshorn, C., Anshelevich, A. and Wangh, L.J. (2005) Rapid, single-tube method for quantitative preparation and analysis of RNA and DNA in samples as small as one cell. *BMC Biotechnology*, **5**, 2. Method for nucleic acid quantification without first extracting RNA.

50. Auer, H., Lyianarachchi, S., Newsom, D. *et al.* (2003) Chipping away at the chip bias: RNA degradation in microarray analysis. *Nature Genetics*, **35**, 292–293.

51. Rupp, G.M. and Locker, J. (1988) Purification and analysis of RNA from paraffin-embedded tissues. *Biotechniques*, **6**, 56–60.

52. Mizuno, T., Nagamura, H., Iwamoto, K.S. *et al.* (1998) RNA from decades-old archival tissue blocks for retrospective studies. *Diagnostic Molecular Pathology*, **7**, 202–208.

53. Klimecki, W.T., Futscher, B.W. and Dalton, W.S. (1994) Effects of ethanol and paraformaldehyde on RNA yield and quality. *Biotechniques*, **16**, 1021–1023.

54. Masuda, N., Ohnishi, T., Kawamoto, S. *et al.* (1999) Analysis of chemical modification of RNA from formalin-fixed samples and optimization of molecular biology applications for such samples. *Nucleic Acids Research*, **27**, 4436–4443.

55. Goldsworthy, S.M., Stockton, P.S., Trempus, C.S. *et al.* (1999) Effects of fixation on RNA extraction and amplification from laser capture microdissected tissue. *Molecular Carcinogenesis*, **25**, 86–91.

56. Godfrey, T.E., Kim, S.H., Chavira, M. *et al.* (2000) Quantitative mRNA expression analysis from formalin-fixed, paraffin-embedded tissues using $5'$ nuclease quantitative reverse transcription-polymerase chain reaction. *Journal of Molecular Diagnostics*, **2**, 84–91.

57. Bock, O., Kreipe, H. and Lehmann, U. (2001) One-step extraction of RNA from archival biopsies. *Analytical Biochemistry*, **295**, 116–117.

58. Specht, K., Richter, T., Muller, U. *et al.* (2001) Quantitative gene expression analysis in microdissected archival formalin-fixed and paraffin-embedded tumor tissue. *American Journal of Pathology*, **158**, 419–429.

59. Cohen, C.D., Gröne, H.J., Gröne, E.F. *et al.* (2002) Laser microdissection and gene expression analysis on formaldehyde-fixed archival tissue. *Kidney International*, **61**, 125–132.

60. Fink, L., Kinfe, T., Stein, M.M. *et al.* (2000) Immunostaining and laser-assisted cell picking for mRNA analysis. *Laboratory Investigation*, **80**, 327–333.

61. Ståhlberg, A., Aman, P., Ridell, B. *et al.* (2003) Quantitative real-time PCR method for detection of B-lymphocyte monoclonality by comparison of kappa and lambda immunoglobulin light chain expression. *Clinical Chemistry*, **49**, 51–59.

62. Tichopad, A., Dilger, M., Schwarz, G. and Pfaffl, M.W. (2003) Standardized determination of real-time PCR efficiency from a single reaction set-up. *Nucleic Acids Research*, **31**, e122.

63. Ramakers, C., Ruijter, J.M., Deprez, R.H. and Moorman, A.F. (2003) Assumption-free analysis of quantitative real-time polymerase chain reaction (PCR) data. *Neuroscience Letters*, **339**, 62–66.

64. Liu, W. and Saint, D.A. (2002) Validation of a quantitative method for real time PCR kinetics. *Biochemical and Biophysical Research Communications*, **294**, 347–353.

65. Pasloske, B.L., Walkerpeach, C.R., Obermoeller, R.D. *et al.* (1998) Armored RNA technology for production of ribonuclease-resistant viral RNA controls and standards. *Journal of Clinical Microbiology*, **36**, 3590–3594.

66. Cloud, J.L., Hymas, W.C., Turlak, A. *et al.* (2003) Description of a multiplex *Bordetella pertussis* and *Bordetella parapertussis* LightCycler PCR assay with inhibition control. *Diagnostic Microbiology and Infectious Disease*, **46**, 189–195.

67. Raengsakulrach, B., Nisalak, A. and Maneekarn, N. (2002) Comparison of four reverse transcription-polymerase chain reaction procedures for the detection of dengue virus in clinical specimens. *Journal of Virological Methods*, **105**, 219–232.

68. Tichopad, A., Didier, A. and Pfaffl, M.W. (2004) Inhibition of real-time RT-PCR quantification due to tissue-specific contaminants. *Molecular and Cellular Probes*, **18**, 45–50.

69. Yenchitsomanus, P.T., Sricharoen, P., Jaruthasana, I. *et al.* (1996) Rapid detection and identification of dengue viruses by polymerase chain reaction (PCR). *Southeast Asian Journal of Tropical Medicine and Public Health*, **27**, 228–236.

70. Lacey, H.A., Nolan, T., Greenwood, S.L. *et al.* (2005) Gestational profile of Na^+/H^+ exchanger and Cl^-/HCO_3^- anion exchanger mRNA expression in placenta using real-time QPCR. *Placenta*, **26**, 93–98.

71. Stangegaard, M., Dufva, I.H. and Dufva, M. (2006) Reverse transcription using random pentadecamer primers increases yield and quality of resulting cDNA. *Biotechniques*, **40**, 649–657.

72. Lewis, F. and Maughan, N.J. (2004) Extraction of total RNA from formalin-fixed paraffin-embedded tissue, In: *A–Z of Quantitative PCR* (ed. S.A. Bustin), IUL, La Jolla, CA, pp. 591–603.

73. Lekanne Deprez, R.H., Fijnvandraat, A.C., Ruijter, J.M. and Moorman, A.F. (2002) Sensitivity and accuracy of quantitative real-time polymerase chain reaction using SYBR green I depends on cDNA synthesis conditions. *Analytical Biochemistry*, **307**, 63–69.

74. Stanley, K.K. and Szewczuk, E. (2005) Multiplexed tandem PCR: gene profiling from small amounts of RNA using SYBR Green detection. *Nucleic Acids Research*, **33**, e180. Protocol for multiplexing using non-probe-based chemistry.

75. Lutfalla, G. and Uze, G. (2006) Performing quantitative reverse-transcribed polymerase chain reaction experiments. *Methods in Enzymology*, **410**, 386–400.

76. Afonina, I.A., Reed, M.W., Lusby, E. *et al.* (2002) Minor groove binder-conjugated DNA probes for quantitative DNA detection by hybridization-triggered fluorescence. *Biotechniques*, **32**, 940–949.

77. Reynisson, E., Josefsen, M.H., Krause, M. and Hoorfar, J. (2005) Evaluation of probe chemistries and platforms to improve the detection limit of real-time PCR. *Journal of Microbiological Methods*, **66**, 206–216.

78. Tyagi, S. and Kramer, F.R. (1996) Molecular beacons: probes that fluoresce upon hybridization. *Nature Biotechnology*, **14**, 303–308. Description of molecular beacons.

79. Huggett, J., Dheda, K., Bustin, S. and Zumla, A. (2005) Real-time RT-PCR normalisation; strategies and considerations. *Genes & Immunity*, **6**, 279–284.

80. Rutledge, R.G. (2004) Sigmoidal curve-fitting redefines quantitative real-time PCR with the prospective of developing automated high-throughput applications. *Nucleic Acids Research*, **32**, e178.

81. Schefe, J.H., Lehmann, K.E., Buschmann, I.R. *et al.* (2006) Quantitative real-time RT-PCR data analysis: current concepts and the novel "gene expression's CT difference" formula. *Journal of Molecular Medicine*, **84**, 901–910.

82. Ke, L.D., Chen, Z. and Yung, W.K. (2000) A reliability test of standard-based quantitative PCR: exogenous vs endogenous standards. *Molecular and Cellular Probes*, **14**, 127–135.

83. Pfaffl, M.W. and Hageleit, M. (2001) Validities of mRNA quantification using recombinant RNA and recombinant DNA external calibration curves in real-time RT-PCR. *Biotechnology Letters*, **23**, 275–282.

84. Hocquette, J.F. and Brandstetter, A.M. (2002) Common practice in molecular biology may introduce statistical bias and misleading biological interpretation. *Journal of Nutritional Biochemistry*, **13**, 370–377.

85. Stephen, B., Vladimir, B., Jeremy, G. *et al.* (2009) The MIQE Guidelines: Minimum Information for Publication of Quantitative Real-Time PCR Experiments. *Clin Chem*, **55**, 4.

7

Gene Expression in Mammalian Cells

Félix Recillas-Targa, Georgina Guerrero, Martín Escamilla-del-Arenal and Héctor Rincón-Arano
Instituto de Fisiología Celular, Departamento de Genética Molecular, Universidad Nacional Autónoma de México, Apartado, Mexico

7.1 Introduction

Understanding of the sophisticated processes of gene expression in mammalian cells necessitates manipulation of the eukaryotic genome. Such manipulation has greatly contributed to the progress in our knowledge of multicellular organisms, albeit having forced amelioration of our experimental strategies. There are new and varied strategies for gene transfer and sequence manipulation with improved methodologies that facilitate the acquisition of results. Recent experimental data has shown that, for convenient stable transgene expression analysis, the influence of chromatin structure should be seriously taken into consideration. Novel regulatory and structural chromatin elements have been proposed as necessary for appropriate and sustained gene expression. These chromatin elements are facing a new era in transgenesis, and we are probably beginning to see a new generation of gene and cancer therapy vectors. In this chapter, we first briefly discuss and describe several strategies for gene transfer and genome manipulation, and then two alternative methods are presented for the study of gene expression in mammalian cells. In the first method, particular attention is paid to the genomic integration context of transgenes, while the second is designed to maintain chromosomal integrity containing the genomic sequences under study.

Starting with recombinant DNA technology, a large spectrum of applications has emerged in gene manipulation area. In particular, intense efforts have been made to transfer genetic information to cell lines, primary cell cultures, diverse organisms and tissues, and in the generation of genetically modified organisms. Without a doubt, gene transfer has been critical in the advancement of our knowledge related to general phenomena such as gene

Genomics: Essential Methods   Edited by Mike Starkey and Ramnath Elaswarapu
© 2011 John Wiley & Sons, Ltd.

regulation, post-translational events, chromatin composition, recombinant protein production and gene therapy, among others. One particular fact to be taken into consideration is that there is no general methodology for gene transfer and, therefore, gene expression studies. Each cell type and organism needs prior careful characterization to ensure optimal transfer conditions to reach the highest efficiencies and reproducibility. In the great majority of experimental biological systems we are constantly confronted with two variables. The first one has to do with the great inconsistency in the expression levels of transgenes and the second with a less-studied phenomenon, namely the progressive extinction of transgene expression, which is present in the great majority of the cases [1, 2]. In addition, a less-clear general phenomenon has to do with multiple copies of the same transgene which, once integrated into the host genome, induces a phenomenon called co-suppression in plants, which causes gene expression silencing when multiple copies of transgenes are integrated in tandem [3].

The lack of accessible *in vivo* systems compels us to search for alternatives to gene expression in eukaryotic cells. Over the years a large list of gene transfer applications has arisen, and the constant appearance of new methodologies has made it more accessible and reproducible for research scientists [4]. One of the most common applications for gene transfer is the study of gene expression patterns. Such kinds of studies can be supported by the use of primary cell cultures from various organisms and tissues, but we can also take advantage of transformed cell lines derived from viral infections or even different types of tumor. Overexpression of gene products can be an alternative to defining gene function, to interfere with and search for a particular signal transduction pathway, or even to titrate post-translational modifications of histones and/or endogenous peptides. At the present time, gene transfer methodologies for gene expression are more reliable as they are based on better knowledge of the parameters involved, even though some problems remain unsolved. Study of gene regulation is probably the most widely used method to understand the activity of eukaryotic regulatory elements. The list of such elements is still growing, including classic examples such as promoters, enhancers, locus control regions (LCRs) and, more recently, insulators [2, 4–7]. All of these studies are based on two main components: (i) the use of measurable reporter genes and (ii) subsequent transfer to cells in a transient or stable way. Plasmids carrying different reporter genes, like chloramphenicol acetyl-transferase (CAT), and more recently luciferase (LUC) genes, β-galactosidase and the green fluorescence proteins (GFPs), are all commercially available, which are efficient and save time in studying the activity of control elements [4].

It is very important to mention that transient transfection experiments, even though they are very instructive, may give inconsistent results, particularly when compared with the same reporter vectors now integrated into the genome of the host cells. In other words, the chromatin environment of an integrated reporter plasmid can give totally different results compared with episomal vectors. In our experience, transient assays reproducibly determined a silencer activity associated with non-coding sequences located at the 3'-end of the chicken α-globin domain. When the same sequences were tested in an integrated context, the silencer contributed positively to the adjacent enhancer elements [8]. This clearly illustrates two opposing activities, depending on whether the regulatory elements under study are located on a chromatin context or not.

Thus, it is critical to understand, without any doubt, the different gene expression profiles within a cell. However, this is not a simple task because of the different networks and redundancies used in nature to reach a highly regulated specific pattern of gene expression.

If we add the complexity and participation of cell nucleus compartments and their associated chromatin structures, it becomes clear that a lot of effort in experimental design and development of new tools is needed for manipulating the genome [9, 11].

7.1.1 Artificial chromosomes and transgenesis

In the course of the last two decades, remarkable efforts have been made on the development of new strategies for gene expression that incorporate large genomic sequences. This is mainly because there is more and more evidence showing that genes require a sophisticated set of proximal (promoters) and distal regulatory elements to achieve relevant expression [12, 13]. Therefore, the use of artificial chromosomes from yeast and bacteria has turned out to be an attractive alternative, not only to recapitulate endogenous gene expression patterns, but also, even more interestingly, to manipulate large genomic regions by taking advantage of homologous recombination strategies [14–16]. Thus, transgenesis with artificial chromosomes has proven to be useful not only in the study of a variety of regulatory and developmental processes, but also for biomedical and biotechnological applications [15]. Two classes of artificial chromosome vector types are commonly used with large cloning capacities: the yeast artificial chromosomes (YACs) and the bacteria-derived bacterial artificial chromosomes (BACs) or P1-artificial chromosomes (PACs). YAC and BAC/PAC vectors permit the incorporation of genomic inserts ranging from 100 kb to more than 1 Mb. The insertion of such large genomic sequences facilitates inclusion of all the regulatory sequences needed for gene expression. Furthermore, the incorporation of all the elements required for gene expression certainly contributes towards ensuring positional independence, copy-number dependence and optimal levels of transgene expression. In addition, the most attractive and useful feature of artificial chromosomes is the apparent unlimited capacity to generate a large variety of modifications that can be incorporated into such vectors, including target disruption of specific sequences, inversion or even insertions [17–19].

7.1.2 Gene transfer and expression problems

Gene transfer expression faces two obstacles: the first consists of frequent variability in transgene expression levels and the second, which is a less-studied phenomenon, is the progressive extinction of expression or silencing of the transgenes. In both cases, the main component responsible for such effects, also known as chromatin position effects, is chromatin structure [1, 2].

7.1.3 Position effects and chromatin

There are two types of position effect: chromatin position effects caused by different integration sites and position effect variegation induced by a rearrangement and subsequent silencing of an active gene, frequently due to its inactivation because of its proximity to heterochromatin [20–22]. Generally, position effect variegation has been defined as a stochastic and heritable silencing of gene expression. Such an effect is particularly accentuated when the integration of the transgene occurs close to heterochromatin, where the silencing pressure is even stronger.

On the other hand, chromatin position effects are basically considered to be the variability in gene expression due to random insertion of each transgene in diverse chromatin environments in the genome. As a consequence, studies associated with chromatin remodeling mechanisms have, in particular, contributed to the understanding of such variability on gene expression. This has led to designing of novel strategies for counteracting the silencing effect of chromatin on random integrated genes, resulting in sustained and homogeneous transgene expression.

7.1.4 Tissue-specific regulatory elements

For any successful gene transfer and expression analysis, the tissue- or cell-specificity should be determined carefully. In addition, the strength of associated regulatory elements represents a complementary factor. Position effects of chromatin had been overcome by the use of strong and dominant regulatory elements. Certain viral enhancer–promoter combinations are largely sufficient to overcome position effects, like the cytomegalovirus (CMV) promoter–enhancer elements (Rincón-Arano and Recillas-Targa, unpublished observations). Unfortunately, these elements inherently possess a large spectrum of cell-type activity, making transgene expression too non-specific. With the discovery of LCR in the human β-globin locus, particular excitement was generated in opening up the possibility of overcoming position effects when those sequences were included in transgene vectors [13, 23]. The β-globin LCR provides strong erythroid-specific gene expression, and, in its absence, these genes are subject to strong chromatin position effects. When linked to a reporter gene, there is a copy-number-dependent gene expression in erythroid cells, independent of the integration site [13, 23]. In other words, the presence of LCR had a positive and dominant effect on transgene expression independent of the chromatin integration context [7, 13, 24]. To date, a significant number of LCRs have been discovered, all of them being tissue specific [25]. Based on their properties, the LCRs have been incorporated into transgene designs in retroviral vectors and in the generation of transgenic mice [26, 27]. The principal restriction for the use of LCRs in recombinant gene expression is their tissue-specificity.

7.1.5 Sustained expression and chromatin insulators

A more recent and attractive alternative for transgene expression is the use of insulator or boundary elements [2, 6]. Initially discovered in *Drosophila*, insulators have emerged in different chromatin domains. At the moment, the best-characterized insulator is the chicken β-globin insulator [2, 28]. Insulators are functionally defined based on two experimental properties: (i) they are able to interfere with enhancer–promoter communication exclusively when located between them and (ii) they have the capacity to protect a transgene, when located on each side of the vector, against chromatin position effects, independent of the genomic integration site [2, 6, 28]. Usually, insulators co-localize with constitutive DNase I hypersensitive sites and, in general, they behave as neutral elements; that is, they are not activators or repressors of transcriptional activity. All these features, and particularly the ability to shield transgenes against chromatin position effects and progressive extinction of expression, show the real potential of insulators in transgenesis and gene therapy. Thus, the use of chromatin insulators, in combination with tissue-specific regulatory elements, is getting closer to becoming a real means of protecting against position effects with direct consequences in the expression patterns of transgenes and gene therapy vectors.

In conclusion, there are several methodologies that can be used for the study of gene expression. Here, we first describe the method associated with transgene interchange, known as Cre recombinase-mediated cassette exchange (RMCE), which targets the transgene into a predetermined chromosomal integration site. Second, we describe the use of the chicken B-cell line, DT40, for the generation of microcell fusion for the homologous recombination modification of genomic regions maintaining the chromosome integrity for subsequent studies.

7.2 Methods and approaches

7.2.1 Site-specific chromosomal integration in mammalian cells

Owing to the complexity of regulatory elements, their study requires the comparison of different portions and/or the analysis by point mutations in a chromatinized context. However, random chromosomal integration in a number of uncontrolled sites and transgene copies generally cannot be predicted or reproduced with precision because of chromosomal position effects. The inability to control the site of integration, the number of integrated copies and the level of expression of transgenes has impeded progress in studies of both gene expression and the physiological effects of transgenes. The RMCE Protocols 7.1 and 7.2 represents, when cell lines are used, a direct and reproducible strategy for transgene integration to predetermined chromosomal loci through the use of sequence-specific Cre recombinase [29–31]. This method is based on the establishment of independent, single-copy integrants of the stably transfected pL1-HYTK-L2 vector containing the *CMV-HYTK* gene. These can subsequently be used for the generation of recipient stable cell lines enabling positive selection with hygromycin. Once characterized by Southern blot and single-copy integrants are confirmed, they can be subjected to a negative selection with gancyclovir to select those cells where the cassette exchange has occurred. This protocol relies on the use of two opposing *loxP* sequences which induce cassette inversion and test cassette exchange (see Figure 7.1a). Cassette exchange will depend on the efficiency of Cre recombinase expression and probably the genomic site of integration. It is worth mentioning that the cassette would integrate in one orientation in half of the clones and in the reverse orientation in the other half.

One central aspect of RMCE methodology is that, by determining the genomic insertional site, we can eliminate chromosomal position effects, allowing reliable comparisons of multiple transgenes individually integrated at the same genomic site [8]. Additionally, independent receptor clones can be generated which are expected to be integrated in random genomic locations, which offers the possibility of discarding or confirming the results obtained for a given integration site. An alternative potential use of the RMCE assay is to drive the integration of the recombination cassette to a particular and neutral (a deserted) chromosomal location via homologous recombination; for example, on chicken DT40 cells (see below) or embryonic stem (ES) cells. One of the more valuable advantages of this procedure is that, after the selection with gancyclovir, the exchange cassette does not need to have a positive selectable marker, favoring a more reliable expression of the test transgene and avoiding any undesired regulatory influence of the selectable gene frequently located nearby. The RMCE protocol can also be used for the functional characterization of regulatory elements or even for chromatin structure studies [31, 32]. When an optimal integration site is defined, we can produce polypeptides in a controlled, sustained and reproducible manner for use in different applications [33, 34]. DT40 technology is well established and widely used by researchers.

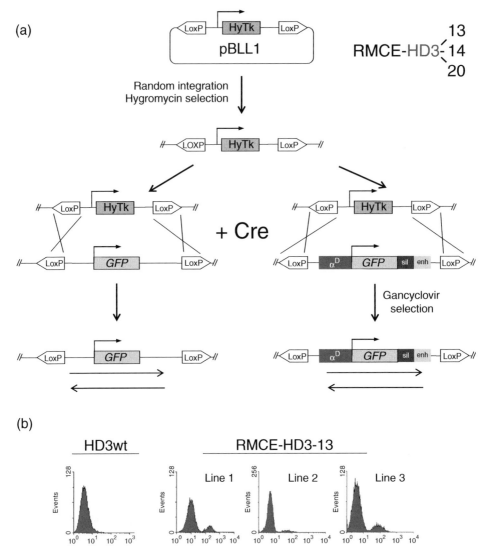

Figure 7.1 Site-specific chromosomal integration in mammalian cells. (a) Schematic representation of recombinase-mediated cassette exchange procedure. Stable and single-copy integrants
are generated and isolated in the chicken erythroblast HD3 cell line (RMCE-HD3-13, -14 and
-20) that are randomly integrated in the HD3 genome. The test vector is co-transfected with
an expression vector for Cre-recombinase. Cells are subjected to gancyclovir and subsequently
individual monoclonal stable lines are isolated. (b) Isolation of individual clones through flow
cytometry analysis. Cells were grown sequentially for 3 days in the presence of gancyclovir,
3 days without selection and again 3 days with gancyclovir selection. Despite all these pre-
cautions, cell sorting remains the most suitable complementary strategy. The diagram shows a
non-transfected wild-type (HD3wt) FACS profile and three independent clones associated with
the same integration site (RMCE-HD3-13) [30, 31, 34].

There are two aspects that require particular attention when performing RMCE. First, high levels of Cre-recombinase gene expression are needed. For this, the use of Cre-GFP fusion vectors allows isolation of GFP-positive cells by cell sorting, increasing the probability of isolating clones in which the cassette exchange has occurred. Second, the enrichment of cell population is required in which the cassette exchange takes place. These goals can be attained by taking advantage of GFP as a reporter gene (whenever possible) to perform one or even two rounds of cell sorting. This is feasible owing to the fact that positive reporter-expressing cells will be favored and this particular cell population will be enriched. For this reason, alternative selection methods are recommended (see Section 7.3).

In conclusion, the RMCE protocol is very efficient, enabling chromatin- and function-related studies with reproducible results for the analysis of regulatory elements and generation of varied types of transgenes in different cell lineages, including ES cells [30, 31, 34].

7.2.2 Plasmid requirement

1 We used the *CMV-HYTK* gene that can be positively selected by hygromycin and negatively by gancyclovir. A plasmid (pL1-HYTK-L1) containing the *CMV-HYTK* gene flanked by inverted *loxP* sites is stably transfected and selected with hygromycin B [30].

2 Linearized plasmids should be dephosphorylated beforehand to ensure the isolation of single-copy integrants.

PROTOCOL 7.1 Recombinase-Mediated Cassette Exchange

Equipment and reagents

- Fluorescent flow cytometer (Becton Dickinson)
- Dulbecco's modified Eagle's medium (DMEM; GIBCO)
- Fetal bovine serum (FBS) (Wisent)
- Gancyclovir (Sigma)
- Hygromycin B (Roche)
- Methocel (Fluka)
- Salmon sperm DNA (Sigma)
- Lipofectamine (Invitrogen).

Method

1 Mix slowly pL1-HYTK-L1 plasmids, in 1 : 1 or 1 : 2 ratio with Lipofectamine in 50 μl total volume.[a]

2 Incubate with cells for 4–6 h.

3 After 24–48 h post-lipofection, replace culture media by DMEM with 750 μg/ml of hygromycin B.

4 Carry out three hygromycin selections for a period of 3–5 days, to allow cell recovery.

5 Select stable lines incorporating the *CMV-HYTK* gene in DMEM plus 2% Methocel, containing hygromycin.

6 Pick individual clones after 2–3 weeks and grow in a medium containing hygromycin.

7 Check the integrity of transgene by polymerase chain reaction (PCR) amplification and Southern blotting using GFP as probe.

8 Identify clones containing a single copy[b] of the L1-HYTK-L1 cassette by Southern blots.[c]

Notes

[a]Cells can also be electroporated using the Gene Pulser II cell electroporator.

[b]Linearization of plasmid is recommended for more efficient isolation of single-copy integrants, followed by dephosphorylation prior to transfection.

[c]Alternatively, it is also possible to determine the site of integration by linear amplification-mediated-polymerase chain reaction (LAM-PCR) [35].

PROTOCOL 7.2   Isolation of Exchanged Transgenes

Equipment and reagents

- Cre recombinase expression plasmid pBS185 (GIBCO)
- DMEM (Gibco)
- FBS (Wisent)
- Fluorescent flow cytometry (Becton Dickinson)
- PCR machine (MasterCycler, Eppendorf)
- Methocel (Fluka)
- Ganyclovir (Sigma)
- Hygromycin B (Roche)

Method

1 For erythroid cell lines, co-transfect the linearized plasmids (the test and Cre expression plasmids) with Lipofectamine for 4 h, in the presence of DMEM cell culture media without serum (Figure 7.1a).[d]

2 Culture the clones (containing the *CMV-HYTK* gene) in selection medium with 0.75 mg/ml of hygromycin B.[e]

3 Three days after the transfection, perform the negative selection over a period of 10 days, in the presence of 50 ng/ml of gancyclovir.[f]

4 In order to increment cell survival without losing the effects of the negative selection, perform three rounds of negative selection for 3 days combined with 3 days of cell recovery (Figure 7.1b).*f*

5 Sort the cells individually into 96-well plates, expanded in the absence of any selection*g* and test for positive RMCE events by fluorescent flow cytometry (fluorescence-activated cell sorting, FACS) (Figure 7.1b).*h* In order to decrease any selection bias against positive expressing clones, perform Southern blotting against the *GFP* reporter gene.*i*

Notes

*d*Invested *loxP* sequences (Cre-recombinase target site) are placed flanking the transgene to be targeted to the same genomic insertion site to induce cassette exchange through Cre-recombinase transient expression. Non-recombinant clones are discarded by negative selection with gancyclovir

*e*Cells should be cultured in selection medium 10 days before the electoporation or Lipofectamine transfection.

*f*Different amounts of Cre-expression plasmid pBS185 were treated and we found that 3 : 1 ratio with respect to the test transgene (the cassette to be exchanged) increases the frequency of recombination and, thus, optimal gancyclovir selection.

*g*The Cre-RMCE procedure seems to have variable degrees of efficiency. We suggest that cells that have been successfully transfected, and where the Cre-mediates cassette exchange has occurred, can be enriched by two or more rounds of FACS, in particular when the *GFP* expression gene was incorporated to the exchange plasmid.

*h*We found a percentage of positive RMCE positive clones near to the 10%.

*i*It is important to outline that such a system allowed us to eliminate the influence of selection gene marker.

7.2.3 Chromosome transfer

Owing to the central role of chromatin structure in the regulation of gene expression, genetic studies need to be performed in the context of intact chromatin environments [7–10]. Homologous recombination and transgenesis have been demonstrated as powerful tools for modifying and manipulating mammalian gene loci. The great majority of such studies are performed by micro-injection of transgenes into the pro-nucleus of a mouse fertilized oocyte, or by the use of mouse ES cells where specific mutations or deletions can be generated by homologous recombination. In both cases we are confronted by two main problems. First, both approaches are limited by their low frequencies and large screening procedures are needed. Second, they are restricted to selected mammalian species, particularly mouse, limiting the manipulation of human genome in a more natural chromatin environment. To circumvent these problems, intact chromosomes could be transferred from one cell type to another, allowing their genetic manipulation [36–38]. Complementary to such chromosomal transfer, for erythroid cell lines analysis, we can manipulate the transferred chromosome in the chicken B-cell line, DT40, which is an avian leucosis virus (ALV) transformed cell line. This cell line is useful for obtaining a very high transformation efficiency by homologous recombination, ranging from 10 to 80% (in our experience we reached a reproducible 40%) [39]. Unfortunately, not much is known about the molecular mechanisms for this unusually high efficiency in this cell line.

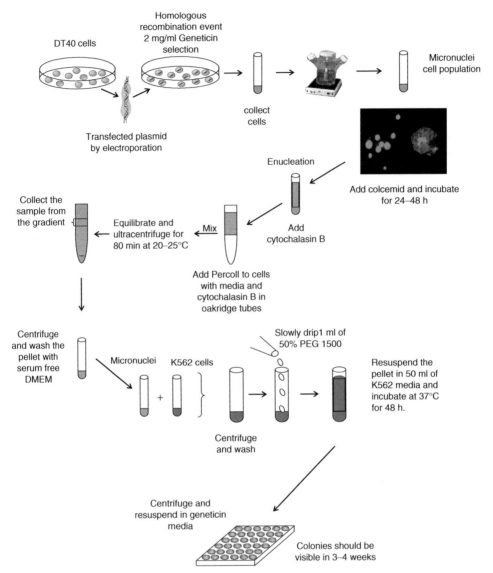

Figure 7.2 A flow diagram describing experimental procedure for the generation of microcell hybrids and modification of chromosomal DNA using homologous recombination in chicken pre-B cell line, DT40. Homologous recombination events are selected and transferred into a mammalian cell line (K562 cells). The cell donor may be obtained from either mammalian cells or directly from chicken DT40 cells. For example, the first donor cells could be from human origin, transferred and modified in DT40 cells by the process of homologous recombination, and eventually the modified human chromosome is transferred a second time for analysis in an appropriate mouse cell line. (See Plate 7.2.)

The critical point when using DT40 cell line is that chromosomes from other species can be transferred into these cells, where targeting efficiencies of the transferred chromosomes are often of significantly higher magnitude than in other vertebrate cells. Thus, genetic manipulation can be done in DT40 cells in the presence of appropriate selectable markers, and then the modified chromosome can again be transferred into another cell line with an appropriate or convenient genetic background where gene expression studies can be performed (Figure 7.2).

Thus, a typical experiment takes into account two steps that involve microcell fusion of particular human chromosomes containing the genetic domain of interest and a previously integrated selectable marker gene. However, it is important to emphasize that, for each microcell fusion event, it is extremely important to verify the integrity of the transferred chromosomes, since during these procedures the loss of entire chromosomal regions had been observed with some frequency [38, 40]. To this end, systematic PCR analysis on entire chromosome with previously defined genetic markers, complemented with fluorescent *in situ* hybridization (FISH), is necessary to confirm chromosome integrity. Microcell fusion procedures coupled with homologous recombination techniques have been successfully applied to chromatin studies, particularly to the human and mouse β-globin loci [38, 40]. In summary, microcell fusion coupled with the DT40 cells' capacity to perform homologous recombination represents a clear alternative to knock-out mice and, more broadly, the spectrum of cell types that can be analyzed.

Microcell fusion protocol takes advantage of the ability of donor cells (DT-40 chicken lymphoid cell line) to become micronucleated. Protocols 7.3, 7.4 and 7.5 are general protocols for microcell transfer into K562 cells (human erythroleukemic cell line). The frequency for this fusion is 30–40%. As DT40 × human microcell fusions are relatively inefficient, a large number of donor microcells is required for successful transfer [36, 37].

PROTOCOL 7.3   Microcell Fusion: Micronucleation of Donor Cells

Equipment and reagents

- DMEM (Gibco)

- FBS (Wisent)

- Chicken serum (Gibco)

- 10% tryptone phosphate buffer (Sigma)

- Penicillin/streptomycin (Invitrogen)

- Geneticin (Invitrogen)

- Colcemid, demecolcine (Sigma), powder is stored desiccated and protected from light at −20 °C. Stock solution of 1 mg/ml in saline is stable for at least 6 months at −20 °C if protected from light.

- Hoechst dye33258, bisbenzimide (cat. B-2883, Sigma) powder is stored at room temperature. Saline stock solution at 50 mg/ml is stable indefinitely at room temperature.

- 500 ml spinner bottle

- Cell electroporator (Gene Pulser II, BioRad)

- Cell culture stirrer (Thermolyne, type 45600 cellgro stirrer)

- Bench-top centrifuge (Labofuge 400R, Heraeus)

- ALV-induced bursal lymphoma DT40 cell line from chicken (ATCC CRL-2111)

Method

1 Generate recombinant plasmids with homologous sequences of the genomic region under study on each side of a selectable marker gene.[j]

2 Transfect the recombinant plasmid by electroporation into 1×10^7 DT40 cells to generate stable clones and select them further with 2 μg/ml of geneticin.[k]

3 After stable DT40 cells are obtained, check the recombination event and the integrity of the recombinant vector by PCR and Southern blot.

4 For micronucleation of recombinant DT40 cells, seed 250 ml of DMEM containing 10% FBS with 50 ml of DT40 cells ($1-3 \times 10^7$ cells /ml) in a 500 ml spinner bottle.

5 Incubate at 37 °C for 24 h.

6 Add Colcemid[l] at 0.5 μg/ml and incubate the cells at 37 °C for 24 h to induce micronucleation.

7 Harvest the micronucleate cells by collecting in 50 ml sterile, polypropylene tubes. Leave one aliquot (5 ml) to count with a hemocytometer and another one for visualization of micronucleate cells.[m]

8 Centrifuge the micronuclei cells at 1000g in a bench-top centrifuge and pool the pellets in 50 ml of DMEM/10% FBS.

Notes

[j]The neomycin-resistant gene should be flanked on each side by *loxP* site-specific recombinant motifs and homologous sequences should be of at least 1.5 kb in length and devoid of repetitive sequences.

[k]25 μg of linearized plasmid is needed to obtain satisfactory transfection efficiencies in DT40 cells [37].

[l]Colcemid is used to induce micronucleation interfering with the microtubule formation. Thus, chromosomes are scattered in the cell and nuclear envelope is formed around one or several chromosomes forming micronuclei. The micronuclei containing the chromosome of interest are then isolated in the presence of geneticin selection.

[m]For visualization, pellet the cells and fix them in at least three changes of 3 : 1 (v/v) methanol : acetic acid. Place a drop of the suspension on a clean, dry microscope slide and air dry. Overlay with Hoechst solution (∼0.5 μg/ml) and stain for 1–3 min. Rinse with water and mount a cover slip with mounting medium. Micronucleate cells can be visualized under UV illumination (excitation 365 nm, emission 480 nm); they are easily distinguished from mononucleate cells (Figure 7.2).

PROTOCOL 7.4   Enucleation of Micronucleate Cell Populations

Equipment and reagents

- Percoll (GE; supplied as a sterile suspension in aqueous solution).
- Cytochalasin B (Sigma) stock solution of 2 µg/ml in DMSO.
- Oakridge, 30 ml polycarbonate tubes (Oakridge Tubes Beckman).
- High-speed refrigerated centrifuge (Beckman XL-90 ultracentrifuge).
- Polycarbonate 50 ml sterile disposable centrifuge tubes (Corning tubes).
- FBS (Wisent)

Method

1 Prepare Percoll by adding 92 ml of Percoll, 3 ml of 5 M NaCl and 5 ml of 1 M Hepes pH 7.0. Add 15 ml of equilibrated Percoll to 2 × 30 ml Oakridge tubes.

2 Prepare DMEM, supplemented with 10% FBS, adding cytochalasin B to a final concentration of 20 µg/ml.

3 Pellet donor cells by centrifugating at 1000g and resuspend in 30 ml DMEM, 10% FBS + cytochalasin B to a density of <3 × 10^7 cells/ml. Clumps of cells should be broken by trituration.

4 Divide the cell suspension into several tubes with Percoll, filling each to capacity (approximately 15 ml), and mix well by inverting the tubes.

5 Centrifuge the tubes at 31 000g in an ultracentrifuge using a 45Ti rotor for 80 min at 20 °C.[n]

6 After centrifugation, two or more bands should be visible. A sample from about 2 cm below the top of the tube[o] down to the region just above the Percoll pellet is collected into 50 ml polypropylene tubes.

7 Centrifuge the tubes again at 2000g for 10 min and remove the supernatant

8 Resuspend and pool the pellets in 50 ml of serum-free DMEM and reserve an aliquot for quantification.

9 Centrifuge pooled pellets at 2000g for 10 min and rinse with serum-free DMEM at least three times to remove Percoll.

Notes

[n] It is recommended to fix a low temperature (∼20 °C) during ultracentifugation to avoid overwarming that can cause massive cell death.

[o] The enucleation process results in extrusion of the nucleus or micronuclei, surrounded by a small amount of cytoplasm and plasma membrane [36, 37].

PROTOCOL 7.5   Fusion of Microcell to Recipient K562 Cells

Equipment and reagents

- Polyethylene glycol (PEG 1500, ICN): Crystals are stored at room temperature.
- Polypropylene 50 ml sterile disposable centrifuge tubes (Corning)
- DMEM (Gibco)
- FBS (Wisent)
- Penicillin/streptomycin (Invitrogen)
- Hemacytometer (Marienfeld)
- Erythroleukemic human transformed cell line K562 (ATCC CCL-243)
- High-speed refrigerated centrifuge (Beckman XL-90 ultracentrifuge)
- Culture plates (150 mm and 96-well) (Corning)

Method

1 Plate 100 ml of K562 cells (seeded at 2×10^5/ml) in 100 mm culture plates on the day prior to the fusion, in DMEM supplemented with 10% FBS and 1% of penicillin/streptomycin.

2 Prepare 1 ml of 50% (w/w) PEG in serum-free DMEM, dissolve at 37 °C, and sterilize by filtration.

3 Count an aliquot of recipient K562 cells and pellet 2×10^6 cells by centrifugation.

4 Resuspend in a volume of 10 ml of DMEM medium in a 50 ml tube.

5 Rinse the cell pellet once and resuspend in 10 ml serum-free DMEM medium.

6 Resuspend the microcell pellet (at least 2×10^7 microcells) in 10 ml of serum-free DMEM. Vigorously mix the suspension by pipetting up and down to resuspend any clumps of cells.

7 Add the resuspended microcells to the recipient K562 cell[p] suspension and mix well. Let the cell/microcell mixture settle for 10 min at room temperature and then centrifuge at 2000g for 10 min.

8 Aspirate the media and slowly drip 1 ml of 50% PEG w/w while gently dispersing the pellet for 1 min. Immediately add 1 ml of serum-free DMEM in a dropwise fashion, while gently swirling for 1 min. Incorporate an additional 1 ml of serum-free DMEM under the same conditions.

9 Add 7 ml serum-free DMEM in a dropwise fashion, while gently swirling over 2 min.

10 Centrifuge the fusion mixture at 1000g for 10 min and gently rinse the pellet with three changes of serum-free DMEM. Centrifuge the fusion mixture each time.

11 Resuspend the pellet in 50 ml non-selective K562 medium, place in 150 mm culture plates and incubate for 48 h at 37 °C.

12 Resuspend the fusion in selective K562 plating medium and divide among 8 × 96-well plates approximately with 0.2 ml/well. Hybrid colonies^q should be visible in 3–4 weeks.^{r}

Notes

^{p}For microcell fusion to recipient cells it is very important to have at least five times more micronuclei in relation to the amount of recipient cells. This is because micronuclei are not stable.

^{q}Hybrid cells, which in an average experiment represent 10% of the total fused cells, can be stored at this point in liquid nitrogen in 10% DMSO and FBS.

^{r}The integrity of the transferred chromosome should carefully be monitored by PCR amplification using several primers designed for the chromosome of interest. Verification by Southern blotting and *in situ* hybridization is also recommended [36, 37].

7.3 Troubleshooting

- For the RMCE protocol one of the critical aspects is the selection of the recombination event. This procedure is highly dependent on the levels of Cre recombinase gene expression.

- Careful gancyclovir incubation conditions should be established to avoid massive cell death or poor selection.

- Another aspect that needs to be taken into consideration, in particular for gene expression studies, is that even though we can compare the effect of different transgenes integrated at the same genomic site, progressive gene expression extinction can occur over time [1, 2].

- For avoiding any kind of FACS selection bias, Southern blot analysis is recommended immediately after gancyclovir selection.

- Stirring of the cells should be performed with specific cell culture stirring plate, since low speed is critical for the integrity of micronuclei formation.

- Before starting the microcell fusion procedure it is essential to establish the optimal Colcemid concentration, since it varies for different cell types.

- It is recommended that micronuclei formation be verified through fluorescence microscopy before proceeding to the protocol (Figure 7.2).

Acknowledgments

We thank Mayra Furlan-Magaril for critical reading of the manuscript. This work was supported by the Dirección General de Asuntos del Personal Académico–Universidad Nacional Autónoma de México (IN209403, IX230104 and IN209403) and Consejo Nacional de Ciencia y Tecnología (CONACyT) (42653-Q and 58767).

References

1. Pikaart, M.J., Recillas-Targa, F. and Felsenfeld, G. (1998) Loss of transcriptional activity of a transgene is accompanied by DNA methylation and histone deacetylation and is prevented by insulators. *Genes & Development*, **12**, 2852–2862.

2. Recillas-Targa, F., Valadez-Graham, V. and Farrell, C.M. (2004) Prospects and implications of using chromatin insulators in gene therapy and transgenesis. *BioEssays*, **26**, 796–807.

3. Garrick, D., Fiering, S., Martin, D.I. and Whitelaw, E. (1998) Repeat-induced gene silencing in mammals. *Nature Genetics*, **18**, 56–59.

4. Recillas-Targa, F. (2006) Multiple strategies for gene transfer, expression, knockdown, and chromatin influence in mammalian cell lines and transgenic animals. *Molecular Biotechnology*, **34**, 337–354.

5. Blackwood, E.M. and Kadonaga, J.T. (1998) Going the distance: a current view of enhancer action. *Science*, **281**, 60–63.

6. Valenzuela, L. and Kamakaka, R.T. (2006) Chromatin insulators. *Annual Review of Genetics*, **40**, 107–138.

7. Bulger, M. and Groudine, M. (1999) Looping versus linking: toward a model for long-distance gene activation. *Genes & Development*, **13**, 2465–2477.

8. Escamilla-Del-Arenal, M. and Recillas-Targa, F. (2008) GATA-1 modulates the chromatin structure and activity of the chicken alpha-globin 3′ enhancer. *Molecular and Cellular Biology*, **28**, 575–586. This is an example of the RMCE protocol used from the chromatin and transcriptional perspective. In this work, endogenous chromatin configuration is restored in an independent chromatin integration site, and different point mutations can then be compared in the same chromatin environment.

9. Recillas-Targa, F. and Razin, S.V. (2001) Chromatin domains and regulation of gene expression: familiar and enigmatic clusters of chicken globin genes. *Critical Reviews in Eukaryotic Gene Expression*, **11**, 227–242.

10. Chakalova, L., Debrand, E., Mitchell, J.A. *et al.* (2005) Replication and transcription: shaping the landscape of the genome. *Nature Reviews Genetics*, **6**, 669–677.

11. Schneider, R. and Grosschedl, R. (2007) Dynamics and interplay of nuclear architecture, genome organization, and gene expression. *Genes & Development*, **21**, 3027–3043.

12. Kosak, S.T. and Groudine, M. (2004) Form follows function: the genomic organization of cellular differentiation. *Genes & Development*, **18**, 1371–1384.

13. Dean, A. (2006) On a chromosome far, far away: LCRs and gene expression. *Trends in Genetics*, **22**, 38–45.

14. Giraldo, P. and Montoliu, L. (2001) Size matters: use of YACs, BACs and PACs in transgenic animals. *Transgenic Research*, **10**, 83–103.

15. Copeland, N.G., Jenkins, N.A. and Court, D.L. (2001) Recombineering: a powerful new tool for mouse functional genomics. *Nature Reviews Genetics*, **2**, 769–779.

16. Ristevski, S. (2005) Making better transgenic models: conditional, temporal, and spatial approaches. *Molecular Biotechnology*, **29**, 153–163.

17. Peterson, K.R., Navas, P.A., Li, Q. and Stamatoyannopoulos, G. (1998) LCR-dependent gene expression in beta-globin YAC transgenics: detailed structural studies validate functional analysis even in the presence of fragmented YACs. *Human Molecular Genetics*, **7**, 2079–2088.

18. Calzolari, R., McMorrow, T., Yannoutsos, N. *et al.* (1999) Deletion of a region that is a candidate for the difference between the deletion forms of hereditary persistence of fetal hemoglobin and $\delta\beta$-thalassemia affects β- but not bold γ-globin gene expression. *EMBO Journal*, **18**, 949–958.

19. Tanimoto, K., Liu, Q., Bungert, J. and Engel, J.D. (1999) Effects of altered gene order or orientation of the locus control region on human bold β-globin gene expression in mice. *Nature*, **398**, 344–348.

20. Robertson, G., Garrick, D., Wu, W. *et al.* (1995) Position-dependent variegation of globin transgene expression in mice. *Proceedings of the National Academy of Sciences of the United States of America*, **92**, 5371–5375.

21. Henikoff, S. (1996) Dosage-dependent modification of position-effect variegation in *Drosophila*. *BioEssays*, **18**, 401–409.

22. Wakimoto, B.T. (1998) Beyond the nucleosome: epigenetic aspects of position-effect variegation in *Drosophila*. *Cell*, **93**, 321–324.

23. Grosveld, F., van Assendelf, G.B., Greaves, D.R. and Kollias, B. (1987) Position-independent, high-level expression of the human beta-globin gene in transgenic mice. *Cell*, **51**, 975–985.

24. Festenstein, R. and Kioussis, D. (2000) Locus control regions and epigenetic chromatin modifiers. *Current Opinion in Genetics & Development*, **10**, 199–203.

25. Bonifer, C. (2000) Developmental regulation of eukaryotic gene loci: which cis-regulatory information is required? *Trends in Genetics*, **16**, 310–315.

26. Pannell, D. and Ellis, J. (2001) Silencing of gene expression: implications for design of retrovirus vectors. *Reviews in Medical Virology*, **11**, 205–217.

27. Neff, T., Shotkoski, F. and Stamatoyannopoulos, G. (1997) Stem cell gene therapy, position effects and chromatin insulators. *Stem Cells*, **15**, 265–271.

28. Burgess-Beusse, B., Farrell, C., Gaszner, M. *et al.* (2002) The insulation of genes from external enhancers and silencing chromatin. *Proceedings of the National Academy of Sciences of the United States of America*, **99** (Suppl. 4), 16433–16437. This is a short manuscript that describes the properties of chromatin insulators with particular reference to chicken cHS4 β-globin insulator.

29. Capecchi, M.R. (1989) Altering the genome by homologous recombination. *Science*, **244**, 1288–1292.

30. Feng, Y.-Q., Seibler, J., Alani, R. *et al.* (1999) Site-specific chromosomal integration in mammalian cells: highly efficient CRE recombinase-mediated cassette exchange. *Journal of Molecular Biology*, **292**, 779–785. This is the original publication describing the theoretical and practical aspects of RMCE assay. There is a clear description of the recombination procedure, the recombinant plasmids and the reagents required.

31. Baer, A. and Bode, J. (2001) Coping with kinetic and thermodynamic barriers: RMCE, an efficient strategy for the targeted integration of transgenes. *Current Opinion in Biotechnology*, **12**, 473–480.

32. Goetze, S., Baer, A., Winkelmann, S. *et al.* (2005) Performance of genomic bordering elements at predefined genomic loci. *Molecular and Cellular Biology*, **25**, 2260–2272.

33. Wong, E.T., Kolman, J.L., Li, Y.-C. *et al.* (2005) Reproducible doxycycline-inducible transgene expression at specific loci generated by Cre-recombinase mediated cassette exchange. *Nucleic Acids Research*, **33**, e147.

34. Toledo, F., Liu, C.-W., Lee, C.J. and Wahl, G.M. (2006) RMCE-ASAP: a gene targeting method for ES and somatic cells to accelerate phenotype analyses. *Nucleic Acids Research*, **13**, e92.

35. Schmidt, M., Schwarzwaelder, K., Bartholomae, C. *et al.* (2007) High-resolution insertion-site analysis by linear amplification-mediated PCR (LAM-PCR). *Nature Methods*, **4**, 1051–1057.

36. Killary, A.M. and Lott, S.T. (1996) Production of microcell hybrids. *Methods*, **9**, 3–11.

37. Dieken, E.S. and Fournier, R.E.K. (1996) Homologous modification of human chromosomal genes in chicken B-cell x human microcell hybrids. *Methods*, **9**, 56–63. This is a reference manuscript in which we can find a detailed protocol for the microcell fusion.

38. Dieken, E.S., Epner, E.M., Fiering, S. *et al.* (1996) Efficient modification of human chromosomal alleles using recombination-proficient chicken/human microcell hybrids. *Nature Genetics*, **12**, 174–182. This manuscript represents one of the most appealing examples of chromosome transfer and describes a targeted modification of the human β-globin locus by homologous recombination using chicken pre-B cell lines, DT40.

39. Buerstedde, J.M. and Takeda, S. (1991) Increased ratio of targeted to random integration after transfection of chicken B cell lines. *Cell*, **67**, 179–188.

40. Epner, E., Reik, A., Cimbora, D. *et al.* (1998) The beta-globin LCR is not necessary for an open chromatin structure or developmentally regulated transcription of the native mouse beta-globin locus. *Molecular Cell*, **2**, 447–455.

8

Using Yeast Two-Hybrid Methods to Investigate Large Numbers of Binary Protein Interactions

Panagoula Charalabous, Jonathan Woodsmith and Christopher M. Sanderson
Department of Physiology, School of Biomedical Sciences, University of Liverpool, Liverpool, UK

8.1 Introduction

The yeast two-hybrid (Y2H) system is a well established procedure which detects direct or 'binary' protein–protein interactions. Since its development in 1989 [1], the classic Y2H assay has undergone a series of adaptations to increase stringency and reduce the occurrence of false-positive interactions [2–5]. In recent years, new procedures have also been developed to facilitate high-throughput analysis of many thousands of potential protein interactions [6–10]. Although global 'interactome' projects often incorporate automated robotic procedures, it is possible to perform thousands of Y2H assays, using relatively inexpensive manual techniques. In the post-genomic era there is an increasing demand for established procedures which can be used to provide a more extensive insight into the organization and complexity of biological and pathological processes. In this chapter we describe a series of reagents and procedures which can be used in any laboratory to perform large-scale manual Y2H experiments.

The classical Y2H system [1] utilizes the inherent properties of the GAL4 transcription factor, which is composed of two functionally distinct regions: a DNA binding domain (BD) and a transcriptional activation domain (AD). The BD region binds to GAL4-specific sequences in the promoter region of particular target genes. In contrast, the AD domain drives transcription of adjacent downstream genes. In the classic Y2H assay the GAL4 protein is fragmented into its component domains. Conventionally, the BD region is fused to the N-terminus of a protein or domain of interest to generate a 'bait' protein, while the AD

Genomics: Essential Methods   Edited by Mike Starkey and Ramnath Elaswarapu
© 2011 John Wiley & Sons, Ltd.

domain is fused to the N-terminus of a second 'prey' protein (Figure 8.1). Once fragmented, the two parts of the GAL4 protein can no longer drive transcription of GAL4-dependent genes. However, if the BD and AD domains are fused to proteins that interact, then the GAL4 BD and AD domains are brought into close enough proximity to drive transcription of an adjacent downstream reporter gene. In most contemporary Y2H systems, a combination of biosynthetic and enzymatic reporter genes are used to minimize the possibility of promoter-specific false-positive results [5]. In the Y2H system described in this chapter, three reporter genes are used to detect positive interactions. These include two biosynthetic reporter genes (ADE2 and HIS3) and one enzymatic (lacZ) reporter gene, each of which is under the control of different Gal4-induced promoters (Gal2p, Gal1p and Gal7p respectively).

Although the basic principles of all Y2H methods are similar, several different Y2H systems have now been developed. Therefore, it is important to emphasize that vectors and host strains from different Y2H systems may not be compatible. Care should be taken to check both the genotype of the host strain(s) being used and the characteristics of different bait and prey vectors. There can be considerable variations in the amount of bait or prey fusion protein expressed in different GAL4-derived Y2H systems. These differences can have dramatic effects on the strength and number of interactions detected under equivalent selection conditions. For example, the pGAD prey vector has a less efficient promoter than that found in pACT vectors. As such, fewer interactions will be detected with a pGAD vector. However, the high stringency of the pGAD system means that interactions that are detected may be strong. At times lower stringency may be required to detect very weak or transient interactions. In this case pACT-derived vectors may be a more appropriate choice if appropriate controls are used. The key point is to fully understand the inherent characteristics of the system being used.

8.2 Methods and approaches

The procedures described in this chapter have been developed to provide sensitive medium or high-throughput assays, which, in combination with stringent controls, can provide large amounts of high-confidence binary protein interaction data. In essence, they utilize the PJ69-4A yeast strain developed by James *et al.* [5], a set of bait and prey recombination vectors developed by Semple and Markie [11] and a selection of high-throughput adaptations inspired by the pioneering work of Walhout and Vidal [12]. Adaptations used in the procedures described in this chapter have the advantage that a single polymerase chain reaction (PCR) product can be used to generate either bait or prey clones, there is no need for the use of expensive cloning enzymes, and vectors are compatible with both Gateway™ format inserts and a large collection of commercially available Y2H libraries (Clontech).

When performing medium- or large-scale Y2H studies, three considerations must be taken into account: the production of large numbers of bait and/or prey constructs; the rapid screening of potential interactions; and the use of appropriate controls to eliminate false-positive data. These criteria are equally valid for both targeted 'matrix-style' Y2H studies or the screening of high-complexity Y2H libraries.

8.2.1 Producing large numbers of bait or prey clones

The process of producing large numbers of bait and prey clones can be expedited by the use of *in vivo* Gap repair cloning methods [13, 14]. This approach is based on the simple

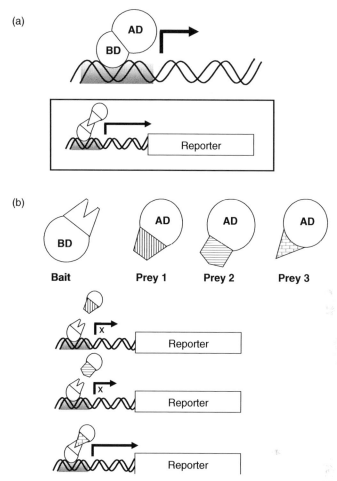

Figure 8.1 Principles of the classical Y2H system. (a) The Gal4 transcription factor is composed of two functionally distinct domains: the DNA binding domain (BD), which recognizes specific sequences in the promoter region of GAL4-dependent genes; and a transcriptional activation domain (AD), which drives transcription of adjacent downstream genes. In the Y2H system the Gal4 protein is fragmented into its two component domains. The DB domain is fused to a protein of interest and the AD domain is fused to a potential binding partner. If the two proteins interact, the DB and AD domains are brought into close proximity and are capable of driving transcription of a downstream reporter gene. (b) When potential binding partners of a protein of interest are not known, bait constructs can be used to screen a library of potential interaction partners. In this method, yeast containing a specific bait protein is mated against many thousands of yeasts containing different prey proteins. Only when bait and prey proteins interact will biosynthetic reporter genes be turned on. This enables yeast containing interacting proteins to grow on selective media. Sequencing of DNA encoded in the prey vector enables the identity of novel interaction partners to be identified.

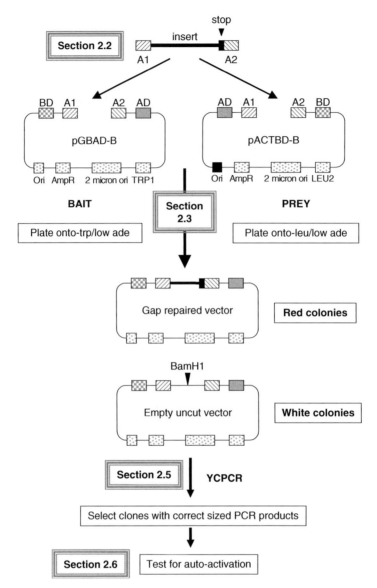

Figure 8.2 Use of *in vivo* Gap repair cloning methods to construct bait or prey clones for use in Y2H studies. The first stage of gap repair cloning is the production of PCR-generated inserts containing the protein coding sequence or domain of interest flanked by different 5′ and 3′ recombination sequences. This PCR product is then co-transfected into an appropriate yeast strain together with a linearized bait or prey vector containing compatible recombination sequences. Following transfection *in vivo*, homologous recombination occurs, resulting in the formation of bait or prey vectors containing in-frame fusions with proteins of interest. When the protein of insert encodes an in-frame stop codon, the downstream Gal4 domain is not expressed. As such, yeast containing vectors with in-frame inserts produce pink/red colonies on selective media containing limiting amounts of adenine. This facilitates rapid detection of bait or prey constructs with potentially correct inserts. Verification of insert size and sequence and auto-activation potential can then be determined prior to use in protein interaction studies.

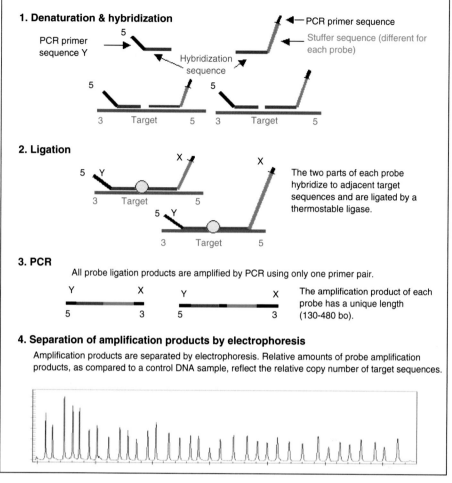

1. Denaturation & hybridization

PCR primer sequence Y

PCR primer sequence

Stuffer sequence (different for each probe)

Hybridization sequence

Target

Target

2. Ligation

The two parts of each probe hybridize to adjacent target sequences and are ligated by a thermostable ligase.

Target

Target

3. PCR

All probe ligation products are amplified by PCR using only one primer pair.

The amplification product of each probe has a unique length (130–480 bo).

4. Separation of amplification products by electrophoresis

Amplification products are separated by electrophoresis. Relative amounts of probe amplification products, as compared to a control DNA sample, reflect the relative copy number of target sequences.

Plate 1.3a Schematic of MLPA. The MLPA reaction is performed using four steps. Genomic DNA is denatured, whereafter the MLPA probes are added and incubated for 16 h, allowing complete hybridization adjacent to all target sequences. Probes completely hybridized to sequences either side of each target region are subsequently ligated to each other, enabling their exponential PCR amplification and final detection, and quantification by capillary electrophoresis.

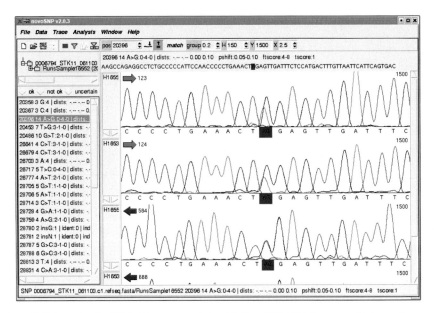

Plate 2.1 SNP detection in ABI 3730 sequence data with NovoSNP. SNP discovery efforts can be organized by projects, each with its own reference sequence (top left). Raw sequence files are basecalled and aligned to the reference sequence, after which a list of candidate sequence changes (left) is generated. Each prediction can be manually reviewed by visualizing the traces as they align to the reference sequence (right).

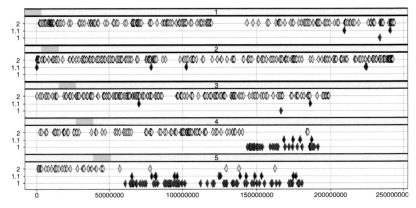

Plate 3.2 Spotfire Genotype and LOH visualization of a single tumor sample relative to a paired normal sample. Five panels (1–5) for chromosomes 1–5 are shown. For each panel, on the x-axis the position of each SNP is depicted in base pairs from the p-telomere to the q-telomere of the chromosome. An SNP that is heterozygous in both the normal sample and paired tumor is represented in yellow diamonds on the 2-line on the y-axis. SNPs that are heterozygous in the normal sample but homozygous in the tumor sample are represented in red diamonds on the 1-line, while SNPs that are heterozygous in the normal but with a quality ratio below 0.8 in the paired tumor sample are represented in blue diamonds on the 1.1 line. LOH is called in regions (relative to their base pair positions on the x-axis), marked by more red and blue markers (SNPs) than yellow markers.

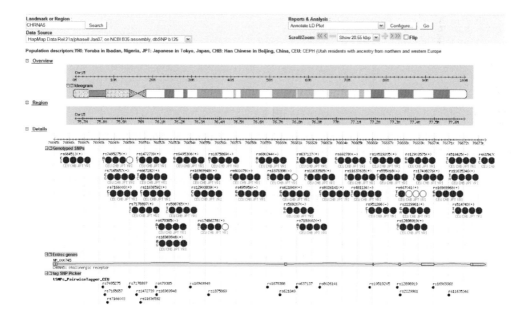

Plate 4.1 An example of the browser display from www.hapmap.org, accessed by clicking on the 'HapMap Genome Browser (B35 – full data set)' link. The gene *CHRNA5* was entered into the 'Landmark or Region' field. The 'Scroll/Zoom' box indicates that the display is showing 28.55 kbp. The 'Overview' panel, or track, indicates the full chromosome on which this gene lies, and the chromosomal region blown up under the 'Region' panel. The 'Details' panel shows the SNPs genotyped by HapMap in the selected region, and also displays a pie chart of the allele frequencies for each SNP in each of the four HapMap population samples: CEU (Centre de Polymorphisme Humaine, CEPH; Utah residents with ancestry from northern and western Europe), YRI (Yoruba in Ibadan, Nigeria), CHB (Han Chinese in Beijing, China) and JPT (Japanese in Tokyo, Japan). The last two tracks display the gene and also tag SNPs which have been selected according to the default settings: tags represent r^2 bins where all bin members satisfy $r^2 > 0.8$ with at least one tag in the CEU population. The parameters for tag SNP selection may be modified using the 'Reports and Analysis' drop-down menu, which is currently set on 'Annotate LD Plot': click on the arrow to the right, select 'Annotate tag SNP Picker' and choose the desired parameters.

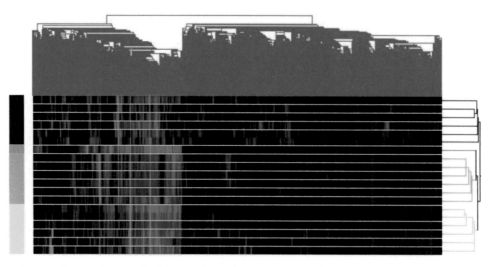

Plate 5.4 Reproducibility and reliability of amplification strategies. HeLa and Stratagene Universal Human Reference RNA were run on 19 000-element cDNA microarrays. 10 µg of each of the RNAs was used as a control condition (gray). RNA was then amplified by T7-amplification (blue), NuGen Ovation™ (orange) or Global-RT-PCR (yellow).

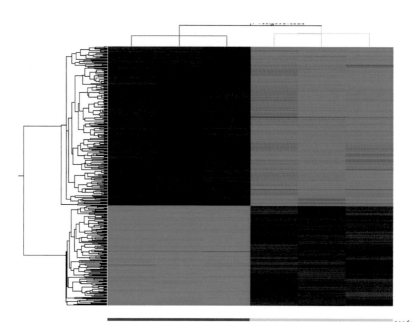

Plate 5.5 Single-cell profiling by Global-RT-PCR. Three individual cells from each of two groups were obtained and the RNA was amplified and profiled on Agilent 44k Whole Human Genome arrays. After a *t*-test was performed to identify a list of 358 genes which distinguished between the two groups, the gene panel was subjected to hierarchical clustering. Reproducible results from each of the cells used in each of the groups were obtained and provided a strong identifier panel.

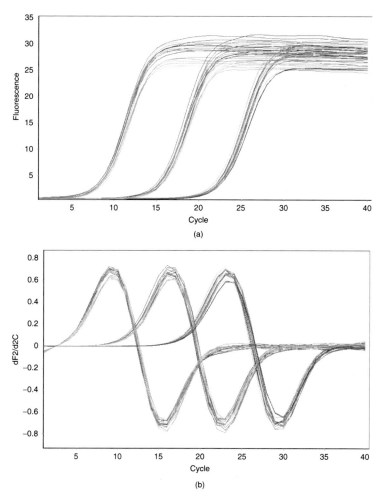

Plate 6.3 Comparative quantification. This procedure provides an alternative to the more usual dilution curve analysis for determining the fold change of a gene of interest relative to a calibrator sample. (a) Three amplification plots derived from a 100-fold serial dilution series of an RNA template used to optimize a new primer set. Note that there is no threshold line, which would be used to determine the C_q using the usual dilution curve method. (b) How the relative concentration of a sample is derived from the amplification plots shown in (a). (1) The second derivatives of the three amplification plots are calculated. These produce peaks corresponding to the maximum rate of fluorescence increase in the reaction denoted by 1, 2 and 3. (2) The 'takeoff' points (labelled 4, 5 and 6) are determined for each curve. A takeoff point is defined as the cycle at which the second derivative is at 20 % of the maximum level, and indicates the end of the noise and the transition into the exponential phase. (3) The average increase in signal four cycles following the takeoff point (denoted by three bars labeled a, b and c) is used to calculate a slope, which provides a measure of the amplification efficiency for each curve. A 100 % efficient reaction should double the signal in the exponential phase. So, if the signal was 10 at cycle 15, then went to 11 at cycle 16, it should go to 13 fluorescence units at cycle 17. (4) All of the amplification values for each sample are averaged to give a mean efficiency of a group of cycling curves for each sample (three in this example). The more variation there is between the estimated amplification values of each sample, the larger the confidence interval will be. In this example, the average amplification is 1.68 ± 0.02 for the neat template and 1.76 ± 0.01 and 1.76 ± 0.02 for the two dilutions. (5) The same procedure is carried out for the calibrator sample and a fold change can then be calculated according to the formula Fold change = Efficiency$^{(\text{Calibrator takeoff} - \text{Target sample takeoff})}$.

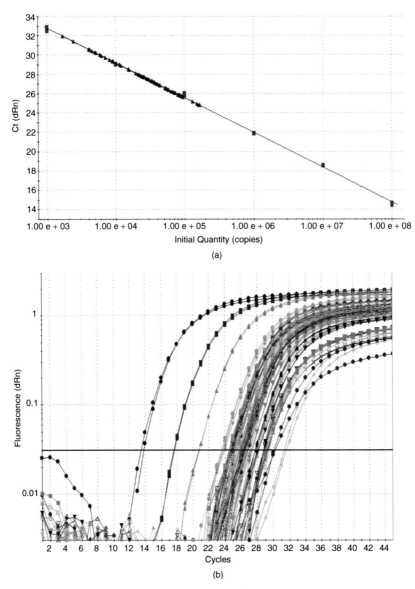

Plate 6.4 (a) Typical standard curve used to quantitate target mRNA from colonic biopsies. All the C_q quantification data from the test samples (blue triangles) in the upper picture are contained within the dynamic range of the standard curve, which is demarcated by the two outermost points of the standard derived from samples of a defined concentration and represented by red squares. This allows accurate quantification of the corresponding mRNAs. (b) Typical amplification plot obtained using a SYBR Green I assay. A single transcript has been quantified in a number of test samples and a serial dilution of standard material using SYBR Green I as the reporter. The two replicates for the three most concentrated standard samples (traces on left of the graph colored blue, red and green) illustrate a good standard of pipetting. The slopes of all the amplification plots are identical, indicating that the amplification efficiencies of every sample are the same. The high relative fluorescence (ΔR_n) value is typical of SYBR Green I assays.

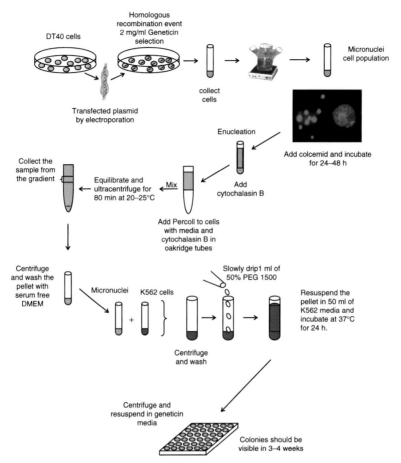

Plate 7.2 A flow diagram describing experimental procedure for the generation of microcell hybrids and modification of chromosomal DNA using homologous recombination in chicken pre-B cell line, DT40. Homologous recombination events are selected and transferred into a mammalian cell line (K562 cells). The cell donor may be obtained from either mammalian cells or directly from chicken DT40 cells. For example, the first donor cells could be from human origin, transferred and modified in DT40 cells by the process of homologous recombination, and eventually the modified human chromosome is transferred a second time for analysis in an appropriate mouse cell line.

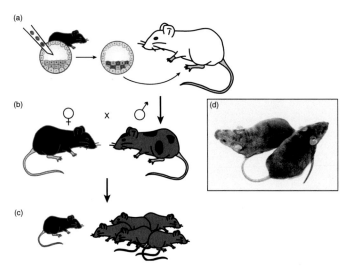

Plate 10.5 ES cell injection and chimera breeding. (a) Blastocysts are obtained from a C57BL/6 mouse (black coat color) and microinjected with targeted R1 ES cells (129 genetic background: agouti coat color). Injected blastocysts are implanted in a pseudopregnant female (white mouse). (b) The resulting black and brown male chimeras are then mated with a C57 black female to obtain brown and black F1 offspring (c). Brown pups are then screened by PCR or Southern blot for germline transmission of the targeted gene. (d) Two adult mouse chimeras generated through gene targeting in ES cells and injection into a C57BL/6 blastocyst. The mouse on the left has a higher percentage of agouti coat color; the probability of germline transmission of the targeted gene to the offspring of this mouse is higher than that for the mouse on the right.

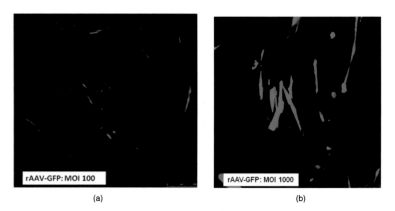

Plate 12.1 AAV-GFP transduction. The green fluorescent cells indicate the *gfp* expression analyzed by confocal microscopy (×250) (Leica Microsystems-TCS SP5). (a) Transduction efficiency was 10% for MOI 100. (b) Transduction efficiency was 40% for MOI 1000. A vector **pUF[11]** (an AAV-based plasmid vector) consisting GFP (535 bp) was transduced in human fetal fibroblasts (IMR-90). IMR-90 cells were plated in six-well plates (5×10^4 cells/well) and were cultured in minimum essential medium (MEM Engle) supplemented with 10% heat-inactivated fetal bovine serum (FBS), 0.1 mm non-essential amino acids and 1.0 mm sodium pyruvate. The cells were cultured for 24 h at 37 °C, 5% CO_2. When cells were approximately 70–80% confluent, the cells were washed in PBS and transduced with AAV-GFP vector.

principle of homologous recombination. By introducing common 5′ and 3′ specific recombination sequences downstream of the BD or AD domains in both bait and prey vectors, it is possible to generate a single PCR product which will directionally recombine with a series of compatible vectors to generate in-frame bait or prey constructs (Figure 8.2). Conventionally, bait and prey vectors are assembled in PJ69-4A MATa [11] or PJ69-4αMATα [5] host strains respectively. This facilitates mating of bait and prey clones during subsequent interaction studies.

Procedures described in this chapter involve the use of bait and prey vectors containing the *att*B1 and *att*B2 recombination sites, which are also used in GATEWAY™ cloning systems. Therefore, PCR products generated for standard GATEWAY™ BP reactions (Invitrogen) can also be used to generate bait or prey clones by *in vivo* gap repair when co-transfected with either pGBAD-B/pACTBD-B or pGBAE-B/pACTBE-B vectors [11]. This series of vectors can also facilitate the rapid detection of bait or prey clones containing in-frame inserts (Figure 8.2).

Unlike conventional Y2H vectors, which encode either DB or AD domains, the pGBAD-B/pACTBD-B and pGBAE-B/pACTBE-B series of vectors encode both DB and AD domains separated by an in-frame linker containing the A1 and A2 recombination sequences (Figure 8.2). In the intact empty vector the DB and AD domains are expressed as one in-frame fusion protein. As a result, Y2H reporter genes are constitutively turned on, allowing yeast to grow efficiently, producing white colonies on low adenine selection media. In contrast, vectors containing an insert with an in-frame stop codon will only express the 5′ Gal4 domain. Fusion proteins expressed from these vectors will not independently drive transcription of biosynthetic reporter genes. Consequently, yeast containing inserts with in-frame stop codons produce colonies with a characteristic pink/red color on media containing low amounts of adenine [11].

8.2.2 Generating recombination-compatible inserts for gap repair cloning

PCR products used for gap repair cloning are composed of three regions: the 5′ upstream recombination sequence; the protein coding region; and the 3′ downstream recombination sequence (Figure 8.2). In this procedure the 5′ recombination sequence termed the 'A1 tag' includes the *att*B1 sequence, which is in-frame with upstream BD or AD domains (see Protocol 8.1). Reverse primers are designed to include the 'A2 tag' (including the *att*B2 sequence) downstream of an in-frame stop codon. Inserts or domains lacking a stop codon must be used in combination with the pGBAE-B (bait) or pACTBE-B (prey) vectors, which use a frame-shift strategy to prevent read-through transcription of downstream AD or BD domains. Significantly, the pGBAE-B and pACTBE-B vectors still allow clones with in-frame inserts to be selected by pink/white color selection [11].

PROTOCOL 8.1   Primer Design and PCR Amplification of Inserts for Use in Gap Repair Reactions

Equipment and reagents

- PCR tubes
- PCR machine

- DNA mini gel
- Standard submarine gel tank and power pack
- Hot-start KOD polymerase (Novagen)
- UV spectrophotometer
- Pipettes (p2, p10 and p200).

Method

1 Set up 25 μl PCR reactions as follows (μl/reaction):

Forward primer (10 μM stock)[a]	1.5 μl
Reverse primer (10 μM stock)[a]	1.5 μl
10× KOD buffer	2.5 μl
dNTPs (provided with KOD)	2.5 μl
MgSO$_4$ (provided with KOD)	1.5 μl
Hot-start KOD	0.3 μl
∼100 ng of PCR product[b]	1.0 μl
$_d$H$_2$O	14.2 μl

PCR reactions should be performed using the following program:

i. 98 °C for 30 s

ii. 55–68 °C for 30 s (dependent on primer T_m)

iii. 70 °C for 1 min

iv. (extension period should be increased if insert size is larger than 3 Kb)

v. Repeat steps i–iii 29 times

vi. Hold at 15 °C.

Run out 5 μl of the PCR reaction on a 1% agarose gel to confirm product size and assess the specificity of the PCR.[c]

Notes

[a]Primers used to generate inserts for gap repair reactions consist of a standard recombination tag (A1 or A2), which is identical to the 5′ recombination sequence in the vector and a stretch of approximately 20 3′ gene-specific nucleotides. Forward primers may encode the ATG 'start codon' of the protein of interest, but this is not essential as inserts will be expressed as in-frame fusions with upstream DB or AD domains. However, it is important to remember that a 'stop codon' is required at the end of the gene-specific segment when using the pGBAD-B or pACTBD-B vectors, to facilitate color selection of clones with inserts. Examples of forward and reverse recombination primers are shown below. Tags to be added to protein or domain of interest:

A1 Tag: 5′ GAA TTC ACA AGT TTG TAC AAA AAA GCA GGC TGG *ATG XXX XXX XXX*3′

A2 Tag: 5′ GTC GAC CAC TTT GTA CAA GAA AGC TGG GTG *CTA XXX XXX XXX 3′*

Sequences in bold represent insert specific sequences (X). Other stop codons can be used as required; 18–21 gene-specific nucleotides must be added in-frame with forward and reverse tag sequences. When generating PCR inserts from clones in the pDONR223 vector the following primer pairs should be used:

Forward primer: 5′ GAATTCACAAGTTTGTACAAAAAAGCTGGCATG 3′

Reverse primer: 5′ GTCGACCACTTTGTACAAGAAAGCTGGG 3′

[b]DNA concentration can be estimated by measuring absorbance at A^{260} on a UV spectrophotometer or by comparison with the intensity of band in molecular weight markers.

[c]If a single band of the correct size is obtained, the PCR product can be used directly in gap repair reactions, without further purification. However, if multiple bands are produced, or if bands are very weak, it is advisable to purify and concentrate the required band. Alternatively, PCR reactions containing multiple bands can be used directly in gap repair reactions, as yeast containing vectors with the correct sized insert can be identified when performing subsequent yeast colony (YC) PCR reactions (Sections 8.2.4 and 8.2.5 below). Annealing temperatures and extension times may need to be optimized for each template in accordance with instructions provided with the hot-start KOD polymerase.

8.2.3 Performing gap repair reactions

Gap repair reactions require host cells to be transfected with a combination of linearized vector and complementary recombination-compatible PCR products (see Protocol 8.2).

PROTOCOL 8.2 Gap Repair Reactions

Equipment and reagents

- PCR tubes
- PCR machine
- Appropriate selective (low ade) plates
- 1 M LiOAc
- QIAquick PCR purification kit (250)
- Filter-sterilized 50% (w/v) PEG 3350
- Sterile water
- Salmon testis carrier DNA (heat denatured at 99 °C for 5 min before use)[d]

- DNA purification kit

- Shaking incubator

- Heat block

- Yeast host strains: PJ69-4A (MATa) for baits or PJ69-4α (MATα) for preys.[e]

- YPAD full rich media for growth of untransfected host strain. Containing: 10 g/l yeast extract, 100 μg/l adenine hemisulfate, 20 g/l peptone and 20 g/l D-glucose. *This media is only used when no selection is required.*

- SD media: 6.7 g/l yeast nitrogen base without amino acids, 20 g/l D-glucose and 20 g/l agar. This provides a minimal media which can be supplemented in accordance with required selection conditions. When using the PJ69-4A/α strains in combination with pGBAD-B, pGBAE-B, pACTBD-B or pACTBE-B vectors for gap repair reactions, low adenine conditions are required. Therefore, SD media should be supplemented with arginine and methionine (20 μg/ml), isoleucine and lysine (30 μg/ml), phenylalanine (50 μg/ml), valine 150 and 20 μg/ml adenine hemisulfate (as opposed to the standard 100 μg/ml that would normally be used in non-selective media). This low level of adenine is essential for the development of pink/red colonies, which indicate the presence of vectors with in-frame inserts. In addition, leucine should be added at 100 μg/ml, while uracil, tryptophan,[f] and histidine should be added at 20 μg/ml, as required for appropriate selection of bait or prey plasmids. Yeast containing bait vectors (pGBAD-B or pGBAE-B) contain the TRP1 gene and, therefore, will grow on media lacking tryptophan, while prey vectors (pACTBD-B or pACTBE-B) contain the LEU2 gene, which facilitates growth on media lacking leucine.

- To prepare *Bam*HI linearized plasmid DNA: digest 2 μg of plasmid DNA with 20 U of *Bam*HI in 1× digestion buffer (NEB buffer 3 with 100 μg/ml BSA) for 1 h at 37 °C. Following digestion, clean up DNA using a QIAquick PCR purification kit and adjust final concentration to 20 ng/μl.

Method

1. Inoculate 2 ml of YPAD broth with fresh yeast PJ69-4A (MATa) for baits or PJ69-4α (MATα) for preys and grow overnight at 30 °C with shaking (~220 rpm).

2. Next day add an additional 8 ml of fresh YPAD broth and continue to incubate at 30 °C with shaking for 5 h.

3. Harvest yeast by centrifugation for 5 min at 2300 rpm (~700g).[g]

4. Resuspend yeast in 5 ml of 100 mM LiOAc and transfer 1.5 ml of the suspension to a microcentrifuge tube. Pellet the cells as above.

5. Wash cells again in 100 mM LiOAc and harvest them by brief centrifugation in a microfuge at 2300 rpm (~700g).

6. After final wash, remove supernatant and add 320 μl of the following master mix: 2.3 ml 50% (w/v), PEG 3350, 350 μl 1 M LiOAc, 450 μl sterile distilled water, 90 μl of 10.5 mg/ml heat-denatured salmon testis DNA as a carrier and 10 μl of 20 ng/μl of *Bam*HI linearized plasmid DNA. This provides sufficient master mix for 10 reactions.

7. Mix well before transferring 32 μl into individual PCR tubes.

8 Add 4 µl of a specific PCR reaction to each tube of suspended yeast and mix by repeated pipetting. Also add 4 µl of H$_2$0 into one control tube to assess background. This should be zero.

9 Use a PCR machine to incubate reactions at 30 °C for 30 min, followed by 25 min at 42 °C and 1 min at 30 °C.

10 Add 100 µl of sterile water to each tube and mix well before spreading on selective plates. SD low adenine media (- trp) for bait vectors or (- leu) for prey vectors.

Notes

[d]Carrier DNA (salmon testis DNA) does not need to be sonicated. However, it must be heat denatured (95 °C for 5 min) and placed on ice prior to use.

[e]Genotype of the PJ69-4A host strain: MATa trp1-901 leu2-3,112 ura3-52 his3-200 gal4Δ gal80Δ LYS2::GAL1-HIS3, GAL2-ADE2 met2::GAL7-lacZ.

[f]Tryptophan solutions should be *filter sterilized* (not autoclaved) and stored in the dark.

[g]Cells should not be chilled during these procedures.

8.2.4 Identifying positive transformants

After 3–5 days' growth at 30 °C, colonies should be visible on transformation plates. Pink/red colonies indicate the presence of an insert with an in-frame stop codon. White colonies probably result from either auto-activating host cells or the presence of residual uncut vector. A selection of red colonies (around six) should be picked for each construct and each colony should be analyzed to check insert size and the potential for auto-activation.

8.2.5 Yeast colony PCR

The gap repair cloning methods described in Protocol 8.3 are directional. Therefore, inserts should always be introduced with the correct orientation and reading frame. However, it is important to ensure that the insert size is as expected. This can easily be checked using primers that flank the recombination sites.

PROTOCOL 8.3   Using Yeast Colony PCR to Confirm Insert Size

Equipment and reagents

- PCR tubes

- PCR machine

- Six positive colonies for each intended bait or prey construct

- Sterile wooden tooth picks

- Primers for YC PCR reactions: The following primers can be used to check insert sizes in both bait (pGBAD-B or pGBAE-B) or prey (pACTBD-B or pACTBE-B) vectors as they contain common 5′ and 3′ recombination sequences.[h]

  Forward primer: 5′ GAATTCACAAGTTTGTACAAAAAAGCAGGC 3′

  Reverse primer: 5′ GTCGACCACTTTGTACAAGAAAGCTGGGTG 3′

  To avoid amplification of residual PCR insert from gap repair reactions, it is advisable to use a combination of the forward primer above and a vector-specific reverse primer. This is particularly important when screening colonies directly from gap repair plates.

  Forward primer: 5′ GAATTCACAAGTTTGTACAAAAAAGCAGGC 3′

  Reverse pGBAD-B or pGBAE-B primer: 5′ GCCAAGATTGAAACTTAGAGGAG 3′

  Reverse pACTBD-B or pACTBE-B primer: 5′ GTCGGCAAATATCGCATGCTTGTTC 3′

Method

1 Pick a small amount of 5–10 colonies using a clean sterile toothpick for each colony.

2 Dip the toothpick into 3 µl of 0.02 M NaOH. Do not remove all yeast. Resuspend the remaining yeast in 20 µl of sterile H_2O. These suspensions can be used for both auto-activation and YC PCR (see below).

3 To preserve colonies, plate out 4 µl of yeast suspension onto selective plates and allow to grow for 3–5 days.

4 To each of the yeast NaOH suspensions prepared in step 1 add 12 µl of the following master mix. Amounts indicated provide enough master mix to perform 10 reactions.

Forward primer (10 µM stock)	7.5 µl
Reverse primer (10 µM stock)	7.5 µl
dNTPs: (10 µM stock)	4.5 µl
NH_4 Buffer (10×)	15 µl
$MgCl_2$ (10 mM stock)	7.5 µl
DMSO	3 µl
$_dH_2O$	73.5 µl
Taq polymerase	1.5 µl

5 PCR reactions should then be performed using the following program:

   i. 95 °C for 5 min

   ii. 95 °C for 1 min

   iii. 55–68 °C for 1 min (dependent on primer Tm)

   iv. 72 °C for 1–3 min (use 1 min per Kb)

v. Repeat steps ii–iv 34–39 times

vi. 72 °C for 5 min

vii. Hold at 15 °C.

6 Run out 5–10 µl of each PCR product on a 1% agarose gel to determine the size of individual PCR products.

Notes

[h]If further verification of inserts is required, PCR products can be directly sequenced to generate either sequence tags or full primary sequence, depending on insert size.

8.2.6 Bait and prey auto-activation tests

It is essential to test both bait and prey clones to ensure that they do not independently activate Y2H reporter genes (see Protocol 8.4). There can be several reasons for auto-activation. Common causes include: the occurrence of spontaneous host mutations, which allow cells to grow on selective media in the absence of bait/prey interactions; bait proteins that have inherent transcriptional activity or are able to interact with other factors, which are capable of activating transcription. Also, prey proteins that bind DNA may non-specifically drive transcription of two hybrid reporter genes.

PROTOCOL 8.4   Testing for Auto-Activation of Bait and Prey Clones

Equipment and reagents

• Tooth picks

• SD (− trp − his + 2.5 mM 3AT) plates and SD (− trp − ade) selective plates (for analysis of bait constructs)

• SD (− leu − his + 2.5 mM 3AT) plates and SD (− leu − ade) selective plates (for analysis of prey constructs)

• Bait and prey clones containing vectors with verified size inserts.

Method

1 Spot 3 µl of the resuspended yeast from Protocol 8.3 (step 2) onto SD plates lacking either tryptophan (baits) or leucine (preys). At the same time, spot the same volume onto bait auto-activation plates, SD (− trp − his + 2.5 mM 3AT) plates and SD (− trp − ade) or SD (− leu − his + 2.5 mM 3AT) and SD (− leu − ade) selective plates for preys.[i]

2 Incubate at 30 °C for an equivalent time to that being used to investigate potential interactions. This may range from 7 to 14 days.

3 Ideally, strong growth should be visible on plates lacking only tryptophan or leucine but no growth should be visible on auto-activation plates.[j]

4 Once non-auto-activating bait and prey clones have been identified, colonies can be resuspended in SD media (− leu or − trp) containing 25% sterile glycerol and stored at −80 °C for future use. It is advisable to archive multiple clones of each construct.[k]

Notes

[i]Normally, the histidine antimetabolite 3-aminotriazole (3-AT) is added to (− his) media (at 2.5 mm) to reduce background growth. The concentration of 3-AT can be increased to eliminate weak auto-activation on (− his) plates.

[j]When auto-activation does occur, there are two typical phenotypes. If the bait or prey construct induces auto-activation, strong growth will normally be seen across the entire area of the spot. However, if auto-activation results from spontaneous mutations within the host, this will appear as isolated colonies growing within the area of the spot. Examples of this type of background growth can be seen in Figure 8.3c. If this kind of background occurs, individual colonies can be isolated from the vector selection plate and retested for auto-activation. Alternately, a new transformation can be performed using a fresh culture of host cells.

Some constructs may induce auto-activation on (− his) plates but not on more stringent (− ade plates). If this occurs, it may still be possible to perform screens under (− ade selection). However, in this case it is important to also perform βGal assays to ensure a second independent reporter is being activated.

If either PJ69-4A or AH109 (MATa) haploid bait strains are mated with the Y187 (MATα) haploid prey strain the lacZ and HIS3 reporters cannot be scored independently as they both have a common (GAL1) promoter.

[k]Clones should not be maintained on selective plates for long periods to prevent the accumulation of auto-activating mutations.

8.2.7 Targeted 'matrix'-style Y2H screens

The purpose of matrix-style Y2H assays (see Protocol 8.5 and 8.6) is to test for binary interactions between a predefined set of bait or prey proteins. This approach can be performed in a 96-well format using 12- or 8-tip multi-channel pipettes.

PROTOCOL 8.5   Performing Matrix Style Y2H Mating Assays

Equipment and reagents

- 96-well plate (round-bottomed or 'v'-shaped wells are best for this procedure)
- Racked tips (1–20 µl and 1000 µl)
- Multichannel pipettes (1–20 µl and 200 µl)
- Sterile tooth picks
- YPAD (rich media) plates
- SD plates − trp − leu (diploid selection plates)

- SD plates — trp — leu — his (+2.5 mM 3-AT) (HIS3 reporter selection plates)
- SD plates — trp — leu — ade (ADE2 reporter selection plates)
- Sterile velvets
- Replicating block.

Method

1 Pipette the required amount of sterile water (4 μl per mating reaction) into the required number of wells of a 96-well plate.

2 Using a sterile toothpick, pick a small sample of a bait colony and resuspend the yeast in the required well. It may help to leave the toothpick in the well until finished. This avoids cross-contamination between samples or wells.

3 When all bait clones are resuspended, mix with tips of a multichannel pipette before spotting out 3 μl of each suspension onto a *dry* YPAD plate (150 mm vented plates are required for standard 96-well format arrays).

4 Allow spots to spread and dry.

5 While bait spots are drying on the YPAD plate, resuspend prey clones as for baits.

6 Pipette 3 μl of prey clones directly on top of appropriate bait spots on the YPAD plate.

7 Allow spots to dry; then invert plate and transfer to 30 °C incubator.

8 Allow cells to mate on YPAD plate for 12–16 h at 30 °C.

9 Pre-dry SD (− trp − leu) diploid selection plates in a sterile hood.

10 Stretch a sterile velvet over a replication block and invert the YPAD plate onto the velvet.

11 Apply even gentle pressure with finger tips to the back of the plate.

12 Remove YPAD mating plate and replace it with the pre-dried SD (− trp − leu) diploid selection plate. Again, apply gentle pressure to the back of the plate. Remove plate and replace lid.

13 Incubate the SD (− trp − leu) plate at 30 °C for 2 days until solid even growth is observed on spots.

14 Invert the SD (− trp − leu) plate onto a sterile velvet and apply gentle even pressure. The color of the colonies will change slightly as they are transferred to the velvet. Too much pressure at this stage will result in too much yeast being transferred and results will be less clear.

15 Place a thick (75 ml of media per 150 mm plate) − trp − leu diploid selection plate onto the velvet and apply even gentle pressure. Remove plate and replace lid.

16 Without changing the velvet place an SD (− trp − leu − ade) plate on to the velvet and again apply gentle even pressure. Remove plate and replace lid.[l]

17 Using the same velvet place an SD (− trp − leu − his + 2.5 mM 3-AT) plate on to the velvet and again apply gentle even pressure. Remove plate and replace lid.

18 Incubate plates at 30 °C for up to 10 days. Selective growth can normally be seen within 3–5 days. However, weak interactions may take longer to develop (Figure 8.3).

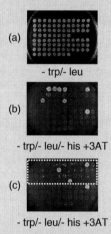

(a)

- trp/- leu

(b)

- trp/- leu/- his +3AT

(c)

- trp/- leu/- his +3AT

Figure 8.3 An example of a targeted matrix style Y2H study. In this study a bait clone containing a human E3-RING protein has been mated against a collection of human E2 Ubiquitin conjugating enzymes. (a) Growth of diploid yeast on (− trp − leu) plates following mating of bait and prey clones. (b) Growth of the same colonies after transfer to selective reporter plates (− trp − leu − his +2.5mM 3AT). On these plates it is possible to see increased growth of some colonies, indicating positive bait–prey interactions. (c) The top section shows an example of a selective reporter plate showing signs of background growth, which is typical of spontaneous mutations in the host yeast. Significantly, because of the colony spotting method used in these screens, it is still possible to score positive interactions above background. This would not be the case if diploids had been patched out onto selective plates. In this case growth would have been more uniform throughout the patch, thereby preventing accurate scoring of true positive interactions.

19 Record growth at regular intervals by photographing plates.

Notes

[i]It is important to replicate onto (− ade) selection plates before (− his) plates. If (− his) plates are used first, too much yeast may be transferred, producing high background growth, which makes positive interactions harder to detect.

PROTOCOL 8.6   Performing *β*-Gal Reporter Assays

Equipment and reagents

- Clean 150 mm Petri dish
- (− trp − his) diploid selection plates

- Circular filter paper (to fit selection plate)

- Liquid nitrogen in a small insulated container.

- Forceps

- Sealable container

- Z buffer: 60 mM Na_2HPO_4 (8.5 g/l), 40 mM NaH_2PO_4 (4.8 g/l), 10 mM KCl (0.75 g/l), 1 mM $MgSO_4$ (0.12 g/l)

Method

1 Using the ($-$ trp $-$ leu) diploid selection plate prepared in Protocol 8.5 (step 15).

2 Incubate the plate for 4–5 days at 30 °C.

3 After the incubation, check the plate for even growth, then place a dry, round filter paper onto the surface of the plate and apply even pressure to transfer cells onto the paper.

4 Using forceps, slowly peel back the filter paper and then immerse in liquid nitrogen (~10 s). Repeat twice.[m,n]

5 Place the filter paper (colony side up) into a Petri dish on top of two filter papers which have been saturated with β-Gal assay mix (6 ml Z buffer, 100 μl of 100 mg/ml X-gal, 11 μl β-mercaptoethanol). 6 ml of reaction mix is sufficient in a 150 mm Petri dish.

6 Incubate at 37 °C for 3–5 h until blue color develops.[o,p]

Notes

[m]Care should be taken at this stage as the filter may become brittle when frozen.

[n]Care should also be taken when handling liquid nitrogen. Although only small quantities are being used in this procedure, appropriate training is essential and institutional safety procedures should be followed at all times.

[o]The β-Gal assay mix is light sensitive and should be incubated in the dark.

[p]As the β-Gal assay mix contains β-mercaptoethanol it is also advisable to make and dispense solutions in a fume hood. Also, Petri dishes should be incubated in a well-sealed container.

8.2.7.1 Streamlining Y2H library screens

In many instances the function of a protein of interest may not be known. Also, most researchers will not have access to large collections of individual prey clones. Therefore, the only way to identify novel interaction partners is to screen high-complexity prey libraries. Although this remains a valuable approach, library screening can be a daunting task. However, with a few simple adaptations, the process can be significantly simplified.

Historically, the main bottleneck in the process of library screening was the identification of interacting prey inserts and the reconfirmation of observed interactions. Previously, this process required the isolation of prey vectors from positive diploid colonies. Once isolated, prey vectors were then reintroduced into fresh yeast, which would then be retested against either the original bait or an irrelevant control. In addition, the isolated prey plasmid would be sequenced to identify the encoded interaction partner. This process can be significantly streamlined by incorporating gap repair cloning and YC PCR procedures (Sections 8.2.3–8.2.5). The procedure consists of four sequential processes.

- Construction of bait clones in PJ69-4A (MATa) (as described in Sections 8.2.2 and 8.2.3).

- Mating of bait clones against a high-complexity prey library. Many good commercial libraries are now available in Matα yeast strains, such as the Matchmaker libraries (Clontech). However, if pretransformed libraries are being used, remember the potential complications of using libraries in the Y187 host strain (Note j, Protocol 8.4).

- Removal of false positive interactions. This involves testing for the activation of multiple reporter genes and reconfirming bait specificity in fresh yeast. By repeating the process in fresh yeast, it is possible to eliminate false positives that arise from spontaneous host mutations.

- Identification of true positive prey proteins. This requires the sequencing of the cDNA insert contained within the prey vector.

As the sequences that flank the inserts in prey libraries are known, it is possible to design PCR primers that include ~30 nucleotides that flank each side of the multiple cloning site. When these primers are used to perform YC PCR reactions on positive diploid colonies, the resulting PCR product will include the encoded prey protein flanked by 5′ and 3′ vector-specific sequences. As such, this PCR product can then be used to perform a gap repair reaction when co-transfected with an appropriately linearized empty prey vector. Using this approach, prey vectors can be rapidly regenerated in fresh yeast using the sequential protocols described in Sections 8.2.1–8.2.6. Once prey clones have been regenerated by gap repair, they can be mated against the original bait clone to reconfirm the interaction, or against a non-specific bait to establish partner specificity (follow procedures outlined in Section 8.2.7).

An additional benefit of this approach is that prey-specific PCR products can be directly sequenced to establish the identity of interacting prey proteins. Using these adaptations, a single person can perform and process 5–10 library screens at a time, which provides a suitable capacity for most biologically focused research projects.

8.3 Troubleshooting

- The key to any good Y2H screen is the inclusion of a full spectrum of controls. In targeted matrix Y2H experiments, the range of different interactions being investigated often provides good internal controls for both background auto-activation and partner specificity. However, in library screens this is not possible and it is essential that all positive interactions must be retested in fresh yeast using the original bait and an irrelevant partner as control.

- We have noticed that, in our hands, YC PCR reactions become more problematic as yeast colonies age. We have also found that when yeast is maintained on plates for several days, multiple bands may be observed in YC PCR reactions. Therefore, whenever possible, perform YC PCR reactions on fresh (~2–3-day-old) colonies.

- To reduce the occurrence of spontaneous auto-activating mutations, it is advisable to use fresh stocks of host yeast when generating bait and prey clones. It is also advisable to produce multiple glycerol stocks of each verified bait and prey clone. In addition, avoid perpetuating clones on selective plates without rechecking auto-activation patterns before screening.

- When using the PJ69-4A strain in combination with either pGBAD-B/pACTBD-B-, pGBAE-B/pACTBE-B- or pACT-based clones; we have occasionally observed 'interactions' which strongly activate the ADE2 reporter, but not the HIS3 or lacZ reporters. These interactions appear to be false positives. This phenotype emphasizes the need for using multiple reporters to score Y2H data.

- The original manuscript describing the PJ69-4A yeast strain suggested that the use of the liquid-nitrogen snap freezing procedure when performing β-Gal assays could be problematic, leading to auto-activation of the lacZ reporter gene. In our experience using the pGBAD-B or pACTBD-B series of vectors, we have not found this to be the case.

References

1. Fields, S. and Song, O. (1989) A novel genetic system to detect protein–protein interactions. *Nature*, **340**, 245–246. The first description of the classical Y2H method.

2. Bartel, P., Chien, C.T., Sternglanz, R. *et al.* (1993) Elimination of false positives that arise in using the two-hybrid system. *Biotechniques*, **14**, 920–924.

3. Vidalain, P.-O., Boxem, M., Ge, H. *et al.* (2003) *Methods*, **32**, 363–370.

4. Koegl, M. and Uetz, P. (2007) Improving yeast two-hybrid screening systems. *Briefings in Functional Genomics & Proteomics*, **6**, 302–312.

5. James, P., Halladay, J. and Craig, E.A. (1996) Genomic libraries and a host strain designed for highly efficient two-hybrid selection in yeast. *Genetics*, **144**, 1425–1436.

6. Uetz, P., Giot, L., Cagney, G. *et al.* (2000) A comprehensive analysis of protein–protein interactions in *Saccharomyces cerevisiae*. *Nature*, **403**, 623–627.

7. Ito, T., Chiba, T., Ozawa, R. *et al.* (2001) A comprehensive two-hybrid analysis to explore the yeast protein interactome. *Proceedings of the National Academy of Sciences of the United States of America*, **98**, 4569–4574.

8. Rual, J.F., Venkatesan, K., Hao, T. *et al.* (2005) Towards a proteome-scale map of the human protein–protein interaction network. *Nature*, **437**, 1173–1178.

9. Stelzl, U., Worm, U., Lalowski, M. *et al.* (2005) A human protein–protein interaction network: a resource for annotating the proteome. *Cell*, **122**, 957–968.

10. Goit, L., Bader, J.S., Brouwer, C. *et al.* (2003) A protein interaction map of *Drosophila melanogaster*. *Science*, **302**, 1727–1736.

11. Semple, J., Prime, G., Wallis, L. *et al.* (2005) Two-hybrid reporter vectors for gap repair cloning. *Biotechniques*, **38**, 927–934.

12. Walhout, A.J.M. and Vidal, M. (2001) High-throughput yeast two-hybrid assays for large-scale protein interaction mapping. *Methods*, **24**, 297–306.

13. Ma, H., Kunes, S., Schatz, P.J. *et al.* (1987) Plasmid construction by homologous recombination in yeast. *Gene*, **58**, 201–216. Plasmid construction by homologous recombination in yeast.

14. Petermann, R., Mossier, B.M., Aryee, D.N. *et al.* (1998) A recombination based method to rapidly assess specificity of two-hybrid clones in yeast. *Nucleic Acids Res.*, **26**, 2252–2253.

9
Prediction of Protein Function

Hon Nian Chua
Data Mining Department, Institute for Infocomm Research, Singapore

9.1 Introduction

Automated protein function prediction (PFP) has gained momentum over the last decade with the proliferation of genomic data. While genomic information becomes available at a rapidly increasing pace, understanding of the functional mechanism of genes and their protein products is relatively less efficient. This has motivated computer scientists and biologists to guide functional discovery by using computational methods to construct functional models and associations using annotated genes in well-studied model organisms and other available biological evidence. With limited success, many computational approaches have been developed to predict gene/protein functions based on a myriad of genomic and experimental evidence, and some of these are readily available for use. In this chapter we provide a concise overview of these *in silico* PFP methods and elaborate the application of a handful of these.

9.2 Methods and approaches

Automated PFP has been, and remains, a popular and important area of research in bioinformatics and computational biology. A considerable number of approaches have been explored, which differ in both the use of computational techniques and biological data. Since a major consideration that guides the choice of an appropriate approach is the availability of data, we group PFP methods according to the type of biological data used as input for inference.

As the focus of the chapter is on the practical application of PFP, we limit the scope to methods that utilize sequence homology [1–3], phylogenetic relationships [4, 5], sequenced-derived functional and chemical properties [6] and protein–protein interaction maps [7–15]. These by no means encompass all available PFP approaches, but are

Genomics: Essential Methods   Edited by Mike Starkey and Ramnath Elaswarapu
© 2011 John Wiley & Sons, Ltd.

chosen because they utilize data that are widely available and usually produce reasonably consistent and relatively good predictions.

Other approaches also exist that use different sources of data, including protein structure [16–18], genomic context [19], gene expression [20, 21], text mining [22, 23] and integration of multiple data sources [24–28]. A comprehensive review on available PFP methods can be found in Hawkins and Kihara [29], while Sharan *et al.* [30] provide an excellent review on the technical aspects of approaches that uses protein–protein interaction networks for PFP. A recent publication also describes a large-scale comparison between many well established gene function prediction methods on *Mus musculus* genes [31].

9.2.1 Annotation schemes

Automated functional prediction is only plausible with the availability of a systematic way of assigning function annotation [32]. Several systems of gene/protein function annotation schemes have been used. One of the earliest standardized schemes is the EC nomenclature [33] developed by the Enzyme Commission of the International Union of Biochemistry and Molecular Biology in the 1950s for classifying enzymes based on their chemical properties. Structural Classification of Proteins (SCOPs) [34] was developed in 1995 to classify proteins based on structure and phylogenetic relationship. The first generalized scheme for classifying protein function was introduced in 1993 for classifying *Escherichia coli* proteins [35]. These classification schemes annotate either a subset of proteins, specific genomes or particular aspects of proteins.

9.2.1.1 FunCat

A more comprehensive functional categorization scheme is the Functional Catalogue (Fun-Cat) [32], developed by the Munich Information Center for Protein Sequences (MIPS) [36]. The FunCat comprises a number of main functional categories (28 in version 2.1, which is the most current at the time of writing) which describe a wide range of general gene functions. Each category consists of a number of gene functional descriptions arranged in a hierarchical structure referred to as a tree in computer science terminology. The annotation term at the top of each hierarchy (or the *root* of the tree) is the most general description of the category, with children terms describing different and more specific forms of their parent term. The trees can span up to six levels in depth. The scheme was originally used for the annotation of the *Saccharomyces cerevisiae* genes, but is generic enough to be extended to other species. A subset of the FunCat annotation scheme is presented in Figure 9.1.

The FunCat scheme can be downloaded from ftp://ftpmips.gsf.de/catalogue/ in both plain text and XML formats. FunCat annotations for a handful of genomes, including *S. cerevisiae*, *Fusarium graminearum* and *Arabidopsis thaliana*, can be downloaded from ftp://ftpmips.gsf.de/catalogue/annotation_data/.

9.2.1.2 Gene ontology

The FunCat annotation scheme is not widely adopted in other databases other than those maintained by MIPS. A more extensively utilized annotation scheme for gene and protein functions is Gene Ontology (GO) [37]. GO was initiated as a collaborative effort in 1998 to

```
# Functional Classification Catalogue    Version 2.1     09.01.2007
01 METABOLISM
01.01 amino acid metabolism
01.01.03 assimilation of ammonia, metabolism of the glutamate group
01.01.03.01 metabolism of glutamine
01.01.03.01.01 biosynthesis of glutamine
01.01.03.01.02 degradation of glutamine
01.01.03.02 metabolism of glutamate
01.01.03.02.01 biosynthesis of glutamate
01.01.03.02.02 degradation of glutamate
01.01.03.03 metabolism of proline
01.01.03.03.01 biosynthesis of proline
01.01.03.03.02 degradation of proline
```

Figure 9.1 A subset of a functional category in FunCat (ftp://ftpmips.gsf.de/catalogue/funcat-2.1_scheme).

address the lack of consistent annotations for gene products in different databases. GO began as a collaboration between FlyBase [38], the *Saccharomyces* Genome Database (SGD) [39] and the Mouse Genome Database (MGD) [40], but has grown to include annotations from a large number of databases, such as WormBase [41], The Institute for Genomic Research (TIGR, now the J. Craig Venter Institute) databases and the Zebrafish Information Network (ZFIN) [42]. In recent years, GO gained popularity quickly and has been used in a large number of studies on function prediction.

PROTOCOL 9.1   Obtaining Gene Ontology Annotations for Function Prediction

Requirements

- Internet access to download data and software
- Programming or scripting language such as C, C++, Perl, Matlab.

Method

1 The GO annotation scheme can be downloaded from http://www.geneontology .org/GO.downloads.ontology.shtml. The ontology can be downloaded in several file formats, including the standard OBO format, SQL database dumps, RDF and OWL.

2 Download GO annotations of genomes of interest from a variety, including *Mus musculus*, *Caenorhabditis elegans* and *Homo sapiens* from http://www.geneontology .org/GO.current.annotations.shtml[a]

3 Each record in an annotation file contains a gene identifier, the GO term identifier and the reference to a database or publication from which the annotation is derived among, other information. See http://www.geneontology.org/GO.format.annotation.shtml for a complete description of the file format.

4 The Qualifier field can contain one or more of 'NOT', 'colocalizes_with' and 'contributes_to'. Of importance is the 'NOT' qualifier, which negates the annotation.

5 The Evidence Code field describes the type of analysis or experiments on which the annotation is based. The types of evidence code are listed at http://www .geneontology.org/GO.evidence.shtml. The type of evidence code has an impact on the credibility of the annotation. In particular, the code 'Inferred from Electronic Annotation' (IEA) is assigned to annotations based on an automatic method without curatorial judgement. The use of such annotations is not recommended and is often avoided.

Notes

[a]Function annotations for a protein may be found in more than one annotation file; for example, the annotation files for Protein Data Bank, UniProt and possibly the organism's specific database.

GO comprises three structured controlled vocabularies, or *ontologies*, for describing molecular function, biological process and cellular component. Each ontology comprises functional terms arranged in a hierarchical structure known as a *directed acyclic graph* (DAG). The DAG differs from a tree mainly in that a term in the former can have multiple parent terms. The relationship between a parent and its children terms is also further specified by two different relationships: *is_a* and *part_of*. Similar to the FunCat, each child term in GO describes a more specific form of its parents. The current version of GO has 26 384 terms. Figure 9.2 illustrates the ancestor terms for 'nucleobase, nucleoside, nucleotide and nucleic acid metabolic process' in the 'biological process' ontology.

Refer to Protocol 9.1 on how to obtain the GO scheme and annotation files. If you wish to use an existing function prediction application for prediction, you may not need to obtain the scheme and annotation data. Many applications are available as online Web services and are already trained on existing annotation data. For such applications, only the features of a protein, such as sequence or structure, are required for prediction. Some applications that

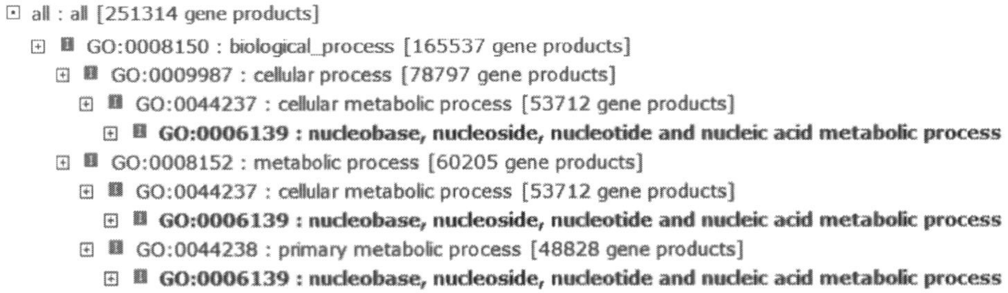

```
⊡ all : all [251314 gene products]
   ⊞ ▣ GO:0008150 : biological_process [165537 gene products]
      ⊞ ▣ GO:0009987 : cellular process [78797 gene products]
         ⊞ ▣ GO:0044237 : cellular metabolic process [53712 gene products]
            ⊞ ▣ GO:0006139 : nucleobase, nucleoside, nucleotide and nucleic acid metabolic process
      ⊞ ▣ GO:0008152 : metabolic process [60205 gene products]
         ⊞ ▣ GO:0044237 : cellular metabolic process [53712 gene products]
            ⊞ ▣ GO:0006139 : nucleobase, nucleoside, nucleotide and nucleic acid metabolic process
         ⊞ ▣ GO:0044238 : primary metabolic process [48828 gene products]
            ⊞ ▣ GO:0006139 : nucleobase, nucleoside, nucleotide and nucleic acid metabolic process
```

Figure 9.2 A subset of a functional category in GO (http://amigo.geneontology.org/cgi-bin/ amigo/term-details.cgi?term=GO:0006139). The term 'nucleobase, nucleoside, nucleotide and nucleic acid metabolic process' has two parent terms 'cellular metabolic process' and 'primary metabolic process.'

can be downloaded and used may require these annotation data to be provided separately. If you wish to implement your own function prediction application based on GO, you will also need to obtain the scheme and annotation data files.

9.2.2 Working with multiple protein identifier systems

To predict functions for proteins, we often need to utilize information from more than one database. For example, to predict functions for proteins from the *S. cerevisiae* genome, we may use functional annotations from GO, protein sequences from the Comprehensive Yeast Genome Database (CYGD) or the SGD [39] and protein–protein interactions from the Biomolecular Interaction Network Database (BIND) [43] or the Biological General Repository for Interaction Datasets (BioGRID) [44]. These databases may refer to the same genes/protein using different identifiers.

The adoption of different naming conventions may stem from various reasons, such as legacy or the nature of the data being referenced (e.g. sequences versus genes). Nonetheless, this poses some problems in PFP when we need to combine data from different sources. Cross-referencing tables are sometimes provided in some of these databases, but these are often incomplete and not up to date. To address the issues of incompleteness and redundancy in cross-referencing genes and proteins, resources such as the International Protein Index (IPI) [45] and the UniProt Universal Protein Resource [46] have been developed. UniProt provides a unique identifier to every distinct protein sequence, while IPI provides a unique identifier for every distinct annotated protein. Efforts have also been made to provide services for cross-referencing genes/proteins between different databases. Here we briefly describe some of these.

MatchMiner [47] provides a set of tools that translates between the different identifiers of a gene. These include interactive lookup for one gene, batch lookup for multiple genes and the merging of two lists of genes under different identifier systems to identify which identifiers refer to the same genes, and can be accessed through the web page at http://discover.nci.nih.gov/matchminer/index.jsp. A command-line version implemented in java is also available at http://discover.nci.nih.gov/matchminer/command.jsp. MatchMiner covers only genes from the *H. sapiens* (human) and *M. musculus* (mouse) genomes.

AliasServer [48] provides translation services between the aliases of a protein under different identifier systems. AliasServer can be accessed through a web interface at http://cbi.labri.fr/outils/alias/ and also provides a web service that can be accessed via the Simple Object Access Protocol (SOAP). Details on how to access the web service are provided at http://cbi.labri.fr/outils/alias/API_SOAP.html, with examples using Perl. At the time of writing, AliasServer covers genes from 29 genomes.

The Protein Identifier Cross-Referencing (PICR) service [49] is another service for translation between different gene identifier systems. One distinct feature of PICR compared with the others described in this section is that the service can take not only gene identifiers, but also protein sequences as input. PICR also does not require the user to specify the type of identifier systems to translate between, which may result in ambiguities when an identifier refers to different proteins in different systems. PICR can be accessed via http://www.ebi.ac.uk/Tools/picr/, and also provides a web service via SOAP. Details on using the PICR web service are available at http://www.ebi.ac.uk/Tools/picr/WSDLDocumentation.do, with examples using the Java API for XML Web Services (JAX-WS). At the time of writing, PICR covers genes from 47 genomes.

Synergizer [50] maintains a database to translate between the different identifiers of bio-logical entities. The translation service can be accessed interactively via the web page at http://llama.med.harvard.edu/synergizer/translate/, or programmatically using a remote pro-cedure call to a web service via the Hyper Text Transport Protocol (HTTP). The web service returns a JSON-encoded object (JavaScript Object Notation), which can be eas-ily decoded for further processing. More details on the JSON format can be found at http://www.json.org/. Details on how to access the Synergizer web service are available at http://llama.med.harvard.edu/synergizer/doc/, with examples using Perl. At the time of writing, Synergizer covers genes from 50 genomes.

9.2.3 Sequence homology

Sequence homology forms the basis of gene/protein function inference in early approaches, and remains very useful and widely used. Using the peptide sequence of an unknown protein to infer its function is not only intuitive (since amino acids form its basic building blocks), but also often necessary, as it is usually the only biological information available for a novel protein, as in the case of a newly sequenced genome.

9.2.3.1 Homologue discovery

The most popular way to get a quick suggestion on the possible characteristics of a protein given its sequence is to search for annotated proteins that have very high levels of sequence similarity to it. Proteins with very similar sequences are likely to be homologous, which means that they originated from the same gene in an ancestor and are conserved during evolution. Since proteins are vital players in the performance of various biological func-tions necessary for the survival of an organism, their sequence is conserved by selective pressure during speciation so that the orthologous proteins in each species retain their abil-ity to function effectively. Paralogous proteins which are homologues in the same species arising from gene duplication events also tend to retain similar sequence and functions, although they are more likely to diverge in these, since only one paralogous gene needs to be conserved to uphold the role of the original gene. However, proteins with high sequence similarity may not necessarily be conserved homologues, but could have arisen by chance during evolution. This is likely when the proteins have very short sequences, but becomes less probable with longer sequences. Hence, this must be taken into account when searching for homologues.

The Basic Local Alignment Search Tool (BLAST) [51] does this very well and very quickly, and has become the run-of-the-mill tool for this purpose for experimental and computational biologists. Using a heuristic combination of exact matching with extension, the tool is able to perform very fast local sequence alignment between a query sequence and a large database of sequences that allows for inexact matching including insertions, deletions and mismatches. BLAST also comes with highly configurable parameters, such as gap initiation and extension penalties, and the choice of substitution matrix [52]. The tool also comes with several scoring metrics, including sequence identity (percentage of sequence with exact match), alignment score and a very useful and widely used statistical score known as the expected value (E-value). The E-value reflects the expected num-ber of sequences in the database that are likely to obtain a similar alignment score with the query sequence by chance. A low E-value indicates that an alignment is more likely to arise from evolutionary conservation, such as that between homologues. The simplest forms of protein function inference given the sequence of a protein would be to search for likely homologues using BLAST and examine the annotations of these homologous proteins. Building on this principle, an array of tools has been developed to extend this concept in different directions and degrees to improve functional inference using sequence homology.

9.2.3.2 Automated function prediction from homologues

GoFigure [2] is one of the earliest among such tools. Given the sequence of a protein, GoFigure first performs a homology search using BLAST to find GO-annotated proteins with similar sequences. The sub-graph of the GO DAG with the greatest depth from the root that includes all GO terms assigned to these proteins is then identified. This graph is termed the minimum covering graph (MCG). Each term in this MCG is then assigned a weighted score derived from alignments to proteins with the term. The amount of contribution of an alignment is inversely related to its E-value. The score for each term in the MCG is then normalized by dividing it by that of the root. Terms with a normalized score of 0.2 and greater are then reported as inferred annotations for the query protein. GoFigure provides a systematic way of assigning weighted GO terms to a query protein based on homology search, giving higher weight to terms annotated to more proteins in the search results, as well as terms associated with more significant alignments. GoFigure was initially available at http://udgenome.ags.udel.edu/frm_go.html/, but is no longer available at the time of writing.

GOblet [1] is another tool that automates GO term inference from BLAST searches. The newest version of GOblet includes statistical analysis on the terms associated with proteins found. Some GO terms are more prevalent than others, and the distribution of such prevalence may differ between species. The observation of a highly prevalent term in the n homologues of a sequence may not be very significant if the probability of observing the same term in a random sample of n sequences from that species is very high. Conversely, the observation of a much less prevalent term is more significant. To quantify the enrichment of a term in the homologues of a protein given prior knowledge of the prevalence of each term, GOblet uses the Fisher exact test with Bonferroni correction to obtain a P-value. This new version of GOblet also includes pathway annotations from MetaCyc [53]. GOblet can be accessed via a web service at http://goblet.molgen.mpg.de.

GOtcha [54] takes a similar approach to GoFigure. Given a protein sequence, a score $R = \max(\log_{10}(E), 0)$ is computed for each alignment, where E is the E-value of the alignment. Each GO term annotated to at least one protein in the BLAST results, as well as its ancestor terms, is assigned a score equivalent to the sum of the scores R of each alignment associated with the term. The score for each term is subsequently normalized by dividing it by the score of the node that is the ancestor term of all scored terms, or the root term. This normalized score is termed the internal score (I-score), and reflects the relative significance of each term in the search results. A second score, termed the C-score, is computed as $\log_e$ of the root node, and reflects the confidence of the search results as a whole. To obtain an intuitive and meaningful score for each prediction, an estimate of the accuracy of various combinations of discretized I-scores and C-scores for each GO term is made using annotated sequences from SwissProt. Each prediction is then assigned a score based on the closest estimated accuracy based on its I-score and C-score. GOtcha is able to provide a more meaningful score than GoFigure that takes into account estimated accuracy based on some characteristics of the search results. Since the estimated accuracy is made separately for each GO term, the approach also accounted for the differences in the background frequency of each term. GOtcha is available as a web service at http://www.compbio .dundee.ac.uk/gotcha/gotcha.php.

GOAnno [55] takes a different approach in the use of sequence homology for function inference. Given a query proteins sequence, PipeAlign [56] is used to search for its homologues and construct a multiple alignment of complete sequences (MACS) that consists of clusters of homologues, each representing a potential functional subgroup. GO terms are then assigned based on three sets of annotations. The first set is the initial protein gene ontology (IPO), which is the set of already known annotations for query gene. The second set, the proximal protein gene ontology (PPO) is the set of GO terms annotated to proteins that share at least 98% sequence identity with the query protein. The last set, the mean subfamily gene ontology (MSO) is the set of GO terms annotated to sequences in the subgroups detected by PipeAlign that fulfill the NorMD [57] multiple sequence alignment score of NorMD > 0.3. Each term is scored by the number of homologous proteins that are annotated with the term or its descendant terms. Some thresholds are also imposed to remove GO branches that are associated with too few proteins. The three sets of annotations are combined to get the final predicted GO terms. GOAnno is available as a web service at http://bips.u-strasbg.fr/GOAnno/GOAnno.html.

GOPET [3] takes a machine-learning approach towards function prediction from sequence homology. A large number of sequences are searched against a database of GO-annotated sequences. For each query, the GO terms annotated to each homologue found are used as training examples; a term is deemed a positive example if it is annotated to the query protein and negative otherwise. Each term is assigned a number of features, such as the E-value, alignment bit scores and sequence identity of the alignment, as well as the background frequency of the term, the evidence codes used for the annotation of these terms, and so on. The training examples are then split randomly into smaller sets that are used to build multiple classifiers using support vector machines (SVMs). To predict functions for a given query protein sequence, homologous proteins are obtained using BLAST and each GO term annotated to these proteins is then scored by building similar features for it and using the classifiers to classify it as positive or negative. The votes from the classifiers are summed to obtain the final score. The authors of GOPET compared the method against GOtcha and found that they performed comparably. GOPET is available as a web service at http://genius.embnet.dkfz-heidelberg.de/menu/biounit/open-husar.

9.2.3.3 Remote homology

PFP [6] (http://dragon.bio.purdue.edu/pfp/) Position-Specific Iterative Basic Local Alignment Tool (PSI-BLAST) improves upon existing sequence-based approaches by extending a sequence homology search beyond sequences with highly similar sequences. Instead of using BLAST, PSI-BLAST [58] is used. PSI-BLAST performs an initial BLAST search using the query sequence and performs multiple sequence alignment on close homologues discovered, using the query sequence as a template. This alignment is then used to create a profile taking into account amino acid variation in specific positions of the profile. The profile, which reflects a model of the homologues found in the BLAST search, is then used to search against sequences in the database with a slightly modified BLAST algorithm.

The homologues found can again be used to modify the profile to get a more representative profile. This process of profile building and homology search is iterated so that the profile becomes more general and remote homologues with lower sequence similarity to the query sequence can be found. Each GO term annotated to the homologues found by PSI-BLAST is assigned a score in a similar manner to GoFigure and GOtcha, but taking into account prior knowledge of the association between GO terms:

$$s(f_a) = \sum_{i \in R} \left\{ [-\log(E(i)) + b] \sum_{j \in F_i} P\left(\frac{f_a}{f_j}\right) \right\}$$

where R is the set of sequences found by PSI-BLAST that is above a threshold, F_i is the set of GO terms annotated to sequence i, $E(i)$ is the E-value of the alignment result associated with sequence i, $P(f_a/f_j)$ is the conditional probability that a protein is annotated with term f_a given that it is annotated with term f_j.

The conditional probability $P(f_a/f_j)$ is computed based on the annotations of a large number of annotated proteins, and the collection of all conditional probabilities between each pair of GO term is collectively termed the function association matrix (FAM). The incorporation of the FAM into the scoring function allows GO terms that are not annotated to homologues found in the PSI-BLAST to be assigned to the query sequence. The use of PSI-BLAST and FAM allow PFP to yield significantly better recall then the other approaches described above while achieving better accuracy than using a standard PSI-BLAST search.

9.2.4 Phylogenetic relationships

Besides using sequence directly, some approaches also explore phylogenetic relationships for functional inference. Genes that participate in similar biological functions tend to be conserved together during speciation. This is intuitive, as proteins do not work alone, but rather form complexes, or interact with each other in biological pathways to perform their functions. This observation forms the underlying principle for using phylogenetic relationships to identify proteins with similar functions, which in turn can be use for function prediction.

9.2.4.1 Phylogenetic profiles

The simplest and probably earliest way to use phylogeny for predicting gene function is proposed by Pellegrini *et al.* [4]. For each gene, an n-bit binary vector known as a phylogenetic profile is constructed. Each index of the vector represents a currently living organism. A value of one is assigned to an index if the corresponding organism has a homologue of the gene; a value of zero is assigned otherwise. The distance between two genes is captured simply by the number of organisms they differ in, or the Hamming distance, and genes with profiles that differ by less than 3 bits are defined as neighbors. Using profiles representing 16 organisms (a 16-bit vector), Pellegrini *et al.* showed that genes with similar profiles tend to be involved in similar biological functions, although some functionally diverse genes still share similar profiles due to the limited resolution of 16-bit vectors. However, since the number of possible profiles for an n-bit profile is 2^n, each additional organism added in a profile effectively doubles the resolution of the profiles. More sophisticated metrics for comparing phylogenetic profiles have also been explored, including hypergeometric distribution [5] and mutual information [59]. Refer to Protocol 9.1 for information on how to construct a phylogenetic profile using BLAST [60] provided an in-depth study on how to select appropriate reference organisms for building effective phylogenetic profiles, and suggested that reference organisms should (i) be sufficiently distant in evolution, (ii) cover the Bacteria, Archaea and Eukarya domains [61] and (iii) be evenly distributed in the fifth level in the evolutionary tree. Refer to Protocol 9.2 on how to create phylogenetic profiles.

9.2.4.2 Phylogeny trees

While the use of a phylogenetic profile provided a simple means for identifying genes with similar functions, it ignores the evolutionary history of organisms, which can have a substantive impact on the effectiveness of the profiles in the identification of functional linkages. For example, genes with profiles similar in the more distant organisms tend to be more reflective of gene conservation due to evolutionary pressure, while genes with profiles similar only in the less distant organisms may appear conserved simply because there is not yet enough time for mutatations to accumulate substantially. To enable a more comprehensive comparison of the phylogenetic relationships between genes, Vert [62] proposed comparing the evolutionary trees of genes instead of their phylogenetic profiles. Phylogenetic profiles reflect only the information associated with the leaves of a phylogenetic tree. Figure 9.3 illustrates a hypothetical phylogenetic tree and the corresponding phylogenetic profiles. While the presence or absence of the homologue of a gene in existing species is known, this is unclear in ancestor species. To model this unknown information, Vert used a Bayesian tree to model the probabilities of each gene existing in ancestor species based on an existing phylogenetic tree [62]. A tree kernel is defined to efficiently compute the inner product of the features representing the Bayesian trees of each pair of genes. The kernel can then be used in a kernel-based method such as an SVM. An SVM can be used to build two classifiers, one using a naive kernel based on the Euclidean distances between phylogenetic profiles and another using the tree kernel, which shows that the classifier using the tree kernel achieved substantially superior performance. Some recent investigations further explore evolutionary models for functional inference [63, 64].

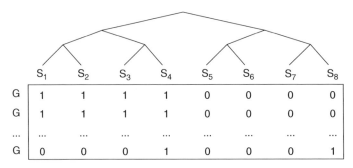

Figure 9.3 A hypothetical phylogenetic tree for eight species and the corresponding profiles for *n* genes.

PROTOCOL 9.2 Computing a Phylogenetic Profile for a Gene[b]

Requirements

- Internet access to download data and software
- Programming or scripting language such as C, C++, Perl, Matlab.

Method

1 Decide on which genomes to use for profile. See Sun *et al.* [60] for guidelines on choosing appropriate genomes.

2 Construct a vector with each column representing a selected genome and initialize the value of each column to zero.

3 Given the protein sequence of the gene of interest, perform a sequence similarity search using BLAST or PSI-BLAST [58] against protein sequences from the selected genomes. Protein sequences can be obtained from SwissProt http://www.ebi.ac.uk/swissprot/.

4 Retrieve all matching proteins below *E*-value threshold as homologues.

5 In Pellegrini *et al.* [4], the *P*-value threshold is computed by $1/nm$, where n is the number of proteins in the genome where the sequence is taken from and m is the number of proteins from other genomes. The *P*-value is defined as $1 - e^{-E}$, where E is the BLAST *E*-value.[c]

6 For each genome in which a homologue is found, the corresponding column in the vector is assigned a value of one.

Notes

[b]This is the original approach described in Pellegrini *et al.* [4]. A simpler method to construct phylogenetic profiles using the COG database is described in Natale *et al.* [65].

[c]More details on the *P*-value and *E*-value, as well as other statistical measures used in BLAST, can be found at http://www.ncbi.nlm.nih.gov/BLAST/tutorial/Altschul-1.html.

Comparing genes based on evolutionary trees better reflects the evolutionary similarities of genes, but is conceptually more complex and computationally more expensive. An appealing approach that avoids the complexity of such methods, but yet can achieve similar sensitivity, is a heuristic approach proposed by some researchers [66]. The heuristic approach considers the underlying phylogeny of the profiles by first ordering reference organisms based on evolutionary distances. Genes that truly co-evolve are likely to be conserved in distant organisms; hence, the matching organisms are less likely to occur in consecutive runs in the sorted profiles. A hypergeometric approach is used to compute the probability that random gene pairs achieve equal or less matching runs than observed between a pair of genes. The larger the numbers of observed runs, the more likely the genes are co-evolved and, hence, share functions.

9.2.4.3 Phylogenomics

Another group of approaches that uses phylogenetic relationships for functional inference involves the reconstruction and in-depth analysis of evolutionary history, commonly referred to as phylogenomics [67–69]. Resampled Inference of Orthologs [69] describes the use of bootstrapped resampled phylogenetic trees to improve orthologue discovery, which can reduce errors in functional inference. Statistical Inference of Function Through Evolutionary Relationships (SIFTER) [68] builds a phylogenetic tree from the homologues of a query protein and annotates speciation and duplication events in the tree. Known functional annotations within the tree are then propagated using a Bayesian approach to assign posterior probabilities of functional annotations to each node. The source code for SIFTER, implemented in Java, is available for download at http://sifter.berkeley.edu/.

9.2.5 Sequence-derived functional and chemical properties

Homology-based methods such as those described above work very well when annotated homologues of the query sequence can be found. However, such approaches are severely limited otherwise. In cases where no or few annotated homologues can be found, it may still be possible to infer a protein's functions from its sequence. A protein's sequence contains vital information that governs its structure and function. For example, a protein involved in signal transduction is likely to have many phosphorylation sites, while a protein involved in DNA binding is likely to be localized to the nucleus [70]. The presence of phosphorylation sites and subcellular localization, as well as many other physical and chemical characteristics of a protein, can be derived or predicted from protein sequences and exploited for functional inference.

ProtFun [71] uses 17 sequence-derived protein features, including predicted post-translational modifications (PTMs), protein sorting signals and secondary structure and physical/chemical properties, calculated from the amino acid composition to characterize each protein. These properties are then used as features to perform supervised learning for function prediction using artificial neural networks. Models are built for each function by learning from labeled examples (annotated proteins). Subsequently, given a protein sequence, similar features can be derived and classified by each model to predict if the protein has the function represented by the model. This approach was shown to work reasonably where homology-dependent approaches fail due to the absence of well-annotated homologues. ProtFun is available as a web service at http://www.cbs.dtu.dk/services/ProtFun/. A similar approach is taken in ProtSVM [72], which uses sequence-derived properties to train SVMs that can assign a protein sequence to 47 enzyme families. ProtSVM has since been updated to include a wide range of functional families, such as lipid transport and immune-response proteins, and can be accessed at http://jing.cz3.nus.edu.sg/cgi-bin/svmprot.cgi.

Lobley et al. [73] propose a model that extends upon ProtFun by introducing new features that encode disordered regions predicted by the DISOPRED server [74]. Disordered regions are regions in proteins that do not have a stable, well-defined tertiary structure in their native states [75]. It was discovered that proteins annotated with different functions exhibit distinguishable bias in the distribution of both the lengths and locations of disordered regions [73]. An SVM classifier is built for each GO term using these features. Based on this approach, an online function prediction server FFPred [70] is made available at http://bioinf.cs.ucl.ac.uk/ffpred/ for the automated prediction of over 300 GO annotations for an input protein sequence.

9.2.6 Protein–protein interaction maps

Protein sequences may encode useful information on the characteristics of the protein but offer little clue on their interaction behaviour. Proteins do not work alone, but interact with DNA, RNA and other proteins in complexes and pathways. Hence, an important source of evidence that can suggest the type of biological processes in which a protein contributes towards is a protein interaction map.

Protein–protein interaction data can be obtained from many databases. The Molecular Interaction database (MINT) [76] at http://mint.bio.uniroma2.it/mint/ contains over 100 000 physical interactions curated from peer-reviewed journals. BioGRID [44] at http://www .thebiogrid.org/ is one of the largest protein–protein interaction databases, with over 200 000 physical and genetic interactions curated from the literature, and has curated the complete set of interactions available in the literature for *S. cerevisiae* and *Schizosaccharomyces pombe*. Other databases from which protein–protein interaction data can be obtained include the Database of Interacting Proteins (DIP) at http://dip.doe-mbi.ucla.edu/ and the Human Protein Reference Database (HPRD) at http://www.hprd.org/. Protocol 9.3 provides details on obtaining protein–protein interaction data from BioGRID.

Although protein–protein interactions are binary relationships (i.e. interact or not), protein–protein interactions in databases vary in reliability. Protein–protein interactions can be observed from many types of experiment, such as two hybrid [77], immuno-precipitation and tandem affinity purification. Two-hybrid experiments are seemingly more susceptible to noise and have been reported to suffer from high false positive rates [13, 78]. Co-purification analysis, on the other hand, tends to be more reliable. Some computational methods on how to reduce noise in protein–protein interaction data are discussed in Chua and Wong [79].

9.2.6.1 Interacting partners

The simplest but yet effective method to infer the function of a protein using protein–protein interactions is to compute the frequency of each function among its interaction partners [80], and is widely referred to as neighbor counting. The function with the highest frequency is assigned to the protein.

Functional terms that are annotated to a larger number of proteins tend to appear in larger numbers in the neighborhood of a protein. Hence, the neighbor counting method may preferentially assign proteins with functions that have significantly high background frequencies. Hishigaki *et al.* [81] identified this problem and proposed using the chi-square statistical measure as a scoring function instead. For each function annotated to the neighbors of a protein, the deviation of its observed occurrence from its expected occurrence is used as a score to assign that function to the protein. Given protein u, function x is assigned to u with

$$S_x(u) = \frac{(f_x(u) - e_x(u))^2}{e_x(u)}$$

where $f_x(u)$ is the observed frequency of x in the neighbors of u and $e_x(u)$ is the expected frequency of x in the neighbors of u based on its background frequency.

PROTOCOL 9.3   Obtaining Protein-Protein Interaction Data from BioGRID for Function Prediction

Requirements

- Internet access to download data and software
- Programming or scripting language such as C, C++, Perl, Matlab.

Method

1 Download the protein–protein interaction data from http://www.thebiogrid.org/downloads.php.

2 Choose between all interaction data and organism-specific data. Three formats are available: Excel tab-delimited text, PSI-MI XML version 1 and PSI-MI XML version 2.5.

3 Each record contains the names of two interacting proteins, the type of experiment system in which the interaction was observed (e.g. two hybrid) and the PubMed identifier of the publication from which the interaction was curated.[d]

4 Convert protein names to the identifier you desire; for example, RefSeq accession numbers, SGD or Ensembl identifiers. You would want to use an identifier system that is consistent with other data, such as GO annotations. This can be done using a tool such as the Synergizer at http://llama.med.harvard.edu/synergizer/translate/.

Notes

[d]The experiment system used in the detection of an interaction reflects the nature and confidence of the interaction. For example, synthetic lethality detects genetic interactions while two hybrid detects physical interactions; interactions detected by affinity purification are more reliable than those detected by two hybrid.

9.2.6.2 Common interacting partners

The above approaches associate proteins through interaction. Some approaches associate proteins based on their sharing of interaction partners. Samanta and Liang [15] computed a distance for each pair of proteins u and v using a hypergeometric score based on each protein's interaction neighbors defined as

$$P(N, u, v, m) = \frac{\binom{N}{m}\binom{N-m}{n_1-m}\binom{N-n_1}{n_2-m}}{\binom{N}{n_1}\binom{N}{n_2}}$$

where N refers to all proteins in the interaction network, N_u refers to the neighbors of protein u, $m = |N_u \cap N_v|$, $n_1 = |N_u|$ and $n_2 = |N_v|$.

The score reflects the likelihood that proteins u and v share m neighbors by chance given N, n_1 and n_2. A smaller P-value implies that u and v are more likely to be biologically associated, and vice versa. Using only protein pairs that have a P-value of 10^{-8} or below, the authors assign function to unannotated proteins based on majority vote, in a similar way to neighbor counting. Other ways of computing association weights between protein pairs for function prediction based on common interaction partners are explored in Brun *et al.* [7] and Chua *et al.* [10].

9.2.6.3 Machine-learning approaches

Some approaches explore machine-learning techniques such as simulated annealing [82], Markov random fields [11] and graph kernels with SVMs [26]. More recent approaches also studied computational methods that detect conserved network patterns [8, 12]. These approaches are more complex and computationally demanding, but usually have better prediction accuracies.

9.3 Troubleshooting

- Hierarchical annotation schemes such as FunCat and GO usually follow the 'true path rule.' This means that a gene or protein annotated with a term is also implicitly annotated with its ancestor terms. However, these ancestor terms are usually not included in annotation files. This should be considered to avoid inconsistent predictions between related terms.

- Owing to differences in the nature of different functions, certain functions have few descendant terms, while others can have a large number of descendant terms. This may be due to many factors, such as the nature of experiments used to determine function and the current level of biological understanding of the mechanisms underlying different functions. As there is some level of overlap between related terms, using all annotations during the evaluation of a prediction method can result in bias towards better understood functions which tend to have more specific terms. A method that can predict well-understood functions better will hence be preferentially evaluated compared with a method that can predict less well understood functions better. Two common ways are used to produce a more balanced evaluation. The first is to evaluate only using terms from a fixed level in the hierarchy (e.g. level 3 in GO) [11, 26, 28, 83, 84]. The other approach is to evaluate only using *informative* terms [10, 21]. A term is defined to be informative if there are at least 30 proteins annotated with it and has no descendant terms annotated with at least 30 proteins. The use of informative terms for evaluation ensures no redundancy (no ancestor or descendant terms of an informative term can be informative), whilst making sure that terms for evaluation are annotated to sufficient proteins.

References

1. Hennig, S., Groth, D. and Lehrach, H. (2003) Automated Gene Ontology annotation for anonymous sequence data. *Nucleic Acids Research*, **31**, 3712–3715.

2. Khan, S., Situ, G., Decker, K. and Schmidt, C.J. (2003) GoFigure: automated Gene Ontology annotation. *Bioinformatics*, **19**, 2484–2485.

3. Vinayagam, A., del Val, C., Schubert, F. *et al.* (2006) GOPET: a tool for automated predictions of Gene Ontology terms. *BMC Bioinformatics*, **7**, 161.

4. Pellegrini, M., Marcotte, E.M., Thompson, M.J. *et al.* (1999) Assigning protein functions by comparative genome analysis: Protein phylogenetic profiles. *Proceedings of the National Academy of Sciences of the United States of America*, **96**, 4285–4288. This is the pioneering work which reported that proteins with similar phylogenetic profiles tend to exhibit conserved function.

5. Wu, J., Kasif, S. and DeLisi, C. (2003) Identification of functional links between genes using phylogenetic profiles. *Bioinformatics*, **19**, 1524–1530.

6. Hawkins, T., Luban, S. and Kihara, D. (2006) Enhanced automated function prediction using distantly related sequences and contextual association by PFP. *Protein Science*, **15**, 1550–1556.

7. Brun, C., Chevenet, F., Martin, D. *et al.* (2003) Functional classification of proteins for the prediction of cellular function from a protein–protein interaction network. *Genome Biology*, **5**, R6.

8. Chen, J., Hsu, W., Lee, M., and Ng, S.-K. (2007) Labeling network motifs in protein interactomes for protein function prediction. Proceedings of the IEEE 23rd International Conference on Data Engineering, pp. 546–555.

9. Chen, Y. and Xu, D. (2004) Global protein function annotation through mining genome-scale data in yeast *Saccharomyces cerevisiae*. *Nucleic Acids Research*, **32**, 6414–6424.

10. Chua, H.N., Sung, W.K. and Wong, L. (2006) Exploiting indirect neighbours and topological weight to predict protein function from protein–protein interactions. *Bioinformatics*, **22**, 1623–1630.

11. Deng, M., Zhang, K., Mehta, S. *et al.* (2003) Prediction of protein function using protein–protein interaction data. *Journal of Computational Biology*, **10**, 947–960.

12. Kirac, M. and Ozsoyoglu, G. (2008) Protein function prediction based on patterns in biological networks. Proceedings of 12th International Conference on Research in Computational Molecular Biology, pp. 197–213.

13. Legrain, P., Wojcik, J. and Gauthier, J.M. (2001) Protein–protein interaction maps: a lead towards cellular functions. *Trends in Genetics*, **17**, 346–352.

14. Letovsky, S. and Kasif, S. (2003) Predicting protein function from protein/protein interaction data: a probabilistic approach. *Bioinformatics*, **19** (Suppl. 1), i197–i204.

15. Samanta, M.P. and Liang, S. (2003) Predicting protein functions from redundancies in large-scale protein interaction networks. *Proceedings of the National Academy of Sciences of the United States of America*, **100**, 12579–12583.

16. Ferre, S. and King, R.D. (2006) Finding motifs in protein secondary structure for use in function prediction. *Journal of Computational Biology*, **13**, 719–731.

17. Laskowski, R.A., Watson, J.D. and Thornton, J.M. (2005) ProFunc: a server for predicting protein function from 3D structure. *Nucleic Acids Research*, **33**, W89–W93.

18. Pazos, F. and Sternberg, M.J. (2004) Automated prediction of protein function and detection of functional sites from structure. *Proceedings of the National Academy of Sciences of the United States of America*, **101**, 14754–14759.

19. Huynen, M., Snel, B., Lathe, W. III and Bork, P. (2000) Predicting protein function by genomic context: quantitative evaluation and qualitative inferences. *Genome Research*, **10**, 1204–1210.

20. Hughes, T.R., Marton, M.J., Jones, A.R. *et al.* (2000) Functional discovery via a compendium of expression profiles. *Cell*, **102**, 109–126.

21. Zhou, X., Kao, M.C. and Wong, W.H. (2002) Transitive functional annotation by shortest-path analysis of gene expression data. *Proceedings of the National Academy of Sciences of the United States of America*, **99**, 12783–12788.

22. Hirschman, L., Yeh, A., Blaschke, C. and Valencia, A. (2005) Overview of BioCreAtIvE: critical assessment of information extraction for biology. *BMC Bioinformatics*, **6** (Suppl. 1), S1.

23. Ray, S. and Craven, M. (2005) Learning statistical models for annotating proteins with function information using biomedical text. *BMC Bioinformatics*, **6** (Suppl. 1), S18.

24. Chua, H.N., Sung, W.-K. and Wong, L. (2007) An efficient strategy for extensive integration of diverse biological data for protein function prediction. *Bioinformatics*, **23**, 3364–3373.

25. Deng, M., Chen, T. and Sun, F. (2004) An integrated probabilistic model for functional prediction of proteins. *Journal of Computational Biology*, **11**, 463–475.

26. Lanckriet, G.R., Deng, M., Cristianini, N. *et al.* (2004) Kernel-based data fusion and its application to protein function prediction in yeast. Pacific Symposium on Biocomputing 2004 (eds R.B. Altman, A.K. Dunker, L. Hunter *et al.*), World Scientific, pp. 300–311.

27. Troyanskaya, O.G., Dolinski, K., Owen, A.B. *et al.* (2003) A Bayesian framework for combining heterogeneous data sources for gene function prediction (in *Saccharomyces cerevisiae*). *Proceedings of the National Academy of Sciences of the United States of America*, **100**, 8348–8353.

28. Tsuda, K., Shin, H. and Schölkopf, B. (2005) Fast protein classification with multiple networks. *Bioinformatics*, **21** (Suppl. 2), ii59–ii65.

29. Hawkins, T. and Kihara, D. (2007) Function prediction of uncharacterized proteins. *Journal of Bioinformatics and Computational Biology*, **5**, 1–30.

30. Sharan, R., Ulitsky, I. and Shamir, R. (2007) Network-based prediction of protein function. *Molecular Systems Biology*, **3**, 88.

31. Pena-Castillo, L., Tasan, M., Myers, C.L. *et al.* (2008) A critical assessment of *Mus musculus* gene function prediction using integrated genomic evidence. *Genome Biology*, **9** (Suppl. 1), S2. This publication provides a comparison between several state-of-the-art computational function prediction methods on the task of predicting functions for *Mus musculus* proteins.

32. Ruepp, A., Zollner, A., Maier, D. *et al.* (2004) The FunCat, a functional annotation scheme for systematic classification of proteins from whole genomes. *Nucleic Acids Research*, **32**, 5539–5545.

33. Barrett, A.J. (1997) Nomenclature Committee of the International Union of Biochemistry and Molecular Biology (NC-IUBMB). Enzyme nomenclature. Recommendations 1992. Supplement 4: corrections and additions (1997). *European Journal of Biochemistry*, **250**, 1–6.

34. Murzin, A.G., Brenner, S.E., Hubbard, T. and Chothia, C. (1995) SCOP: a structural classification of proteins database for the investigation of sequences and structures. *Journal of Molecular Biology*, **247**, 536–540.

35. Riley, M. (1993) Functions of the gene products of *Escherichia coli*. *Microbiological Reviews*, **57**, 862–952.

36. Mewes, H.W., Heumann, K., Kaps, A. *et al.* (1999) MIPS: a database for genomes and protein sequences. *Nucleic Acids Research*, **27**, 44–48.

37. Ashburner, M., Ball, C.A., Blake, J.A. *et al.* (2000) Gene Ontology: tool for the unification of biology. *Nature Genetics*, **25**, 25–29. GO is currently the most widely adopted annotation scheme for protein function.

38. Powell, J.R. (1996) *Progress and Prospects in Evolutionary Biology: The Drosophila Model*, Oxford University Press, New York.

39. Cherry, J.M., Adler, C., Ball, C. *et al.* (1998) SGD: *Saccharomyces* Genome Database. *Nucleic Acids Research*, **26**, 73–79. The SGD database provides comprehensive resources for the model organism *S. cerevisiae*.

40. Bult, C.J., Blake, J.A., Richardson, J.E. *et al.* (2004) The Mouse Genome Database (MGD): integrating biology with the genome. *Nucleic Acids Research*, **32**, D476–D481.

41. Stein, L., Sternberg, P., Durbin, R. *et al.* (2001) WormBase: network access to the genome and biology of *Caenorhabditis elegans*. *Nucleic Acids Research*, **29**, 82–86.

42. Sprague, J., Clements, D., Conlin, T. *et al.* (2003) The Zebrafish Information Network (ZFIN): the zebrafish model organism database. *Nucleic Acids Research*, **31**, 241–243.

43. Bader, G.D. and Hogue, C.W. (2000) BIND – a data specification for storing and describing biomolecular interactions, molecular complexes and pathways. *Bioinformatics*, **16**, 465–477.

44. Breitkreutz, B.J., Stark, C., Reguly, T. *et al.* (2008) The BioGRID Interaction Database: 2008 update. *Nucleic Acids Research*, **36**, D637–D640. BioGRID is an up-to-date repository of protein–protein interactions for a variety of species.

45. Kersey, P.J., Duarte, J., Williams, A. *et al.* (2004) The International Protein Index: an integrated database for proteomics experiments. *Proteomics*, **4**, 1985–1988.

46. Bairoch, A., Apweiler, R., Wu, C.H. *et al.* (2005) The Universal Protein Resource (UniProt). *Nucleic Acids Research*, **33**, D154–D159.

47. Bussey, K.J., Kane, D., Sunshine, M. *et al.* (2003) MatchMiner: a tool for batch navigation among gene and gene product identifiers. *Genome Biology*, **4**, R27.

48. Iragne, F., Barre, A., Goffard, N. and De Daruvar, A. (2004) AliasServer: a web server to handle multiple aliases used to refer to proteins. *Bioinformatics*, **20**, 2331–2332.

49. Côté, R.G., Jones, P., Martens, L. *et al.* (2007) The Protein Identifier Cross-Referencing (PICR) service: reconciling protein identifiers across multiple source databases. *BMC Bioinformatics*, **8**, 401.

50. Berriz, G.F. and Roth, F.P. (2008) The Synergizer service for translating gene, protein and other biological identifiers. *Bioinformatics*, **24**, 2272–2273.

51. Altschul, S.F., Gish, W., Miller, W. *et al.* (1990) Basic local alignment search tool. *Journal of Molecular Biology*, **215**, 403–410. BLAST is an important sequence analysis tool for both experimental and computational biologists.

52. Henikoff, S. and Henikoff, J.G. (1992) Amino acid substitution matrices from protein blocks. *Proceedings of the National Academy of Sciences of the United States of America*, **89**, 10915–10919.

53. Caspi, R., Foerster, H., Fulcher, C.A. *et al.* (2008) The MetaCyc database of metabolic pathways and enzymes and the BioCyc collection of pathway/genome databases. *Nucleic Acids Research*, **36**, D623–D631.

54. Martin, D.M., Berriman, M. and Barton, G.J. (2004) GOtcha: a new method for prediction of protein function assessed by the annotation of seven genomes. *BMC Bioinformatics*, **5**, 178.

55. Chalmel, F., Lardenois, A., Thompson, J.D. *et al.* (2005) GOAnno: GO annotation based on multiple alignment. *Bioinformatics*, **21**, 2095–2096.

56. Plewniak, F., Bianchetti, L., Brelivet, Y. *et al.* (2003) PipeAlign: a new toolkit for protein family analysis. *Nucleic Acids Research*, **31**, 3829–3832.

57. Thompson, J.D., Plewniak, F., Ripp, R. *et al.* (2001) Towards a reliable objective function for multiple sequence alignments. *Journal of Molecular Biology*, **314**, 937–951.

58. Altschul, S.F., Madden, T.L., Schaffer, A.A. *et al.* (1997) Gapped BLAST and PSI-BLAST: a new generation of protein database search programs. *Nucleic Acids Research*, **25**, 3389–3402. PSI-BLAST extends conventional BLAST to retrieve the homologues of a protein with much lower sequence similarity.

59. Date, S.V. and Marcotte, E.M. (2003) Discovery of uncharacterized cellular systems by genome-wide analysis of functional linkages. *Nature Biotechnology*, **21**, 1055–1062.

60. Sun, J., Li, Y. and Zhao, Z. (2007) Phylogenetic profiles for the prediction of protein–protein interactions: how to select reference organisms? *Biochemical and Biophysical Research*, **353**, 985–991.

61. Woese, C.R., Kandler, O. and Wheelis, M.L. (1990) Towards a natural system of organisms: proposal for the domains Archaea, Bacteria, and Eucarya. *Proceedings of the National Academy of Sciences of the United States of America*, **87**, 4576–4579.

62. Vert, J.P. (2002) A tree kernel to analyse phylogenetic profiles. *Bioinformatics*, **18** (Suppl. 1), S276–S284.

63. Barker, D., Meade, A. and Pagel, M. (2007) Constrained models of evolution lead to improved prediction of functional linkage from correlated gain and loss of genes. *Bioinformatics*, **23**, 14–20.

64. Barker, D. and Pagel, M. (2005) Predicting functional gene links from phylogenetic-statistical analyses of whole genomes. *PLoS Computational Biology*, **1**, e3.

65. Natale, D.A., Galperin, M.Y., Tatusov, R.L. and Koonin, E.V. (2000) Using the COG database to improve gene recognition in complete genomes. *Genetica*, **108**, 9–17.

66. Cokus, S., Mizutani, S. and Pellegrini, M. (2007) An improved method for identifying functionally linked proteins using phylogenetic profiles. *BMC Bioinformatics*, **8** (Suppl. 4), S7.

67. Eisen, J.A. (1998) Phylogenomics: improving functional predictions for uncharacterized genes by evolutionary analysis. *Genome Research*, **8**, 163–167.

68. Engelhardt, B.E., Jordan, M.I., Muratore, K.E. and Brenner, S.E. (2005) Protein molecular function prediction by Bayesian phylogenomics. *PLoS Computational Biology*, **1**, e45.

69. Zmasck, C.M. and Eddy, S.R. (2002) RIO: analyzing proteomes by automated phylogenomics using resampled inference of orthologs. *BMC Bioinformatics*, **3**, 14.

70. Lobley, A.E., Nugent, T., Orengo, C.A. and Jones, D.T. (2008) FFPred: an integrated feature-based function prediction server for vertebrate proteomes. *Nucleic Acids Research*, **36**, W297–W302.

71. Jensen, L.J., Gupta, R., Staerfeldt, H.H. and Brunak, S. (2003) Prediction of human protein function according to Gene Ontology categories. *Bioinformatics*, **19**, 635–642.

72. Cai, C.Z., Han, L.Y., Ji, Z.L. *et al.* (2003) SVM-Prot: web-based support vector machine software for functional classification of a protein from its primary sequence. *Nucleic Acids Research*, **31**, 3692–3697.

73. Lobley, A., Swindells, M.B., Orengo, C.A. and Jones, D.T. (2007) Inferring function using patterns of native disorder in proteins. *PLoS Computational Biology*, **3**, e162. This publication describes the use of a large variety of sequence-derived biochemical properties of proteins with SVMs for function inference.

74. Ward, J.J., McGuffin, L.J., Bryson, K. *et al.* (2004) The DISOPRED server for the prediction of protein disorder. *Bioinformatics*, **20**, 2138–2139.

75. Dunker, A.K., Garner, E., Guilliot, S. *et al.* (1998) Protein disorder and the evolution of molecular recognition: theory, predictions, and observations. *Pacific Symposium on Biocomputing '98* (eds R.B. Altman, A.K. Dunker and T.E. Klein), World Scientific, pp. 473–484.

76. Chatr-aryamontri, A., Ceol, A., Palazzi, L.M. *et al.* (2007) MINT: the Molecular INTeraction database. *Nucleic Acids Research*, **35**, D572–D574.

77. Gietz, R.D., Triggs-Raine, B., Robbins, A. *et al.* (1997) Identification of proteins that interact with a protein of interest: applications of the yeast two-hybrid system. *Molecular and Cellular Biochemistry*, **172**, 67–79.

78. Sprinzak, E., Sattath, S. and Margalit, H. (2003) How reliable are experimental protein-protein interaction data? *Journal of Molecular Biology*, **327**, 919–923.

79. Chua, H.N. and Wong, L. (2008) Increasing the reliability of protein interactomes. *Drug Discovery Today*, **13**, 652–658.

80. Schwikowski, B., Uetz, P. and Fields, S. (2000) A network of protein-protein interactions in yeast. *Nature Biotechnology*, **18**, 1257–1261. This is the publication that first explored the use of high-throughput protein–protein interactions for protein function prediction.

81. Hishigaki, H., Nakai, K.,Ono, T. *et al.* (2001) Assessment of prediction accuracy of protein function from protein–protein interaction data. *Yeast*, **18**, 523–531.

82. Vazquez, A., Flammini, A., Maritan, A. and Vespignani, A. (2003) Global protein function prediction from protein–protein interaction networks. *Nature Biotechnology*, **21**, 697–700.

83. Chua, H.N., Sung, W.K. and Wong, L. (2007) Using indirect protein interactions for the prediction of Gene Ontology functions. *BMC Bioinformatics*, **8** (Suppl. 4), S8. This is a recent review on computational techniques to improve the reliability of experimentally derived protein–protein interactions.

84. Gabow, A., Leach, S., Baumgartner, W. *et al.* (2008) Improving protein function prediction methods with integrated literature data. *BMC Bioinformatics*, **9**, 198.

10

Elucidating Gene Function through Use of Genetically Engineered Mice

Mary P. Heyer, Cátia Feliciano, João Peca and Guoping Feng
Department of Neurobiology, Duke University Medical Center, Durham, North Carolina, USA

10.1 Introduction

The development of molecular genetics techniques for the direct manipulation of the mouse genome has revolutionized biological and biomedical research. These techniques are now in widespread use, with increasingly sophisticated genetic alterations possible. The mouse genome is the most easily manipulated of all mammals; thus, genetically engineered mice facilitate detailed *in vivo* analysis of gene function and provide models for a variety of human disorders. Several methods are available for loss or gain of gene function in mice, including overexpression, deletion, point mutation and reporter expression, with additional control through techniques facilitating temporal and spatial restriction of these manipulations. Phenotypic consequences of these heritable genetic alterations can be examined within the rich complexity of the living mouse, yielding insights into gene function in various biological phenomena and disease states.

Transgenic technology is widely used for gene overexpression *in vivo* through random insertion of an exogenous vector DNA sequence after pronuclear injection of mouse oocytes. This method can be adapted for expression of a reporter gene downstream of a promoter of interest, while gene function can also be inhibited through overexpression of dominant-negative mutants. Transgenic technology is relatively straightforward, well established and highly efficient, generating founder animals in a relatively short time period. In some cases, owing to random vector integration, transgene expression patterns and levels may be affected by neighboring sequences. DNA is often inserted as multiple, concatemerized copies, generating variable expression levels. The injected DNA must contain all the

Genomics: Essential Methods Edited by Mike Starkey and Ramnath Elaswarapu
© 2011 John Wiley & Sons, Ltd.

regulatory elements for reliable reproduction of endogenous gene expression – sometimes located at a great distance from the coding sequence. Bacterial artificial chromosome (BAC) transgenic technology is increasingly used to overcome these difficulties [1]. However, transgene insertion may produce unwanted effects on expression of a nearby, endogenous gene. Transgenic methods have been described in detail elsewhere [2, 3].

Loss of gene function in mouse can be achieved through gene trap methods [4, 5]. Gene trap utilizes a reporter gene that is designed to insert randomly into the genome, potentially interrupting expression of a gene at or near the insertion site. The mutated gene and its expression patterns can then be identified by means of the inserted reporter. Gene trap can generate many lines of mice with different disrupted genes; these can be screened based on a particular phenotype or expression pattern of interest. Recently, the DNA transposons *piggyBAC* and *Sleeping Beauty* have been used to inactivate murine genes by random insertional mutagenesis [6–8]. Gene trap can be achieved in this manner by incorporating a reporter gene into the transposon. One advantage of gene trap and transposon-based methods is that they do not require prior knowledge of the affected gene's sequence or structure. While gene trap facilitates serendipitous insights into gene function, mutation events may occur at low frequency, and the investigator has no control over the locus of the mutation.

Gene targeting overcomes many of the limitations of other genetic engineering methods, as mutations are targeted to inactivate or modify a specific, endogenous gene of interest. This precise control stems from the use of homologous recombination at the integration site, combined with negative selection against randomly integrated vector copies. Typically, targeted mutations have been designed to eliminate gene function, resulting in 'knockout' mice. However, more elegant strategies are available for gene overexpression, mutation and deletion, with increasing degrees of temporal and spatial control. These have been facilitated by the availability of the human and mouse genomic sequences. Gene targeting is not restricted to alteration of coding regions; gene regulatory regions, untranslated mRNA regions and noncoding RNAs such as microRNAs can all be specifically targeted in order to investigate their function.

Owing to the diverse applications of genetic engineering and in the interests of space, this chapter focuses on selected gene targeting strategies and methods. The protocols included aim to provide a brief guide to targeting vector design, homologous recombination in mouse embryonic stem (ES) cells and transfer of mutations to the mouse germline. More in-depth consideration of the theory and methodology of genetic engineering and gene targeting in mice can be found in Nagy [9] and Tymms [10].

10.2 Methods and approaches

10.2.1 Principles of targeted gene deletion in mice

Gene targeting is defined as the introduction of site-specific modifications into the genome by homologous recombination. This powerful technique can be applied to almost any biological question or phenomenon. Indeed, the *in vivo* functions of over 10 000 genes have now been analyzed by gene targeting, and over 500 different mouse models of human disorders have been created. The technique of gene targeting in mice was developed as a combination of two major breakthroughs, recognized by the 2007 Nobel Prize in Physiology or Medicine, awarded to Mario Capecchi, Martin Evans and Oliver Smithies. The first advance was the ability to culture mouse ES cells that have the potential to contribute to the germline of mice [11, 12], while the second was the introduction of specific mutations into a chosen gene by homologous recombination in the cultured mouse ES cells [13, 14]. The first reports using homologous recombination in ES cells to generate gene-targeted mice followed thereafter [15–18].

Gene targeting in mice relies on the unique attributes of mouse ES cells, which are derived from the pluripotent, uncommitted cells of the inner cell mass of pre-implantation blastocysts. They can be manipulated in culture, introduced into a pre-implantation blastocyst embryo and allowed to mature in a foster mother. These pluripotent cells have the ability to contribute efficiently to the formation of both somatic and germline tissues after reintroduction into a blastocyst. This capacity is highly dependent on culture conditions that keep ES cells in an uncommitted, undifferentiated state. Differentiation inhibitory signals are provided by mouse embryonic fibroblast (MEF) feeder cells that also act as a matrix for ES cell adherence and by addition of leukemia inhibitory factor (LIF) to the culture medium [19]. While the R1 ES cells described in these protocols do rely on feeder cells, several ES cell lines are available that do not have this requirement, and thus require less labor-intensive cultivation.

Targeted mutations can be introduced in ES cells by homologous recombination of a suitable targeting vector (see Figure 10.1). Homologous recombination in mammalian cells occurs at a very low frequency; therefore, positive–negative selection was developed as a method for enrichment of cells in which homologous targeting has occurred [20]. Incorporation of the targeting vector into the genome is positively selected for by addition of the antibiotic G418/geneticin, due to the presence of a cassette carrying the neomycin resistance gene under the control of a strong promoter. For negative selection, the herpes simplex virus thymidine kinase (TK) gene is found at the end of the linearized targeting construct. Cells undergoing homologous recombination will lose the TK gene, while those with random insertion of the targeting vector are eliminated by applying the toxic nucleoside analog, FIAU [1-(2-deoxy-2-fluoro-β-D-arabino-furanosyl)-5-iodouracil]. Diphtheria toxin A (DT-A) has also been successfully used as a negative selection marker [21].

The following factors have been shown to influence targeting frequency:

1 The recombination rate increases with total length of homology between a vector and its target locus, up to about 10 kb [22]. Here, this is maximized by the use of a 6–10 kb long homology arm (LHA) and an ~1 kb short homology arm (SHA) flanking the neomycin resistance cassette (Figure 10.1). The LHA can be cloned through retrieval of DNA from a BAC, described in Protocols 10.1–10.4, while the SHA is cloned using standard polymerase chain reaction (PCR)-based methods.

2 Targeting frequency increases when the homology regions of the gene of interest are isolated from the same genetic background (isogenic) as the ES cells to be used [23]. R1 ES cells are derived from the 129 genetic background [24]; however, cell lines from other mouse strains such as C57BL/6 and BALB/c are also available and may be used [25, 26].

3 Targeting frequency is locus dependent; thus, while the methods described here maximize the efficiency of homologous recombination, factors such as chromatin structure and accessibility make the overall targeting frequency difficult to predict for a given gene.

Several methods can be used to screen for ES cell clones with the desired recombination event, including PCR and optimized mini-Southern blot. The strategy outlined here takes

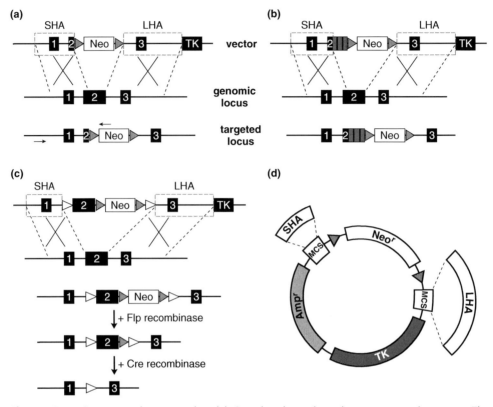

Figure 10.1 Gene targeting strategies. (a) *Gene knockout via replacement targeting vector*. The second exon of a target gene is disrupted by the neomycin resistance cassette (Neo) after homologous recombination with the linearized targeting vector. The thymidine kinase gene (TK) is lost upon homologous recombination. Exons are numbered, black boxes; short and long homology arms are outlined by gray dotted lines. PCR primers P1 and P2 are used to screen for homologous recombination. FRT sites (gray triangles) flank the Neo cassette; if desired, crossing targeted mice with a line of mice expressing FLP recombinase mediates deletion of the Neo cassette. (b) *Gene knockin via replacement targeting vector*. Part of the second exon of a target gene is replaced by a sequence containing point mutations or a reporter gene (gray box with black stripes) in addition to Neo. (c) *Conditional gene deletion*. Two loxP sites (white triangles) flanking the exon to be deleted and the neomycin cassette are inserted by homologous recombination. Crossing targeted mice with a line expressing FLP recombinase mediates deletion of the Neo cassette. Conditional deletion is achieved by crossing these mice with a line expressing tissue- or developmental stage-specific Cre recombinase. (d) *Targeting vector design*. Required components include the TK gene, an antibiotic resistance gene (here, ampicillin resistance, Amp^r), and the neomycin resistance cassette (Neo^r) flanked by FRT sites and multiple cloning sites (MCSs). Short (~1 kb) and long (6–10 kb) homology arms (SHA and LHA) may be cloned into either of the MCSs.

advantage of PCR screening, with one primer derived from the newly introduced neomycin resistance cassette and the other hybridizing to a genomic sequence outside the SHA. The SHA length is restricted to 1 kb to allow for efficient amplification by PCR during screening of ES cell DNA.

The high degree of precision of genetic manipulation through gene targeting in mice comes at some expense. The procedures can be involved, labor intensive, relatively costly and lengthy. Automated approaches to gene targeting in ES cells and subsequent production of gene-targeted mice may be suited to the needs of some laboratories [27, 28].

10.2.2 Strategies for gene targeting in mice

Because of the extensive variety of mutations and sequence manipulations that are possible through gene targeting, this chapter will discuss a selection of the most common targeting strategies as listed below. In all cases, a targeting vector must be designed such that chromosomal sequences are replaced by vector sequences through homologous recombination of the flanking homology regions. A modified pBluescript (pBSK) plasmid may be employed as a suitable targeting vector (Figure 10.1d).

1 **Knockout**: Targeting constructs for generation of null mutants typically consist of the target gene sequence into which a loss-of-function mutation has been engineered (Figure 10.1a). This can be achieved by replacing the initiation codon (ATG) and/or an exon (encoding the functional domain of a protein) with the neomycin resistance cassette. In order to effectively eliminate gene function, this replacement should also incorporate a translational termination codon, while generating a frameshift mutation that precludes alternative splicing. It is important to note that transcription from the strong promoter driving the neomycin resistance gene may interfere with expression of neighboring genes, thus confounding phenotypic analysis [29, 30]. Thus, the neomycin resistance cassette should be flanked by two FLP recombination target (FRT) sites, allowing for FLP recombinase-mediated excision [31]. This is achieved by crossing with a mouse expressing FLP recombinase in the germline [32]. Null mutations may be embryonic lethal, restricting the study of gene function to early development. In other cases, compensation for constitutive gene deletion may yield phenotypes that underestimate gene function. These pitfalls can be avoided through conditional or inducible mutations described below.

2 **Knockin**: Knockin mutations are generated through sequence replacement or insertion at a specific gene locus (Figure 10.1b). Possible applications include the introduction of specific point mutations or the fusion of reporter genes such as green fluorescent protein (GFP) or β-galactosidase. Such reporter gene fusions facilitate accurate analysis of endogenous gene expression patterns, while manipulations such as point mutations allow more subtle dissection of gene function. In addition, overexpression can be achieved by knockin of the gene of interest at the *ROSA26* locus. This locus is regarded as one from which proteins can be expressed ubiquitously at a moderate level [33].

3 **Conditional deletion**: In order to overcome some potential difficulties associated with constitutive deletion and to analyze gene function in a particular cell, tissue and/or developmental stage of interest, gene targeting can be combined with the powerful Cre/loxP system (reviewed in Refs. [34, 35]). Cre recombinase efficiently catalyzes

reciprocal DNA recombination between two loxP sites of the same directionality. Thus, a sequence with flanking loxP sites (floxed) will be excised by site-specific recombination in the presence of Cre, leaving a single loxP site marking the point of excision (Figure 10.1c). By crossing a line of mice with a floxed allele generated by gene targeting and a line of transgenic mice expressing Cre under the control of a tissue-specific promoter, the allele will be deleted specifically in that tissue. In an alternative approach, conditional knockin of point mutations has also been reported [36]. The repertoire of mice expressing Cre recombinase under different endogenous, tissue and developmental stage-specific promoters is continually growing. Viral delivery of Cre to specific tissues may be useful if a suitable Cre line of mice is not available. It is important that the endogenous gene functions normally until recombined by Cre; therefore, targeting modifications must be made without interfering with coding, splicing or regulatory regions. Similar to the generation of null mutants, the loxP sites must be inserted in such a way that recombination will render the gene inactive, avoiding compensatory alternative splicing.

In order to approach as close to normal expression as possible, it is advisable to remove the drug selection marker by designing FRT sites flanking the selection cassette. Transfection of ES cells with an FLP-expressing vector or crossing chimeras with transgenic mice expressing FLP in the germline effects excision of the cassette.

In some cases, deletion of the entire gene is desired, especially in cases of complicated alternative splicing, which may be the case for some large genes. Cre recombinase can also mediate large genomic deletions between two loxP sites located at a great distance. Two separate targeting vectors with different positive selection markers should be used for insertion of two loxP sites by homologous recombination at the 5' and 3' proximal regions of a target gene. At least one TK gene should also be introduced in one of the targeting constructs. Correctly targeted clones are then transfected with a Cre-expressing plasmid, and clones with deletion, and thus TK deletion, are selected for by resistance to FIAU. Alternatively, targeting vectors can be designed that carry two complementary but non-functional fragments of a positive selection marker. Upon Cre-mediated excision, the two fragments are brought together and drug resistance is restored. Using these methods, deletions of several hundred kilobase pairs can be achieved. Further discussion of conditional and inducible techniques is found elsewhere [37, 38].

4 **Inducible deletion**: Inducible mutation strategies using tetracycline or steroid receptor antagonists have been successful in achieving genetic manipulations at a desired stage of development, and can also provide tissue specificity [38]. Tet on/off systems allow switching on and off of gene expression through activation or inhibition of the tetracycline-dependent transactivating (tTA) factor with tetracycline analogs such as doxycycline. tTA binds to the tetO operator sequence placed upstream of a minimal promoter and the gene of interest through gene targeting. This is an appropriate strategy for reversible gene expression changes, but can often have 'leaky' gene expression. The Cre-ER^{T2}-tamoxifen system is a recommended approach for inducible and irreversible gene deletion [39, 40]. When fused to a mutant estrogen receptor ligand binding domain that is activated by tamoxifen but not estrogen, Cre remains in the cytoplasm and cannot catalyze recombination events. On addition of tamoxifen, the specific ligand for ER^{T2}, the Cre-ER^{T2} fusion protein translocates to the nucleus, where it now mediates deletion of floxed alleles. Thus, deletion is controlled temporally by tamoxifen administration and spatially by tissue-specific Cre-ER^{T2} expression. Tamoxifen availability can be limiting,

such that deletion may not reach 100% efficiency, resulting in mosaic expression of the target gene. However, the phenotype is still informative, especially for deletion of cell-autonomous genes.

10.2.3 Retrieval of DNA from BAC by recombineering

Recombineering stands for '*recomb*ination-mediated genetic eng*ineering*.' It is a very efficient and useful technique in providing direct manipulation of large sequences of bacterial DNA. More specifically, it involves targeted homologous recombination in a modified *Escherichia coli* strain. These strains (which include the DY380 strain used in our protocols)

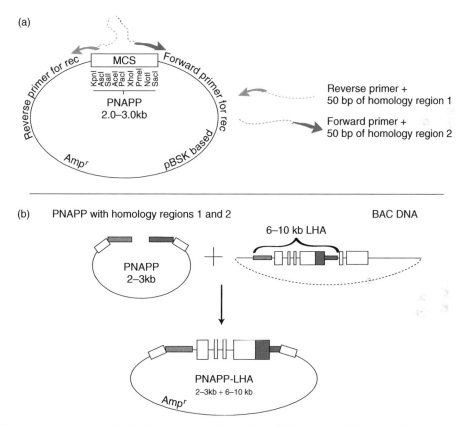

Figure 10.2 DNA retrieval by BAC recombineering. (a) *PCR amplification of a pBluescript (pBSK)-derived retrieval vector (pNAPP).* Standard PCR primers are designed complementary to regions of pNAPP overlapping the MCS (solid light gray and dark gray lines); the primers must have an additional 50 bp sequence at their 3' end that is homologous to one end of the sequence to be retrieved (dashed light gray and dark gray lines). Primers are designed such that the retrieved sequence is flanked by rare cutting restriction enzyme sites, to allow cloning into the final targeting vector (see Figure 10.1). (b) *Recombineering in bacteria.* Recombinogenic bacteria containing BAC DNA are transformed with the pNAPP PCR product. Homologous recombination yields the pNAPP vector containing the retrieved sequence. Positive cells are identified by PCR screening of bacterial colonies.

have been altered through the introduction of exogenous genes which promote homologous recombination, including a defective λ prophage. The phage genes of interest, *exo*, *bet* and *gam*, are transcribed from the λ*P*L promoter. This promoter is temperature sensitive and is repressed at 32 °C and de-repressed at 42 °C. When bacteria are grown at 32 °C, no recombination proteins are produced. However, briefly culturing the bacteria at 42 °C induces expression of all the recombination proteins, which readily promotes the modification of targeted DNA sequences.

These protocols allow the retrieval of a large region of mouse DNA contained in a BAC. The retrieved sequence serves as the LHA for recombination in ES cells (see Figure 10.1 and Figure 10.2). Recombineering bacterial strains can be obtained from The National Cancer Institute at Frederick (NCI-Frederick).

In Ensembl, selection of desired BACs containing DNA from the 129 mouse strain can easily be identified using the mouse genome browser (http://www.ensembl.org/ Mus_musculus/index.html). Here, 129 BACs are used in order to maximize recombination efficiency in the R1 line of ES cells, which has a 129 genetic background. The first step involves searching for the gene or region of interest and selecting the 'graphical view' window. The region of interest should be visible with several other layers of information. By default, BAC maps are not displayed, and in order to have the browser layer them on the 'graphical View' it is necessary to select the 'DAS sources' menu and highlight '**129S7/AB2.2 clones**' and the '**BAC map**' option to view the span and genomic region mapped to each library. Each individual BAC can be easily identified by its specific library identifier and can then ordered online from Geneservice, Ltd, Cambridge, UK (http://www.geneservice.co.uk/products/sanger/order.jsp) (bMQ 129 clones).

PROTOCOL 10.1   BAC Purification Protocol

Equipment and reagents

- Luria–Bertani (LB) media with appropriate antibiotic resistance
- 37 °C incubator
- Reagents included in the plasmid DNA purification Mini Kit (Qiagen)
- 2 ml Eppendorf tubes
- Isopropanol (Sigma)
- 70% ethanol
- Microcentrifuge at 4 °C
- 15 ml tubes.

Method

1 Grow bacteria containing the desired BAC in 6 ml of LB broth with appropriate antibiotic resistance overnight at 37 °C with shaking (220 rpm).

2 Divide bacteria into three tubes and pellet by centrifugation at maximum speed (16 000*g*) for 1 min.

3 Resuspend each pellet in 250 µl of buffer P1.

4 Add 250 μl of buffer P2 to each tube and mix.

5 Add 350 μl of buffer P3 to each tube and mix.

6 Spin tubes for 4 min at maximum speed.

7 Transfer the supernatant to new tubes.

8 Spin for 4 min at the maximum speed to clear the supernatant.

9 Transfer the supernatant to new tubes.

10 Add 750 μl of isopropanol and keep at room temperature for 10 min.

11 Collect the DNA by spinning the tubes for 10 min at maximum speed.

12 Wash the DNA pellet once with 1 ml of 70% ethanol and air dry.

13 Resuspend all pellets in a total volume of 50 μl water (total from three tubes).

14 For electroporation use 1 μl of the BAC DNA.

PROTOCOL 10.2   Transformation of BAC into Recombinogenic Bacterial Strain (DY380)

Equipment and reagents

- Decay-wave ECM 630 electroporator (BTX – Genetronics)
- DY380 cells (National Cancer Institute)
- LB media with appropriate antibiotic resistance
- 15 ml tubes
- Spectrophotometer
- Microcentrifuge at 4 °C
- 0.1 cm gap electroporation cuvette
- 1.5 ml Eppendorf tubes
- LB-agar plates with appropriate antibiotic selection
- 32 °C incubator
- Thermal cycler.

Method

1 Grow DY380 cells in 5 ml of LB broth at 32 °C overnight with shaking.[a]

2 On the next day, check cell growth by spectrophotometer reading ($OD_{600} = 1.2$).

3 Collect the cells by centrifuging at 2000g for 5 min at 4 °C.

4 Resuspend the cell pellet in 1 ml of ice-cold water and transfer to a 1.5 ml Eppendorf tube (on ice).

5 Centrifuge in a microcentrifuge at maximum speed for 15–20 s at 4 °C.

6 Place tubes on ice and aspirate the supernatant.

7 Repeat steps 3–5 two more times.

8 Resuspend the cell pellet in 50 μl of ice-cold water mixed with 1 μl of BAC DNA (from Protocol 10.1).

9 Transfer to a chilled electroporation cuvette.

10 Electroporate once using the following conditions: 1.75 kV, 25 μF and 200 .The time constant should be around 5–6 ms.

11 Add 1 ml of LB to the cuvette, transfer to an Eppendorf tube and incubate at 32 °C for 1 h.

12 Spin down cells and spread the entire volume on one LB-agar plate containing the appropriate antibiotic resistance. Incubate at 32 °C overnight.[b]

Notes

[a]It is important to always grow the recombineering strains at the restrictive temperature of 30–32 °C.

[b]It is critical to screen the DY380 colonies by PCR to ensure that they contain both ends of the BAC DNA or, preferably, to ensure that they contain both ends of the sequence that will be retrieved from the BAC.

PROTOCOL 10.3 Preparation of PCR Product

Equipment and reagents

- Thermal cycler
- *Dpn*I restriction enzyme (New England Biolabs)
- High-fidelity *Taq* or *Pfu* enzyme and buffer (e.g. Stratagene)
- 37 °C water bath
- Zymoclean gel purification kit (Zymo Research)
- 70% ethanol
- Spectrophotometer.

Method

1 Design primers with 50 bp homology to the BAC sequence and approximately 20 bp priming site on target cloning vector (pNAPP or pBSK; see Figure 10.2).

2 Dilute the cloning vector DNA in water to 0.1 ng/μl (approximately 1 : 10 000 dilution of mini prep DNA) and use 1 μl as a template for the PCR reaction.

3 Amplify the vector by high-fidelity PCR in a 50 μl reaction for 40 cycles (see DNA polymerase manufacturer's instructions).

4 Add 1 μl of *DpnI* restriction enzyme directly to the final PCR product and incubate at 37 °C for 1 h to digest the template DNA.

5 Gel-purify the PCR product using the Zymoclean gel purification kit. Perform an extra 70% ethanol wash to remove all salts prior to elution.

6 Elute DNA in 20 μl water and measure the absorption at 260 nm using a spectrophotometer in order to calculate DNA concentration.

PROTOCOL 10.4   Retrieval of Sequence of Interest from BAC DNA

Equipment and reagents

- LB media with appropriate selection
- 32 °C incubator
- Spectrophotometer
- 42 °C water bath
- Autoclaved 250 ml flasks
- 15 ml tubes
- Centrifuge at 4 °C
- LB-agar plates with appropriate antibiotic
- 0.1 cm gap electroporation cuvette
- Spectrophotometer
- Decay-wave ECM 630 electroporator (BTX – Genetronics).

Method

1 Grow DY380 cells containing the BAC in 5 ml of LB broth with appropriate antibiotic at 32 °C overnight with shaking (220 rpm).

2 On the following day, transfer 1 ml of the overnight culture ($OD_{600} = 1.2$) to a 250 ml flask containing 20 ml of LB with antibiotic. Incubate for 2 h at 32 °C with shaking ($OD_{600} = 0.5$).

3 Transfer 10 ml of the cells to a new 250 ml flask and shake in a 42 °C water bath for 15 min.

4 Put both flasks (42 °C treated and control) into wet ice and swirl to make sure that the temperature drops as quickly as possible. Leave on wet ice for 5 min.

5 Transfer the cells to cold 15 ml centrifuge tubes and spin at 2000*g* for 5 min at 4 °C.

6 Resuspend the cells in 1 ml of ice-cold water and transfer to a 1.5 ml Eppendorf tube on ice.

7 Wash three times with ice-cold water as described in steps 3–5 of Protocol 10.2.

8 Resuspend the cell pellet in 50 µl of ice-cold water.

9 Add 300 ng of the PCR product obtained from Protocol 10.3 and electroporate as described in Protocol 10.2, steps 3–11.

10 Plate the electroporated bacteria on LB-agar plates with antibiotic selection for the plasmid (not the BAC) and incubate overnight at 32 °C.

11 Screen colonies by standard *Taq* PCR for the retrieved DNA sequence.

10.2.4 ES and MEF cell culture

The R1 line of ES cells is a well-established line with an impressive track record of germline transmission of targeted genes. The R1 cell line was derived from deliberate outcrossing of 129/*Sv* and 129/*SvJ* strains. The majority of available ES cell lines were derived from 129 substrains, due to the ability of this strain to generate ES cells that efficiently contribute to the mouse germline after extensive manipulation in culture. This cell line is available to the scientific community from the laboratory of Andras Nagy at Mount Sinai hospital (http://www.mshri.on.ca/nagy/) and in our hands has proven a reliable resource for the generation of knockout mice. The following protocols have been specifically adapted for R1 ES cells; culture conditions for other ES cell lines have been successfully developed in many laboratories.

10.2.4.1 MEF cell culture protocols

ES cells must be maintained in an undifferentiated, healthy status throughout the entire gene targeting procedure. When using R1 ES cells, one critical aspect is to culture them on top of a layer of feeder MEF cells. Although it is possible to culture R1 cells without this feeder layer, it is highly recommended to use MEF cells when possible (ref Nagy website). Protocols 10.5–10.10 describe how to produce MEF cells from a transgenic mouse line with a neomycin resistance cassette. This cassette will confer resistance to the antibiotic G418, which is necessary for positive ES cell clone selection. Additionally, after the MEF cells are isolated and expanded, they are γ-irradiated in order to prevent cell division of the feeder cells. One transgenic mouse line expressing the neomycin resistance cassette, C57BL/6J-Tg(pPGKneobpA)3Ems/J, is commercially available from The Jackson Laboratory.

PROTOCOL 10.5   Isolating MEF Cells (from 8 to 12 embryos)

Equipment and reagents

- Incubator at 37 °C with 5% CO_2

- Water bath at 37 °C

- E13–15 embryos from G418 resistant mouse
- Isoflurane (Baxter Healthcare)
- Sterile dissection tools
- 1× Dulbecco's phosphate-buffered saline (PBS)
- 0.25% trypsin with EDTA
- 10 ml glass pipettes
- 50 ml conical tubes
- 600 ml cell culture flasks with air filter caps (Nunc/Nalgene), coated with 0.1% gelatin
- 1000× penicillin/gentamicin (0.59% penicillin (w/v)/8% gentamicin (w/v))
- Dulbecco's modified Eagle's medium (DMEM; high glucose, high bicarbonate, without pyruvate, with L-glutamine, Invitrogen)
- Fetal bovine serum (FBS, heat inactivated, Hyclone)
- MEF medium: DMEM, 1× penicillin/gentamicin, 10% FBS
- 10 cm diameter Petri dishes.

Method

1 Sacrifice the pregnant mouse after isoflurane-induced anesthesia.

2 Dissect out the uterus and place in a large Petri dish with PBS.

3 Use sterile forceps to free the embryos from the uterus and place in a new Petri dish with PBS.

4 Use two sterile forceps to remove and discard all red organs from the thoracic and abdominal cavities of each embryo.

5 After dissection of each embryo, place in a new Petri dish with fresh PBS.

6 When all embryos have been dissected, swirl gently to wash.

7 Transfer the embryos to a 10 cm Petri dish containing 5 ml of 0.25% trypsin/EDTA.

8 Use clean, sterile scissors to chop the embryos into small pieces.[c]

9 Add 5 ml of 0.25% trypsin and incubate at 37 °C for 15 min.

10 Dissociate cells by pipetting the mixture up and down with a 10 ml glass pipette about 20 times. When pipetting down, press the pipette tip to the bottom of the Petri dish to help dissociate tissue pieces.

11 Repeat steps 9 and 10 three more times, but reduce the incubation time to 10 min (37 °C/5% CO_2).

12 Divide the mixture into two 50 ml conical tubes and add equal volumes of MEF medium.

13 Allow tissue pieces to settle down for 3–4 min.

14 Transfer supernatant (containing MEF cells) to new 50 ml conical tubes.

15 Spin down cells at 600g for 5 min and aspirate supernatant.

16 Resuspend pellet in 25 ml of MEF medium.

17 Transfer each 25 ml to a 600 ml flask.

18 Mix well by swirling the flasks gently.

19 Place flasks in incubator (37 °C/5% CO_2). These cells are considered to be at passage zero (P0).

20 Change medium after 24 h and then every 2 days until they are 100% confluent.

21 Monitor cell growth every day for 2–4 days. When cells are 100% confluent they are ready to be passaged. Protocol 10.6 describes how to expand the MEF cells at a 1 : 3 dilution.

Notes

^{c}This is a tiring and time-consuming step, but it is very important in order to obtain a good yield.

PROTOCOL 10.6   Expansion and Passage of MEF Cells

Equipment and reagents

- Incubator at 37 °C with 5% CO_2

- Water bath at 37 °C

- 1× Dulbecco's PBS

- 0.25% trypsin with EDTA

- 10 ml glass pipettes

- 50 ml conical tubes

- 600 ml cell culture flasks with air filter caps (Nunc/Nalgene), coated with 0.1% gelatin

- 1000× penicillin/gentamicin (0.59% penicillin (w/v)/8% gentamicin (w/v))

- DMEM (high glucose, high bicarbonate, without pyruvate, with L-glutamine, Invitrogen)

- FBS (heat inactivated, Hyclone)

- MEF medium: DMEM, 1× penicillin/gentamicin, 10% FBS

- 0.1% gelatin-coated 10 cm Petri dishes.

Method

1 Pre-warm MEF media in a 37 °C water bath.

2 Aspirate medium from 600 ml flasks and gently rinse each flask with 10 ml PBS.

3 Aspirate PBS and add 3 ml 0.25% trypsin/EDTA to each flask. Rock the flask to cover the entire bottom surface.

4 Incubate at 37 °C for 5 min, rocking the flasks twice during the incubation.[d]

5 Add 15 ml MEF media to each flask and rock well to inhibit trypsin.

6 Dissociate cells by pipetting up and down 10 times with a 10 ml glass pipette. When pipetting down, press the pipette tip to the bottom of the Petri dish to help dissociate tissue pieces.

7 Transfer cell suspension to a 50 ml conical tube.

8 Spin for 3 min at 600g.

9 While spinning, aliquot 25 ml MEF medium to new 600 ml flasks.

10 Resuspend MEF cell pellet from each flask in 15 ml MEF medium.

11 Add 5 ml to each new 600 ml flask containing 25 ml MEF medium (for a 1 : 3 passage). These are P1 MEF cells. Incubate at 37 °C/5% CO_2.

12 Monitor growth of the P1 cells, and change medium every 2 days until they are 100% confluent.

Notes

[d] Notice cells detaching from the bottom of the flask.

PROTOCOL 10.7 Harvesting and Cryopreservation of P2 MEF Cells

Equipment and reagents

- Incubator at 37 °C with 5% CO_2

- Water bath at 37 °C

- 1× Dulbecco's PBS

- 0.25% trypsin with EDTA

- 10 ml glass pipettes

- 50 ml conical tubes

- 15 ml conical tubes

- 600 ml cell culture flasks with air filter caps (Nunc/Nalgene), coated with 0.1% gelatin

- 1000× penicillin/gentamicin (0.59% penicillin (w/v)/8% gentamicin (w/v))

- DMEM (high glucose, high bicarbonate, without pyruvate, with L-glutamine, Invitrogen)
- FBS (heat inactivated, Hyclone)
- MEF medium: DMEM, $1\times$ penicillin/gentamicin, 10% FBS
- Dimethyl sulfoxide (DMSO, Sigma)
- 2 ml cryotube vials.

Method

1 Aspirate medium from 600 ml flasks and rinse each flask with 10 ml PBS.

2 Aspirate PBS and add 3 ml 0.25% trypsin/EDTA to each flask. Rock the flask to cover the entire bottom surface.

3 Incubate at 37 °C for 5 min, rocking the flasks twice during the incubation.

4 Add 8 ml of MEF medium into each flask.

5 Dissociate cells by pipetting up and down 10 times with a 10 ml glass pipette. When pipetting down, press the pipette tip to the bottom of the Petri dish to help dissociate tissue pieces.

6 Pool cells from all flasks into 50 ml conical tubes.

7 Spin down cells at 600*g* for 5 min.

8 During this time, prepare MEF medium containing 20% DMSO.

9 Resuspend cells in 1 ml of medium per flask.

10 Add an equal volume of 20% DMSO–MEF medium and mix well.

11 Aliquot cells as quickly as possible to cryotube vials (1 ml per vial) and store in a polystyrene box at −80 °C overnight. These are P2 MEF cells.

12 The following day, transfer the tubes to liquid nitrogen for long-term storage.

13 In the future, to avoid *de novo* generation of MEF cells, proceed as follows:

   (a) Thaw two vials of P2 MEF cells in a 37 °C water bath.

   (b) Add cells to 7 ml MEF medium in a 15 ml tube.

   (c) Spin at 600*g* for 3 min.

   (d) Aspirate supernatant.

   (e) Resuspend pellet in 10 ml MEF medium.

   (f) Add 5 ml cells to the bottom of two 0.1% gelatin-coated 600 ml flasks, and add 20 ml MEF medium to the bottom of each flask for a total of 25 ml per flask. Mix well by swirling flask gently.

   (g) Place flasks in 37 °C incubator. These MEFs are ready to be expanded according to Protocol 10.6 'Expansion and Passage of MEF cells'. Passage the cells at a 1 : 3 ratio until they reach P5 (a total of 54 flasks).

PROTOCOL 10.8   Harvesting of MEF Cells

Equipment and reagents

- Incubator at 37 °C with 5% CO_2

- Water bath at 37 °C

- 1× Dulbecco's PBS

- 0.25% trypsin with EDTA

- 10 ml glass pipettes

- 50 ml conical tubes

- 600 ml cell culture flasks with air filter caps (Nunc/Nalgene), coated with 0.1% gelatin

- 1000× penicillin/gentamicin (0.59% penicillin (w/v)/8% gentamicin (w/v))

- DMEM (high glucose, high bicarbonate, without pyruvate, with L-glutamine, Invitrogen)

- FBS (heat inactivated, Hyclone)

- MEF medium: DMEM, 1× penicillin/gentamicin, 10% FBS.

Method

Perform the following steps using four to five flasks of expanded MEF cells at a time.

1 Aspirate medium from each 600 ml cell culture flask containing MEF cells.

2 Rinse cells with 10 ml PBS.

3 Aspirate PBS and add 3 ml trypsin/EDTA. Place flasks in 37 °C incubator for 5 min, rocking once during this time.[e]

4 Add 6 ml MEF medium to each flask and wash cells from the bottom by pipetting up and down five times.

5 Transfer cells to a 50 ml conical tube (pool four to five flasks in one conical tube).

6 Spin cells down (600g for 3 min) and resuspend in 0.75 ml of MEF medium per harvested flask.[f]

Notes

[e] Notice cells detaching from the bottom of the flask

[f] Keep the conical tube with pooled cells on ice.

PROTOCOL 10.9 γ-Irradiation of MEF Cells

Equipment and reagents

- 50 ml conical tubes

- 1000× penicillin/gentamicin (0.59% penicillin (w/v)/8% gentamicin (w/v))

- DMEM (high glucose, high bicarbonate, without pyruvate, with L-glutamine, Invitrogen)

- FBS (heat inactivated, Hyclone)

- MEF medium: DMEM, 1× penicillin/gentamicin, 10% FBS

- DMSO (Sigma)

- 2 ml cryotube vials.

Method

1 Expose MEF cells harvested in Protocol 10.8 to a dose of 3000 rad (γ-rays).[g]

2 To aliquot cells for storage, prepare 50 ml of MEF medium containing 20% DMSO in a 50 ml conical tube.

3 Mix irradiated cells in a 50 ml conical tube with an equal volume of MEF medium containing 20% DMSO.

4 Mix well and immediately aliquot 1 ml per cryotube. Mix the stock of cells frequently during aliquoting.

5 Remove a small aliquot to count the cells.[h]

6 Store the cryotubes at −80 °C overnight in a polystyrene box.

7 The following day, transfer the tubes to liquid nitrogen.

Notes

[g]The exact procedure for irradiating cells will depend on the irradiator and the procedures in the facility housing it.

[h]Calculate how many cells will be present in each ml.

PROTOCOL 10.10   Plating γ MEF Cells for Culturing ES Cells

Equipment and reagents

- Incubator at 37 °C with 5% CO_2

- Water bath at 37 °C

- 1× Dulbecco's PBS

- 10 ml glass pipettes

- 15 ml conical tubes

- 1000× penicillin/gentamicin (0.59% penicillin (w/v)/8% gentamicin (w/v))

- DMEM (high glucose, high bicarbonate, without pyruvate, with L-glutamine, Invitrogen)

- FBS (heat-inactivated, Hyclone)

- MEF medium: DMEM, 1× penicillin/gentamicin, 10% FBS

- 0.1% gelatin-coated 10 cm Petri dishes.

Method

1 Remove a cryotube of γ-irradiated MEF cells from liquid nitrogen and warm it in a 37 °C water bath.

2 Transfer cells to a 15 ml conical tube with 7 ml of MEF medium.

3 Spin cells at 600g for 5 min.

4 Resuspend cells in 9 ml of MEF medium.[i]

5 Distribute cells to several 10 cm gelatin-coated plates.

6 Add MEF medium to make up a total volume of 10 ml per plate.[j,k]

7 Each plate will be ready to receive ES cells after 24–48 h.

Notes

[i]Usually 1.5 million cells will be sufficient to plate one 10 cm plate.

[j]MEF cells will adhere rapidly so sometimes it is better to have the 10 cm plates ready with several ml of medium.

[k]It is important to make sure that the MEF cells are fully spread out over the plate, with no large gaps. This ensures a sufficient adherence substrate for the ES cells.

10.2.4.2 ES cell culture protocols

Once the targeting vector has been constructed and MEF feeder cells are prepared, genetic manipulations of ES cells in culture may proceed. An overview of the steps involved in achieving successful gene targeting in ES cells is found in Figure 10.3. These steps involve growth of ES cells, transfection of ES cells with targeting vector via electroporation, picking of colonies obtained after positive-negative selection, harvesting of colonies and screening of ES cell DNA by PCR (see Protocols 10.11–10.19). It is imperative to plan ahead meticulously, as feeder cells must be plated at least 48 h before plating, passaging or picking ES cells.

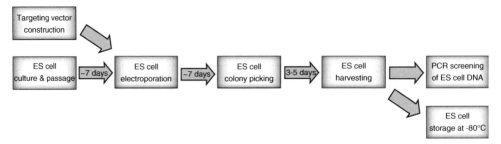

Figure 10.3 Outline of ES cell manipulations for gene targeting. MEF feeder cells must be plated 2 days prior to ES cell plating or passage. Arrows depict approximate time from one step to the next (shaded boxes). ES cell colony picking usually lasts for 5 days; harvesting occurs 3–5 days from the time of individual colony picking. Harvested colonies are screened immediately by PCR for homologous recombination.

PROTOCOL 10.11   Plating ES Cells

Equipment and reagents

- Incubator at 37 °C with 5% CO$_2$

- Water bath at 37 °C

- 10 ml glass pipettes

- 15 ml conical tubes

- 1000× penicillin/gentamicin (0.59% penicillin (w/v)/8% gentamicin (w/v))

- FBS (heat inactivated)

- DMEM (high glucose, high bicarbonate, no pyruvate, with L-glutamine, Invitrogen, cat. no. 11965)

- β-Mercaptoethanol (BME)

- ESGRO leukemia inhibitory factor (LIF; 10^7 units/ml, Chemicon, cat. no. ESG1107)

- ES medium: DMEM, 20% FBS, 1 : 1000 dilution of penicillin/gentamicin, 1 : 100 dilution of BME, 1 : 10000 dilution of LIF

- γ-irradiated MEF cells plated on 0.1% gelatin-coated 10 cm plates 48 h prior to ES cell plating.

Method

1 Pre-warm ES medium in a 37 °C water bath.

2 Aspirate MEF medium from one 0.1% gelatin-coated 10 cm dish previously plated with γ-MEF cells, and add 7 ml of ES medium to the MEF cells.

3 Remove one vial of ES cells from storage in liquid nitrogen and thaw immediately in a 37 °C water bath.

4 Add cells to 5 ml ES medium in a 15 ml tube and spin for 3 min at 600*g*.

5 Aspirate supernatant and resuspend cells in 5 ml of ES medium.

6 Seed ES cells (5 ml) in the MEF cell plate containing ES medium (total volume of 12 ml). Mix well by rocking the plate gently, and transfer to a 37 °C incubator.[

Notes

[Change ES medium daily by aspirating medium and then adding 12 ml of fresh medium against the side of the culture dish so as not to disturb the cells. Two to three days after plating, the ES cell colonies will be large enough to reach 60–70% confluency. At this stage they are ready to be passaged. It is very important to prevent colonies from being in contact with one another. If this occurs, they have become overconfluent and may start to differentiate (see Figure 10.4).

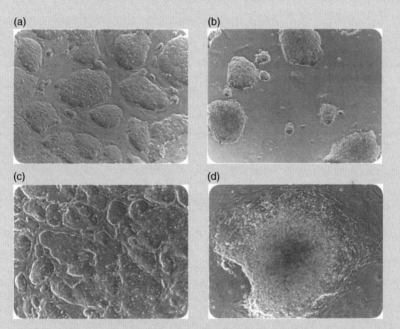

Figure 10.4 ES cell morphology and confluence. (a) Healthy R1 ES cells after 4 days in culture, at a density ready for passage (10× magnification). (b) ES cells plated at a low density. Another passage is required to increase individual colony numbers (10×). (c) Overconfluent ES cells (10×). (d) Highly differentiated ES cells. Note the loss of the smooth ES cell colony border, the formation of cobblestone-like cells and ES cells differentiating into fibroblasts that extend outwards from the colony (25×). a, b, c and d = Images 17C, 17B, 17D, 17F, pp. 35–36, from the Murine Embryonic Stem Cell Culture instruction manual, Millipore, Billerica, MA.

PROTOCOL 10.12   Passaging ES Cells

Equipment and reagents

- Incubator at 37 °C with 5% CO_2

- Water bath at 37 °C

- 10 ml glass pipettes

- 50 ml conical tubes

- 15 ml conical tubes

- 1000× penicillin/gentamicin (0.59% penicillin (w/v)/8% gentamicin (w/v))

- FBS (heat inactivated)

- 0.25% trypsin with EDTA

- DMEM (high glucose, high bicarbonate, no pyruvate, with L-glutamine, Invitrogen, cat. no. 11965)

- BME

- ESGRO LIF (10^7 units/ml, Chemicon, cat. no. ESG1107)

- G418/Geneticin selective antibiotic (50 mg/ml)

- FIAU (1 mg/ml, dissolved 1 : 1 in ethanol/H_2O, = 2.7 mM for 10 000 × stock, Moravek Biochemicals)

- MEF medium: DMEM, 1 : 1000 penicillin/gentamicin, 10% FBS

- ES medium: DMEM, 20% FBS, 1 : 1000 penicillin/gentamicin, 1 : 100 BME, 1 : 10 000 LIF

- ES medium + selection:

  — ES medium plus:

  — 1 : 10 000 FIAU

  — 1 : 167 dilution of G418 (e.g. 3 ml for 500 ml of medium) or 1 : 83.5 (during first day of selection)

- γ-irradiated MEF cells plated on 0.1% gelatin-coated plates 48 h prior to ES cell plating

- 1× Dulbecco's PBS

- DMSO

- Cryotubes.

Method

1 Aspirate ES medium, then rinse plate with 5 ml of Dulbecco's PBS.

2 Aspirate PBS and add 2.5 ml of pre-warmed 0.25% trypsin/EDTA. Rock gently until the entire plate is covered.

3 Place the plate in the incubator for 5 min. Rock the cells once during digestion to encourage dissociation from the plate.

4 Remove the plate from the incubator, add 10 ml of ES medium and rock gently.

5 Dissociate cells by pipetting up and down 15 times in a 10 ml glass pipette while forcing the pipette tip against the bottom of the dish.

6 Spin the ES cells for 3 min at 600g. Aspirate medium.

7 At this stage the cells are ready to be replated or frozen down. To transfer to fresh dishes, proceed to step 8. To freeze cells, proceed to step 11.

8 Aspirate the medium from fresh 10 cm plates of γ-MEF cells and add 10 ml of ES cell medium to each.

9 Resuspend the pellet of ES cells obtained in step 6 in 1 ml of ES medium (containing the appropriate selection if needed) per plate of MEF cells. Ensure complete resuspension by by pipetting up and down at least 10 times.

10 Distribute 1 ml of ES cells to each plate of γ-MEF cells. Place plates in the incubator and change medium daily.[m]

11 Prepare cryotubes with loosened caps so as to allow quick aliquoting following DMSO addition.

12 Make 20% DMSO in ES medium (including G418 if appropriate).

13 Resuspend the pellet of ES cells in 3 ml of ES medium.

14 Mix the cells with 3 ml of ES medium containing 20% DMSO. Pipette up and down several times and aliquot 1 ml per cryotube. Freeze the cryotubes overnight in a polystyrene box at $-80\,^{\circ}$C.

15 The following day, transfer the vials to liquid nitrogen, keeping the cells on dry ice while they are being transferred.

Notes
[m]Usually one 10 cm plate of ES cells trypsinized as above can be re-plated onto six 10 cm plates of γ-MEF cells.

PROTOCOL 10.13   Preparation of Targeting Vector DNA for Electroporation

Equipment and reagents

- Endotoxin-free plasmid maxi kit (Qiagen)
- User-designed plasmid targeting vector DNA
- 37 °C water bath
- Appropriate restriction enzyme and buffers
- 1% agarose gel

- Phenol chloroform
- Chloroform
- 70% ethanol
- Microcentrifuge
- Spectrophotometer.

Method

1 Purify targeting vector plasmid DNA using the Qiagen endotoxin-free plasmid maxi kit in order to obtain a high concentration and purity of DNA.

2 Linearize 80 µg of targeting vector (in order to obtain at least 25 µg after subsequent purification) for 6–8 h in a 37 °C water bath with an appropriate restriction enzyme that has a unique cut site on one end of a homology arm; for example, between the TK cassette and a homology arm. Use 25 µl of enzyme and suitable buffers in a total volume of 400 µl.

3 Before purification, run 2 µl of digested DNA on a 1% agarose gel to verify digestion efficiency. If not completely digested, add more enzyme (total enzyme can be 40 µl).

4 Add an equal volume of phenol chloroform, vortex and then spin for 3 min at maximum speed in a bench-top centrifuge.

5 Draw off aqueous (top) phase (contains DNA) and add to it an equal volume of chloroform (to remove phenol). Vortex and spin down for 3 min at maximum speed.

6 Draw off aqueous phase (contains DNA), and add to it 1/10 volume of 3 M sodium acetate plus two volumes of 100% ethanol. Vortex and freeze at −80 °C for 15 min.

7 Spin for 10 min at 4 °C at maximum speed to pellet the DNA.

8 Suck off the ethanol and wash the pellet with 150 µl of 70% ethanol.

9 Spin for 3 min at maximum speed.

10 Draw off ethanol and air-dry the pellet for 15 min.

11 Resuspend the pellet in 50 µl of deionized H_2O.

12 Measure absorbance of DNA at 260 nm and calculate concentration.

PROTOCOL 10.14   Electroporation of ES Cells with Targeting Vector

Equipment and reagents

- Incubator at 37 °C with 5% CO_2
- Water bath at 37 °C

- 10 ml glass pipettes

- 50 ml conical tubes

- 15 ml conical tubes

- 1000× penicillin/gentamicin (0.59% penicillin (w/v)/8% gentamicin (w/v))

- FBS (heat inactivated)

- 0.25% trypsin with EDTA

- DMEM (high glucose, high bicarbonate, no pyruvate, with L-glutamine, Invitrogen, cat. no. 11965)

- BME

- ESGRO LIF (10^7 units/ml, Chemicon, cat. no. ESG1107)

- G418/Geneticin selective antibiotic (50 mg/ml)

- FIAU (1 mg/ml, dissolved 1:1 in ethanol/H_2O, = 2.7 mM for 10 000 × stock, Moravek Biochemicals)

- MEF medium: DMEM, 1 : 1000 penicillin/gentamicin, 10% FBS

- ES medium: DMEM, 20% FBS, 1 : 1000 penicillin/gentamicin, 1 : 100 BME, 1 : 10 000 LIF

- ES medium + selection:

  — ES medium plus:

  —1× (1:10 000 dilution of) FIAU

  —1× (1:167 dilution of) G418 (e.g. 3 ml for 500 ml of medium) or 2× (1:83.5 dilution of) G418 (during first day of selection)

- γ-irradiated MEF cells plated on 0.1% gelatin-coated plates 48 h prior to ES cell plating

- 1× Dulbecco's PBS

- Decay-wave ECM 630 electroporator (BTX – Genetronics)

- 1× Hank's balanced salt solution (HBSS)

- Hemocytometer

- Trypan blue

- Trypsin/EDTA.

Method

1 Aspirate ES medium, and rinse plate with 5 ml of PBS.[n]

2 Aspirate PBS and add 3 ml of 0.25% trypsin/EDTA. Rock gently until the entire plate is covered.

3 Place the plate in the incubator for 5 min. Rock the cells once during this time to promote dissociation from the plate.

4 Add 9 ml of ES medium and rock gently.

5 Dissociate cells by pipetting up and down 15 times while forcing the pipette tip against the bottom of the dish.[o]

6 Transfer the cells to a 50 ml conical tube and spin for 3 min at 600g.

7 Aspirate the supernatant and thoroughly resuspend cells in 10 ml of PBS.

8 Dilute 5 µl of cell suspension in 20 µl of trypan blue (1 : 5 dilution). Add to a hemacytometer and calculate the number of cells per milliliter.

9 For each electroporation, spin down 20 million cells and resuspend the pellet in PBS with 25 µg of linearized targeting vector DNA in a total volume of 800 µl.

10 Add resuspended cells to a 4 mm sterile, disposable gap cuvette (800 µl capacity).

11 Cap cuvette and let stand for 5 min at room temperature.

12 Apply a single 225 V, 500 µF pulse to each cuvette (ECM 630 – low-voltage mode, resistance = 0). Let the cuvette stand on ice for 10 min.

13 Transfer cells to 9.2 ml of ES cell medium in a 15 ml conical tube. Rinse cuvette with medium to maximize recovery of cells.

14 Take 100 µl of the 10 ml of electroporated cells and calculate survival rate: mix 100 µl of cells with 100 µl of trypan blue, wait 5 min and count number of live cells. Survival rate equals number of live cells after electroporation divided by 20 million. A typical electroporation will have a survival ratio 30–50%.

15 Prepare γ-MEF cells plates by substituting MEF medium with ES medium. Typically, each electroporation will seed five plates.

16 Plate 2 ml of cells (from 10 ml) in each plate and disperse to allow good separation between colonies.

17 After 24 h, change medium from ES medium to ES medium + 2× G418 (1 : 83.5 dilution of G418).[p]

18 48 h after electroporation, change medium to ES medium + 1× G418 + FIAU.[q]

19 Change medium once per day with ES medium + 1× G418 + FIAU.

Notes

[n]Cells will be ready for electroporation approximately four days after being passaged. Each plate typically yields 10–20 million cells.

[o]It is critical that a thorough dissociation be accomplished, so as to allow for efficient electroporation and subsequent plating of individual cells.

[p]ES cells will grow on days 2–6 after electroporation, but most will die off as the cells only transiently transfected are eliminated after gradual loss of the neomycin resistance cassette. By days 7–9, individual colonies should become visible on the surface of the MEF cells and be ready for picking.

^qTo test effectiveness of negative selection, do not add FIAU to one plate. Carefully label this plate and expect it to have a much larger number of colonies than those plates with FIAU. This plate should not be used for picking.

PROTOCOL 10.15   Picking ES Cell Colonies

Equipment and reagents

- Incubator at 37 °C with 5% CO_2

- Water bath at 37 °C

- 96 well plate

- 10 ml glass pipettes

- 50 ml conical tubes

- 15 ml conical tubes

- 1000× penicillin/gentamicin (0.59% penicillin (w/v)/8% gentamicin (w/v))

- FBS (heat inactivated)

- 0.25% trypsin with EDTA

- DMEM (high glucose, high bicarbonate, no pyruvate, with L-glutamine, Invitrogen, cat. no. 11965)

- BME

- ESGRO LIF (10^7 units/ml, Chemicon, cat. no. ESG1107)

- G418/geneticin-selective antibiotic (50 mg/ml)

- FIAU (1 mg/ml, dissolved 1 : 1 in ethanol/H_2O, = 2.7 mM for 10 000× stock, Moravek Biochemicals)

- MEF medium: DMEM, 1 : 1000 penicillin/gentamicin, 10% FBS

- ES medium: DMEM, 20% FBS, 1 : 1000 penicillin/gentamicin, 1 : 100 BME, 1 : 10 000 LIF

- ES medium + selection:

  — ES medium plus:

  — 1× (1 : 10 000 dilution of) FIAU

  — 1× (1 : 167 dilution of) G418 (e.g. 3 ml in 500 ml of medium)

- γ-irradiated MEF cells plated on 0.1% gelatin-coated plates 48 h prior to ES cell plating.

Method

1 Plate MEF cells on 0.1% gelatin-coated 24-well plates 48 h before starting to pick colonies.

2 Pre-warm ES + G418 medium and 0.25% trypsin/EDTA at 37 °C. Thoroughly clean a microscope, 200 µl pipette, pipette tip box and table with 70% ethanol.[r]

3 Change the medium of a 24-well plate of MEF cells to ES + 1× G418 medium.

4 Add 100 µl of trypsin/EDTA to each well in a few rows of a 96-well plate. This should be repeated as necessary during picking.

5 Remove each plate of ES cell colonies from the incubator and circle all visible colonies by holding the plate above your head and marking visible colonies on the bottom.[s,t]

6 Set a 20 µl pipette to 8 µl. Place a plate with circled colonies on the microscope stage. To pick a colony, open the lid of the 10 cm plate with one hand and simultaneously bring the pipette tip immediately adjacent to a colony. To pick the colony, gently nudge the sides to loosen it from its MEF cell moorings and quickly aspirate, bringing the colony with 8 µl of medium.

7 Add the colony to a well of trypsin/EDTA in the 96-well plate and pipette up and down a few times to help break up the colony. Start a countdown timer for 10 min.

8 Repeat steps 6 and 7 as many times as possible inside the 10 min countdown.[u]

9 After 10 min, use a 200 µl pipette to vigorously pipette up and down each colony 10 times and then immediately transfer each trypsinized colony to a well of MEF cells in a 24-well plate (containing ES medium). Take care to also pipette up and down several times at this stage to distribute the cells throughout the well.[v]

10 After picking from a 10 cm plate is complete for the day, replace medium in the plate with fresh ES medium + 1× G418 + 1× FIAU.

11 48 h after picking, replace the medium in the 24-well plate with ES medium + 1× G418 and then daily until harvesting.[w]

Notes

[r] If possible, it is preferable to use a microscope inside a culture hood since this will reduce the possibility of culture contamination.

[s] Different colored markers may be used for each day of picking, to avoid repeated picking of the same colony. Picking will occur over four to five days.

[t] Do not pick colonies that appear to be differentiating (poorly defined edges, very spread out, less rounded cells).

[u] Even if doing only a small number of colonies per picking cycle, a minimum of 3 minutes is necessary to achieve good trypsinization.

[v] Each individual well should be numbered in order to track the number of colonies picked.

[w] Picked ES colonies will generally be ready for harvesting three to five days later, but please note that exact harvesting time must be empirically determined by carefully monitoring cell confluency in the 24 well plates on a daily basis.

PROTOCOL 10.16   Harvesting ES Cell Colonies

Equipment and reagents

- Incubator at 37 °C with 5% CO_2
- Water bath at 37 °C
- FBS (heat inactivated)
- 0.25% trypsin with EDTA
- 1× Dulbecco's PBS
- DMSO
- Cryotubes.

Method

1 Thaw heat-inactivated FBS and pre-warm 0.25% trypsin/EDTA to 37 °C.

2 Select which wells of the 24-well plates containing ES cells that are ready for harvesting.[x]

3 Fully aspirate the medium from 15 wells that are to be harvested. Add 350 µl of 0.25% trypsin/EDTA to each well and put the plate in the 37 °C incubator for 5 min.

4 During the 5 min incubation, label two sets of 15 tubes, both from 1 to 15 (16 to 30 in second round, etc.).[y]

5 After the 5 minute incubation, add 400 µl of FBS to each trypsinized well. Pipette up and down 15 times to dissociate cells.

6 Put 400 µl of cells into a tube for freezing at −80 °C and the remaining 350 µl into the other tube for PCR screening.

7 To each of the tubes destined for −80 °C freezing, add 44 µl of DMSO and mix immediately and thoroughly.[z,aa]

8 Spin the tubes destined for PCR screening at maximum speed for 2 min in a bench-top centrifuge and aspirate the medium. The pellets should then be stored at −80 °C in boxes separate from those containing cells frozen in FBS/DMSO.

9 Repeat steps 3–8 until all colonies are ready to be harvested that day have been frozen.

Notes

[x]A well that is ready will generally have 20–100 medium to large sized colonies per well.

[y]This will allow the preservation of half the cells (first set) for injection and the digestion of the other half (second set) for identification of positive clones by PCR.

[z]These must be stored at −80 °C immediately. The tubes for PCR can be temporarily stored at 4 °C for a few rounds of harvesting.

[aa]It is helpful to write down the last number harvested before beginning the next round so as to correctly number the tubes in the proper, sequential manner.

PROTOCOL 10.17   Screening ES Cell Colonies for Homologous Recombination of Targeting Vector

Equipment and reagents

- Water bath at 55 °C

- PCR reagents and primers

- Digestion buffer (50 mM Tris-HCl, pH 8.0; 1 mM $CaCl_2$; 1% Tween-20 in dH_2O; 10 mg/ml proteinase K).

Method

1 Resuspend harvested ES colony pellets (*for PCR screening*) in 50 µl of digestion buffer.

2 Let digestion proceed at 55 °C for 6 h.

3 Denature proteinase K in each tube by immersing the tubes in boiling water for 10 min.

4 Centrifuge the digestions briefly to collect evaporated droplets and allow to cool to room temperature.

5 Use 2 µl of each digested colony for one PCR reaction.[bb]

Notes

[bb]The exact PCR screening strategy will depend on the targeting vector, primer design and experimental conditions (see Figure. 10.1). Nevertheless, it is generally recommended to perform several screens per colony, including sets of primers to test for:

1 The quality of the DNA prep, by screening an endogenous region not targeted for modification

2 The efficiency of the primers used to screen for correct homologous recombination

3 Random insertion of the targeting construct

4 Homologous recombination of the targeting construct at the targeted locus

PROTOCOL 10.18   Expansion of Correctly Targeted ES Cells

Equipment and reagents

- Incubator at 37 °C with 5% CO_2

- Water bath at 37 °C

- 10 ml glass pipettes

- 15 ml conical tubes

- 1000× penicillin/gentamicin (0.59% penicillin (w/v)/8% gentamicin (w/v))

- FBS (heat inactivated)

- DMEM (high glucose, high bicarbonate, no pyruvate, with L-glutamine, Invitrogen, cat. no. 11965)

- BME

- ESGRO LIF (10^7 units/ml, Chemicon, cat. no. ESG1107)

- G418/geneticin-selective antibiotic (50 mg/ml)

- FIAU (1 mg/ml, dissolved 1 : 1 in ethanol/H_2O, = 2.7 mM for 10 000× stock, Moravek Biochemicals)

- MEF medium: DMEM, 1 : 1000 penicillin/gentamicin, 10% FBS

- ES medium: DMEM, 20% FBS, 1 : 1000 penicillin/gentamicin, 1 : 100 BME, 1 : 10 000 LIF

- ES medium + selection:

  — ES medium plus:

  — 1× (1 : 10 000 dilution of) FIAU

  — 1× (1 : 167 dilution of) G418 (e.g. 3 ml for 500 ml of medium)

- γ-irradiated MEF cells plated on gelatinized plates 48 h prior to ES cell plating

- 1× Dulbecco's PBS

- 0.25% trypsin with EDTA

- DMSO

- Cryotubes.

Method

1 Plate MEF cells on 0.1% gelatin-coated six-well dishes 24–48 h in advance.[cc]

2 Pre-warm ES medium to 37 °C.

3 Fill a 15 ml conical tube with 9 ml of ES medium + 1× G418.

4 Quickly thaw tubes with positive cells in a 37 °C water bath.

5 After thawing, transfer each clone to a conical tube with 9 ml of medium and spin for 3 min at 600g.

6 While spinning down cells, replace MEF medium in the six-well dish with 2 ml of ES medium + 1× G418.

7 Resuspend the pelleted cells in 1 ml ES medium + 1× G418 and plate onto one well of a six-well dish. Change the medium daily.[dd]

8 As soon as colonies are clearly visible, trypsinize and plate half of the cells in a new well of a 6 cm dish containing MEF cells with 5 ml ES medium + G418. Freeze down the other half in ES medium (without G418) plus 10% DMSO.*ee*

9 After the first trypsinization, the cells will grow faster. When cells in the 6 cm dish are ready to be passaged, trypsinize and plate half the cells in a 10 cm dish with MEF cells and freeze the other half as previously described.

10 When colonies have appeared in the 10 cm dish, passage to another 10 cm plate and freeze approximately five vials, depending on colony density.

11 If necessary, propagate to another 10 cm dish for further expansion.*ff*

Notes

*cc*Once a positive clone has been identified by PCR, the appropriately labeled cells at −80 °C should be expanded. To maximize ES cell viability, PCR screening should be completed within one to two weeks.

*dd*It will take several days before colonies grow up. They are often very sparse.

*ee*For each round of freezing, the cryotube should be clearly labeled with the passage number.

*ff*In addition, cells may be passaged to an uncoated plate in order to expand the cells for genomic DNA extraction and subsequent Southern blot analysis to check for [1] single integration of the targeting vector and [2] no unwanted genomic rearrangements.

PROTOCOL 10.19   Preparing Positive ES Cells for Blastocyst Injection

Equipment and reagents

- Incubator at 37 °C with 5% CO_2

- Water bath at 37 °C

- 10 ml glass pipettes

- 50 ml conical tubes

- 15 ml conical tubes

- 1000× penicillin/gentamicin (0.59% penicillin (w/v)/8% gentamicin (w/v))

- FBS (heat inactivated)

- 0.25% trypsin with EDTA

- DMEM (high glucose, high bicarbonate, no pyruvate, with L-glutamine, Invitrogen, cat. no. 11965)

- BME

- ESGRO LIF (10^7 units/ml, Chemicon, cat. no. ESG1107)

- G418/geneticin-selective antibiotic (50 mg/ml)

- FIAU (1 mg/ml, dissolved 1 : 1 in ethanol/H_2O, = 2.7 mM for 10 000× stock, Moravek Biochemicals)

- MEF medium: DMEM, 1 : 1000 penicillin/gentamicin, 10% FBS

- ES medium: DMEM, 20% FBS, 1 : 1000 penicillin/gentamicin, 1 : 100 BME, 1 : 10 000 LIF

- γ-irradiated MEF cells plated on gelatinized plates 48 h prior to ES cell plating

- 1× Dulbecco's PBS

- Injection buffer:

  — 8.3 g/l of DME powder without phenol red, without sodium bicarbonate (Sigma cat. no. D5030)

  — 4.5 g/l D-glucose

  — 25 mM HEPES

  — 584 mg/l L-glutamine.

Method

1 Plate one 0.1% gelatin-coated 10 cm plate with MEF cells 48 h in advance.

2 Prepare ES cell injection buffer.

3 Plate PCR-positive ES cells, using ES medium without selection, from either a fresh passage or from frozen stocks.[gg,hh,ii]

4 Allow cells to grow until colonies are large. This must be timed precisely so that the injections can be performed on a planned day. Often this involves starting 5–7 days prior to blastocyst injection, with one passage in between plating and the injection day.[jj]

5 On the day of injection, select one 10 cm plate of cells with the best morphology and density.

6 Rinse the plate with 5 ml of PBS.

7 Aspirate the PBS and add 3 ml of 0.25% trypsin/EDTA.

8 Add 9 ml of ES medium and thoroughly pipette the cells up and down against the bottom of the plate using a 10 ml glass pipette (15 times).[kk]

9 Spin the cells down (3 min at 600g) and resuspend in 2 ml of injection buffer supplemented with 5% FBS.

10 Place the cells on ice until blastocyst injection.

Notes

[gg]Cells from a fresh passage are preferable, but not necessary.

[hh]If using cells from a frozen stock, it is generally advised to make use of a earlier passage at which the cells are still reasonably concentrated. This is often the third passage after plating from −80 °C.

ii Remember to use ES medium without selection.

jj Use several dilutions in the final passage to ensure that at least one plate will have an optimal density of colonies.

kk Ensure a thorough dissociation of cells in order to allow injection of individual cells.

10.2.5 Mating of chimeras and downstream applications

Homologous recombinant ES cells obtained using the protocols described in this chapter are microinjected into the fluid-filled blastocoele cavity of 3.5-day-old embryos at the blastocyst stage; the injected embryos are then surgically implanted in the uterus of pseudopregnant females (Figure 10.5a). These techniques are described in detail elsewhere and are beyond

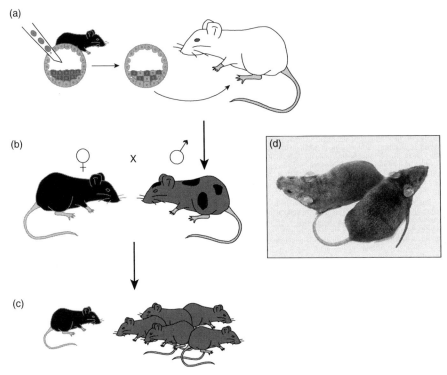

Figure 10.5 ES cell injection and chimera breeding. (a) Blastocysts are obtained from a C57BL/6 mouse (black coat color) and microinjected with targeted R1 ES cells (129 genetic background: agouti coat color). Injected blastocysts are implanted in a pseudopregnant female (white mouse). (b) The resulting black and brown male chimeras are then mated with a C57 black female to obtain brown and black F1 offspring (c). Brown pups are then screened by PCR or Southern blot for germline transmission of the targeted gene. (d) Two adult mouse chimeras generated through gene targeting in ES cells and injection into a C57BL/6 blastocyst. The mouse on the left has a higher percentage of agouti coat color; the probability of germline transmission of the targeted gene to the offspring of this mouse is higher than that for the mouse on the right. (See Plate 10.5.)

the scope of this chapter. The foster mother gives birth 17 days later to chimeric pups derived partly from the injected ES cells and partly from the host embryo.

Chimeras with the potential to transmit the recombinant ES cell genome through the germline are expected to be male, as they are derived from male R1 ES cells, and can be bred with C57BL/6 females in order to monitor germline transmission to the resulting offspring (F1 generation) (Figure 10.5b). Coat color is an indicator of germline transmission; agouti coat color in F1 mice demonstrates germline transmission, as 129-derived cells yield agouti, while C57-derived cells yield black pigmentation. PCR and Southern blot screening of purified tail DNA from agouti F1 animals will determine which F1 animal carries the mutated allele. Heterozygous F1 animals are then intercrossed to generate homozygous mutant and wild-type control littermates. In some instances, genetic background can contribute to the observed phenotype; in this case, animals should be backcrossed to the desired strain (usually C57BL/6) for more than eight generations so that mutants are maintained and studied on a pure genetic background.

10.3 Troubleshooting

1 **BAC retrieval does not work**

- Use three sets of PCR primer pairs to confirm the presence in the BAC of the two ends and middle of the sequence to be retrieved.

- The orientation of the long primers used to amplify the pNAPP vector is critical. Refer to Figure 10.2 for a schematic.

- Multiple freeze–thaw cycles compromise the ability of recombineering bacteria to accept BAC DNA. It may be necessary to obtain a fresh stock of these bacteria.

- In rare cases, BAC DNA may contain mutations that interfere with BAC retrieval. These may arise due to low DNA replication fidelity in bacteria. Therefore, we recommend ordering at least two different BAC clones when performing BAC retrieval.

2 **Poor MEF cell yield or morphology**

- During expansion of MEF cells, make sure the cells are highly confluent before passage. After 100% confluency is reached, wait 1–2 days before passage.

3 **ES cells undergo differentiation**

- Ensure that the MEF feeder cells are plated 48 h before ES cells, and that they form a monolayer over the plate.

- Ensure that LIF is at the correct concentration and is not past its expiration date.

- ES cells should be passaged approximately every 3 days; only undifferentiated cells survive frequent passaging. Refer to Figure 10.4 for images of ES cell confluency and differentiation.

- Culture medium should be less than 4 weeks old, as glutamine by-products may be toxic to ES cells.

- Slowly growing ES cells are more likely to undergo differentiation. Increase serum concentration if cells are growing slowly.

- ES cells have optimal growth when slightly crowded. Plate or passage ES cells at a higher density if it is necessary to increase colony numbers.

4 **Lack of chimeras**

- After identification of targeted ES cell clones, it is important to minimize time in culture. We recommend passaging ES cells no more than one to two times from thawing of frozen stock until blastocyst injection.

- Chimeras are more easily generated from ES cells with low passage number. It is preferable to begin ES cell manipulation with as low a passage number as possible.

- ES cells should be microinjected as soon as possible after collection from culture dishes, not left on ice for an extended period of time.

- If troubleshooting attempts have failed due to undetectable differentiation or defects of the first clone, it may be necessary to prepare a second targeted ES cell clone for blastocyst injection.

References

1. Heintz, N. (2001) BAC to the future: the use of bac transgenic mice for neuroscience research. *Nature Reviews Neuroscience*, **2**, 861–870.

2. Feng, G., Lu, J. and Gross, J. (2004) Generation of transgenic mice. *Methods in Molecular Medicine*, **99**, 255–267.

3. Hofker M.H. (2003) Introduction: the use of transgenic mice in biomedical research. *Methods in Molecular Biology*, **209**, 1–8.

4. Evans, M.J., Carlton, M.B. and Russ, A.P. (1997) Gene trapping and functional genomics. *Trends in Genetics*, **13**, 370–374.

5. Stanford, W.L., Cohn, J.B. and Cordes, S.P. (2001) Gene-trap mutagenesis: past, present and beyond. *Nature Reviews Genetics*, **2**, 756–768.

6. Ding, S., Wu, X., Li, G. *et al.* (2005) Efficient transposition of the *piggyBac* (*PB*) transposon in mammalian cells and mice. *Cell*, **122**, 473–483.

7. Dupuy, A.J., Akagi, K., Largaespada, D.A. *et al.* (2005) Mammalian mutagenesis using a highly mobile somatic *Sleeping Beauty* transposon system. *Nature*, **436**, 221–226.

8. Wu, S., Ying, G., Wu, Q. and Capecchi, M.R. (2007) Toward simpler and faster genome-wide mutagenesis in mice. *Nature Genetics*, **39**, 922–930.

9. Nagy, A. (2003) *Manipulating the Mouse Embryo: A Laboratory Manual*, Cold Spring Harbor Laboratory Press, Cold Spring Harbor, NY. A comprehensive guide to transgenics, gene targeting and other manipulations of the mouse embryo.

10. Tymms, M.J. and Kola, I. (eds) (2001) *Gene Knockout Protocols*, Methods in Molecular Biology, vol. **158**, Humana Press, Totowa, NJ.

11. Bradley, A., Evans, M., Kaufman, M.H. and Robertson, E. (1984) Formation of germ-line chimaeras from embryo-derived teratocarcinoma cell lines. *Nature*, **309**, 255–256. Generation of the first ES cell-derived mouse.

12. Robertson, E., Bradley, A., Kuehn, M. and Evans, M. (1986) Germ-line transmission of genes introduced into cultured pluripotential cells by retroviral vector. *Nature*, **323**, 445–448. Generation of the first mutant line of mice derived from ES cells.

13. Thomas, K.R. and Capecchi, M.R. (1987) Site-directed mutagenesis by gene targeting in mouse embryo-derived stem cells. *Cell*, **51**, 503–512. Demonstration of homologous recombination in ES cells by inserting a mutation into the *Hprt* wild-type locus.

14. Doetschman, T., Gregg, R.G., Maeda, N. *et al.* (1987) Targetted correction of a mutant HPRT gene in mouse embryonic stem cells. *Nature*, **330**, 576–578. Demonstration of homologous recombination in ES cells by repairing the spontaneous *Hprt* mutation.

15. Thompson, S., Clarke, A.R., Pow, A.M. *et al.* (1989) Germ line transmission and expression of a corrected HPRT gene produced by gene targeting in embryonic stem cells. *Cell*, **56**, 313–321.

16. Koller, B.H., Hagemann, L.J., Doetschman, T. *et al.* (1989) Germ-line transmission of a planned alteration made in a hypoxanthine phosphoribosyltransferase gene by homologous recombination in embryonic stem cells. *Proceedings of the National Academy of Sciences of the United States of America*, **86**, 8927–8931.

17. Zijlstra, M., Li, E., Sajjadi, F. *et al.* (1989) Germ-line transmission of a disrupted β_2-microglobulin gene produced by homologous recombination in embryonic stem cells. *Nature*, **342**, 435–438.

18. Thomas, K.R. and Capecchi, M.R. (1990) Targeted disruption of the murine int-1 proto-oncogene resulting in severe abnormalities in midbrain and cerebellar development. *Nature*, **346**, 847–850.

19. Pease, S. and Williams, R.L. (1990) Formation of germ-line chimeras from embryonic stem cells maintained with recombinant leukemia inhibitory factor. *Experimental Cell Research*, **190**, 209–211.

20. Mansour, S.L., Thomas, K.R. and Capecchi, M.R. (1988) Disruption of the proto-oncogene int-2 in mouse embryo-derived stem cells: a general strategy for targeting mutations to non-selectable genes. *Nature*, **336**, 348–352. Application of a general positive–negative selection strategy to enrich for cells in which homologous recombination has occurred.

21. Yagi, T., Ikawa, Y., Yoshida, K. *et al.* (1990) Homologous recombination at c-*fyn* locus of mouse embryonic stem cells with use of diphtheria toxin A-fragment gene in negative selection. *Proceedings of the National Academy of Sciences of the United States of America*, **87**, 9918–9922.

22. Deng, C. and Capecchi, M.R. (1992) Reexamination of gene targeting frequency as a function of the extent of homology between the targeting vector and the target locus. *Molecular and Cellular Biology*, **12**, 3365–3371.

23. Te Riele, H., Maandag, E.R. and Berns, A. (1992) Highly efficient gene targeting in embryonic stem cells through homologous recombination with isogenic DNA constructs. *Proceedings of the National Academy of Sciences of the United States of America*, **89**, 5128–5132.

24. Nagy, A., Rossant, J., Nagy, R. *et al.* (1993) Derivation of completely cell culture-derived mice from early-passage embryonic stem cells. *Proceedings of the National Academy of Sciences of the United States of America*, **90**, 8424–8428. First description of the R1 line of ES cells.

25. Ledermann, B. and Bürki, K. (1991) Establishment of a germ-line competent C57BL/6 embryonic stem cell line. *Experimental Cell Research*, **197**, 254–258.

26. Noben-Trauth, N., Kohler, G., Burki, K. and Ledermann, B. (1996) Efficient targeting of the IL-4 gene in a BALB/c embryonic stem cell line. *Transgenic Research*, **5**, 487–491.

27. Valenzuela, D.M., Murphy, A.J., Frendewey, D. *et al.* (2003) High-throughput engineering of the mouse genome coupled with high-resolution expression analysis. *Nature Biotechnology*, **21**, 652–659.

28. Poueymirou, W.T., Auerbach, W., Frendewey, D. *et al.* (2007) F0 generation mice fully derived from gene-targeted embryonic stem cells allowing immediate phenotypic analyses. *Nature Biotechnology*, **25**, 91–99.

29. Fiering, S., Epner, E., Robinson, K. *et al.* (1995) Targeted deletion of 5–HS2 of the murine β-globin LCR reveals that it is not essential for proper regulation of the β-globin locus. *Genes & Development*, **9**, 2203–2213.

30. Pham, C.T., MacIvor, D.M., Hug, B.A. *et al.* (1996) Long-range disruption of gene expression by a selectable marker cassette. *Proceedings of the National Academy of Sciences of the United States of America*, **93**, 13090–13095.

31. Dymecki, S.M. (1996) Flp recombinase promotes site-specific DNA recombination in embryonic stem cells and transgenic mice. *Proceedings of the National Academy of Sciences of the United States of America*, **93**, 6191–6196.

32. Rodriguez, C.I., Buchholz, F., Galloway, J. *et al.* (2000) High-efficiency deleter mice show that FLPe is an alternative to Cre-*loxP*. *Nature Genetics*, **25**, 139–140.

33. Zambrowicz, B.P., Imamoto, A., Fiering, S. *et al.* (1997) Disruption of overlapping transcripts in the ROSA βgeo 26 gene trap strain leads to widespread expression of β-galactosidase in mouse embryos and hematopoietic cells. *Proceedings of the National Academy of Sciences of the United States of America*, **94**, 3789–3794.

34. Kuhn, R. and Torres, R.M. (2002) Cre/*loxP* recombination system and gene targeting. *Methods in Molecular Biology*, **180**, 175–204. Detailed overview of the use of Cre/*loxP* recombination for conditional gene targeting in mice.

35. Sauer, B. (1998) Inducible gene targeting in mice using the Cre/*lox* system. *Methods*, **14**, 381–392.

36. Skvorak, K., Vissel, B. and Homanics, G.E. (2006) Production of conditional point mutant knockin mice. *Genesis*, **44**, 345–353.

37. Wirth, D., Gama-Norton, L., Riemer, P. *et al.* (2007) Road to precision: recombinase-based targeting technologies for genome engineering. *Current Opinion in Biotechnology*, **18**, 411–419.

38. Lewandoski, M. (2001) Conditional control of gene expression in the mouse. *Nature Reviews Genetics*, **2**, 743–755.

39. Wunderlich, F.T., Wildner, H., Rajewsky, K. and Edenhofer, F. (2001) New variants of inducible Cre recombinase: a novel mutant of Cre-PR fusion protein exhibits enhanced sensitivity and an expanded range of inducibility. *Nucleic Acids Research*, **29**, E47.

40. Li, M., Indra, A.K., Warot, X. *et al.* (2000) Skin abnormalities generated by temporally controlled RXRα mutations in mouse epidermis. *Nature*, **407**, 633–636.

11

Delivery Systems for Gene Transfer

Charlotte Lawson[1] and Louise Collins[2]
[1]*Veterinary Basic Sciences, Royal Veterinary College, London, UK*
[2]*Department of Clinical Sciences, Kings's College London School of Medicine, James Black Centre, London, UK*

11.1 Introduction

Insertion of genetic material into the cells of an individual to treat a disease or correct a hereditary condition has evolved from the distant dream of molecular biologists to a viable treatment option, with a number of successful clinical trials published to date. Historically, advancement within the gene therapy field has required development in other areas of molecular biology. The therapeutic potential of being able to harness the power of viruses that integrate their DNA into mammalian cells was recognized long before the recombinant DNA technology was available to isolate and clone the genes to be delivered [1]. Safety has always been a large consideration when designing 'viral vectors' for gene therapy, from the point of view of toxicity of the introduced genetic material or its gene products, the possibility of recombination events occurring that could lead to production of mobilizable viruses, and insertional mutagenesis. In view of this there is also a large literature on the development and use of 'non-viral' vectors for gene therapy applications. Since before the first clinical trials were carried out to treat severe combined immunodeficiency (SCID) patients in the 1990s there has been significant debate about viral versus non-viral gene transfer. Here, we will give a brief overview of some of the more commonly used transfer strategies together with protocols for viral and non-viral gene transfer into mammalian cells.

Genomics: Essential Methods   Edited by Mike Starkey and Ramnath Elaswarapu
© 2011 John Wiley & Sons, Ltd.

11.2 Methods and approaches

11.2.1 The ideal gene therapy vector

Efficient transfer of naked DNA into cells and tissues can sometimes be achieved (see later); however, introduction of genetic material into most sites requires the use of a vector to deliver the DNA to cells efficiently . Current strategies to develop vectors for gene therapy can be divided into viral and non-viral vector-mediated gene delivery, with the ultimate goal being to develop the ideal gene therapy vector with the following properties:

1 **Efficient delivery**: Since naked DNA is not efficiently taken up by cells, many strategies have been applied to improve gene delivery. Two approaches have been taken, namely the use of modified viruses (viral vectors) or non-viral delivery with charged polymers.

2 **Safety**: Safety vectors must not be pathogenic or toxic to patients. Non-viral delivery systems are generally relatively safe, but there are a number of safety issues when viral vectors are used (see Table 11.1).

   (a) Recombination events may occur between endogenous viral elements and transduced viral vectors, leading to the formation of a pathogenic replication-competent virus.

   (b) Integration of introduced genetic material into the host cell genome, whilst being desirable since it leads to longer term expression of the transgene, also carries the risk of activation of oncogenes due to insertional mutagenesis.

   (c) Plasmid DNA (either naked or packaged) will be contaminated with lipopolysaccharide (LPS) and cannot be completely removed from plasmid preparations. Thus, LPS-mediated toxicity may be triggered [2].

Table 11.1 Advantages and disadvantages of viral and non-viral vectors.

	Advantages	Disadvantages
Viral vectors	High efficiency	Oncogenic activation
	Long-term expression through integration (retrovirus)	Replication-competent virus formation
	Integral intracellular trafficking properties	Immunogenicity and viral protein overload
		Helper virus carryover (AAV)
		Limited cell tropism
		Batch-to-batch variation
Non-viral vectors	Absence of viral proteins	Low efficiency
	Low or no immunogenicity	Transient gene expression
	Unlimited size of DNA insert	Some cellular toxicity (PEI, liposomes)
	Targeting options for cell specificity	Inflammation due to unmethylated CpG DNA sequences
	Homogeneous, standardized and stable reagents	Additional agents needed to overcome intracellular barriers

(d) Selectable markers incorporated into vectors for *ex vivo* propagation could be antigenic if they enter the hemopoietic cell lineage. For example, if the protein were to be expressed, it could lead to humoral immune responses, which would be detrimental to patients [3].

3 **Specificity**: Ideally a vector needs to be capable of targeting and entering a specific pre-empted cell type or tissue, depending on the intended clinical application. This can be done by adding targeting ligands or by using tissue-specific promoters, as discussed later. Unpredictable side effects due to the ectopic expression of the transgene in normal tissues should be avoided at all costs.

4 **Regulation**: This is highly desirable, since it allows for

(a) activation of transgene when needed

(b) maintenance of transgene expression within a therapeutic window

(c) possibility to silence the gene if necessary; there has been some success *in vitro* and in animal models using antibiotic-responsive promoters (e.g. TetON [4]), or use of a hypoxia switch [5].

5 **Delivery of genes of any size or function**: There is often a limit to the amount of foreign DNA that can be inserted into viral vectors depending on the size of the wild-type viral genome and the number of viral genes that can be deleted whilst retaining infectivity. Inserts much larger than the endogenous DNA that has been removed will not be efficiently packaged by viral structural proteins. In contrast, there is in theory no limit to the size of DNA that can be delivered by non-viral methods of gene transfer.

6 **Long term and elevated expression**: Viral transfer of DNA is more efficient than non-viral transfer, and viruses have evolved strategies to avoid degradation within the host cell. However, duration of transgene expression could be limited by the immunogenicity of viral genes. DNA delivered by non-viral methods is sometimes rapidly removed from cells by lysosomal degradation, although there are now a number of different strategies to overcome this (see below).

7 **Cost effective to produc**: Vectors should be cheap to produce in large quantities. Vectors for non-viral transfer of DNA can be prepared rapidly with less stringent quality control assays required before use, and usually have less batch-to-batch variation in potency. On the other hand, complex purification procedures and quality control measures are often required during production of sufficient quantities of viruses for gene therapy applications. Viral vectors and non-viral vectors vary greatly in their design and ability to deliver DNA to a cell. There are advantages and disadvantages to both systems, which are summarized in Table 11.1.

11.2.2 Plasmid design

11.2.2.1 Promoters

It is essential when optimizing a vector for gene therapy that the therapeutic gene is expressed with maximum efficiency at the right time and in the right cells; thus, a great deal of research has been carried out into the best promoters to use. Many strategies have employed viral promoters. Endogenous viral promoters (for example, the long terminal repeat (LTR) promoters in retroviral vectors such as the pBABE series) may be exploited [6]. Many early vector plasmids used the SV40 promoter; however, in recent years the Rous sarcoma virus (RSV) promoter and the cytomegalovirus (CMV) immediate early promoter have been more often utilized since they have been shown to drive a higher level constitutive expression, although with different temporal characteristics [7]. Where tissue-specific expression is desirable, a number of tissue-specific promoters have been employed with some success. For example, therapeutic levels of factor VIII were achieved after gene therapy with a liver-specific high-capacity adenoviral vector [8] in murine and canine models of hemophilia; and a number of cell-type specific promoters have been described for most other tissues, including endothelium, cardiac, skeletal, smooth muscle, epithelia, and skin with varying results (reviewed in Sadeghi and Hitt [9]). One drawback of using tissue-specific promoters can be that they are not as strong expressers as the viral promoters; therefore, the use of viral enhancer elements has also been employed [10].

It is often desirable to be able to regulate gene expression, and the tetracycline (Tet)-regulated expression system has been extensively used *in vitro* and *in vivo* [11, 12]. These elements have also been incorporated into a number of gene therapy vectors (e.g. [13, 14]).

11.2.2.2 Other features

Viral vector plasmids usually lack some or all of the genes required for propagation, but a packaging signal is maintained, together with other essential virally encoded regulatory sequences; this is discussed in more detail below.

Expression of a selectable marker and/or a second therapeutic gene may be desirable depending on the gene therapy strategy. These genes may be under the control of a separate promoter [15] or an internal ribosome entry site (IRES) may be included for multi-cistronic transcription [16]. As with plasmids designed for other applications, it is also desirable that the plasmids developed for use in gene therapy applications contain sequences required for efficient propagation of the vector in bacteria and a multi-cloning site (MCS) for cloning of insert DNA. For viral vector plasmids these genes may be encoded on more than one separate plasmid to further minimize the potential for replication-competent virus production.

11.2.3 Viral vectors

The life cycle of a generic virus comprises attachment of the virus via specific host cell surface receptors, followed by entry into the cell. The virus then becomes uncoated to allow release of viral genetic material into host cells (infection) followed by expression of viral proteins and assembly of new viral particles (replication). Gene therapy vectors have been developed to exploit these life cycles, using modified genomes that carry the therapeutic gene cassette in place of large amounts of the viral genome. This is sometimes called 'transduction,' which is defined as non-replicative or abortive infection to introduce functional genetic information expressed from recombinant vectors into target cells [17, Figure 11.1].

Within the virus genome are genes required for replication and infection as well as *cis*-acting regulatory sequences. Removal of most of the viral genes and regulatory sequences is advantageous since it improves safety by minimizing the risk of reconstitution into productive viral particles by recombination events, and increases the size of insert DNA that can be incorporated. However, at least some of the excised viral genes are still essential for propagation of vectors. Such genes are expressed on separate plasmids

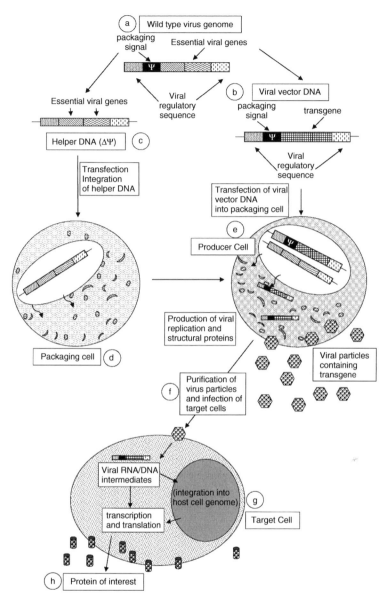

Figure 11.1 Strategy for viral vector production. (Modified from Lawson [17].) (a) Typical viral genome with essential viral genes, regulatory sequences and packaging signal. In order to produce replication-deficient viral vectors the genetic material is separated onto (b) plasmid encoding viral packaging signal and minimum viral regulatory sequences and (c) helper plasmids encoding essential viral genes required for viral replication and packaging. (d) Production of packaging cell line by transfection of helper plasmids. (e) Production of producer cell line by transfection of packaging cell line with viral vector plasmid. (f) Isolation and purification of virus particles from producer cell culture supernatants (e.g. retrovirus vectors) or lysates (e.g. Ad5 vectors). (g) Infection of target cells with viral vectors and (h) expression of protein of interest in target cells (can be secreted, cell surface or intracellular protein, or could be siRNA to inhibit expression of endogenous protein).

in trans using engineered 'helper' or 'packaging' cells to express the viral proteins needed, limiting remobilization and ensuring stability [18]. Usually the packaging cell lines are stably transfected with plasmids encoding viral proteins. Typically, packaging cells express the viral proteins required for packaging of vectors but lack a packaging signal. In contrast, viral vector plasmids contain a packaging signal together with other essential virally encoded regulatory sequences, together with a strong constitutive (or inducible) promoter (which may be a viral promoter) and polyadenylation signals. Genes may be encoded on more than one separate plasmid to minimize replication-competent virus production. Helper viruses are not used owing to the likelihood that a replication-competent virus could be generated through high-frequency recombination [19].

Several different virus families have been exploited for gene therapy applications to take advantage of their efficiency of infection and cell tropism. A brief overview of some of the more commonly used viral vector systems is outlined below and in Lawson [17].

11.2.3.1 Retroviruses

Retroviruses are RNA viruses that replicate through an integrated DNA intermediate (see Coffin *et al.* [20] for review). Retroviral particles encapsidate two copies of the full-length viral RNA with each copy containing the full genetic information needed for virus replication, including the *gag* (group-specific antigen), *pro* (protease), *pol* (polymerase) and *env* (envelope) genes. Retroviruses can be classified into simple and complex retroviruses. Complex viruses encode the essential viral genes above as well as several accessory genes. A further classification of retroviruses is into oncoretroviruses (mostly simple retroviruses, e.g. murine leukemia virus; MLV), lentivirus (complex retroviruses, e.g. human immunodeficiency virus-1; HIV-1) and spumavirus (complex retroviruses, e.g. human foamy virus (FV)). Currently, all three types are being exploited as gene therapy tools.

11.2.3.2 Oncoretroviruses

The earliest gene therapy trials using retroviral transfer used vectors based on MLV, an amphotropic (able to infect human cells) oncoretrovirus. The oncoretrovirus genome is relatively simple and can be easily rearranged to generate replication-defective recombinant viral vectors. In general, the retroviral LTR sequences are retained, together with a minimal packaging signal and the gene of interest. Transfection of oncoretroviral vectors (e.g. pMFG; [21], pBAbe series [6]) into a suitable eukaryotic packaging cell line (e.g. Omega E; GP+E; GP EnvAm12 [6, 22]) results in production of recombinant replication-defective particles. Although straightforward to propagate, the main drawback of using these vectors is that only dividing cells can take up the particles and integrate the DNA into the host cell genome for long-term expression of the transgene. Therefore, their usefulness is limited clinically, since many targets for gene therapy are non-dividing cells [23].

11.2.3.3 Lentiviruses

Recently, there has been some progress in the development of vectors based on lentivirus, in particular HIV-1. The lentiviruses have a more complex replication cycle than oncoretroviruses and, therefore, a more complex genome. The major advantage of use of vectors based on lentiviruses for gene therapy applications is that they are able to infect non-dividing and terminally differentiated cell types, which is a significant improvement over oncoretrovirus vectors [23]. The development of lentivirus vectors and packaging cell lines has been more difficult than for oncoretrovirus vector delivery systems. The earliest HIV-1-based vectors were nearly intact viral genomes with the *env* gene deleted and substituted in *trans*, enabling efficient targeting of CD4-expressing cells. However, targeting to other cells was limited and viral titers were low. Targeting to other cell types was improved by pseudotyping (use of envelope glycoproteins/capsid proteins from a virus that differs from the source of the genome and replication apparatus of the virus by substitution of the amphotropic MLV envelope glycoprotein), whilst inclusion of vesicular stomatitis virus-G-protein (VSV-G) also improved vector titer and stability of virus particles (for a review, see Sanders [24]).

VSV-G pseudotyped HIV-1 vectors can infect cell-cycle-arrested cells *in vitro* and *in vivo*, with stable expression being reported for several months [25]. The latest generation of packaging systems has been designed so that either transient transfection of four separate plasmids is required or packaging cell lines are stably transfected with all the packaging constructs requiring only transfection of the lentiviral expression vector [26]. So-called self-inactivating (SIN) vectors have also been developed where the transcriptional capacity of the LTR is lost once they are integrated into the target cell. These vectors are reported to show improved performance *in vivo* [26].

Owing to concerns of over safety with vectors derived from HIV-1, vectors have also been developed from other less-pathogenic lentiviruses, including HIV-2, simian immunodeficiency virus (SIV) and feline lentivirus (feline immunodeficiency virus; FIV). Chimeric lentivirus vector systems have also been developed, and pseudotyping with VSV-G combined with strong promoters improves cell tropism and transgene expression [27].

11.2.3.4 Spumavirus

Spumaviruses are also known as FVs, since they induce a cytopathic foam effect when in culture. They are complex retroviruses that can persist indefinitely in their hosts, even in the presence of antibodies directed against FV proteins [28]. They are innocuous in their natural primate hosts, and appear to be absent in humans. Accidental infection of humans appears to be apathogenic and there have been no reports of horizontal transmission. In culture they can infect non-dividing cells and exhibit a wide cell tropism [28].

FV vectors have been developed that contain only the minimal viral sequences necessary for efficient gene transfer. This vector has been shown to transduce human hematopoietic CD34+ cells and human mesenchymal stem cells *in vitro* using a four-plasmid packaging cell system [29]. Recently, an FV vector has been successfully used to treat canine leukocyte adhesion deficiency [30]. FVs are a promising prospect for future gene therapy applications for both human and veterinary medicine.

11.2.3.5 Adeno-associated viruses

Adeno-associated virus (AAV) is a small non-enveloped, single-stranded DNA virus belonging to the Parvoviridae family (for review, see Berns and Giraud [31]). All the *cis*-acting elements required for replication, packaging and integration of AAV are located within inverted terminal repeat (ITR) sequences at each end of the genome, and the two open reading frames (ORFs) can be deleted. Heparan sulfate proteoglycan (HSPG), fibroblast growth factor receptor 1 (FGF-R1) and $\alpha v \beta 5$ have been identified as primary and co-receptors for AAV [32].

Use of AAV vectors as gene therapy vectors is attractive owing to their large packaging capacity of up to 4.9 kb [33] and the apparent lack of induction of cellular immune responses [34]. However, humoral responses to AAV have been detected in animal models, and the presence of neutralizing antibodies greatly reduces the success of vector readministration [35, 36]. Since up to 90% of the population is seropositive for AAV and a significant percentage could have preformed neutralizing antibodies to AAV-2 [37], this could limit the usefulness of AAV for gene therapy applications [32].

AAV requires the presence of a helper virus for propagation, usually adenovirus (Ad) or herpesvirus. In the absence of help, AAV integrates into a particular locus on q13.3qter human chromosome 19 and becomes latent. Although this process requires AAV rep proteins which are deleted in AAV vectors, transgene expression has been reported for months and up to several years in some *in vivo* models, possibly due to long-lived double-stranded episomal rAAV genomes or by random integration into the host cell genome For a recent review of AAV vector developments, see Buning *et al.* [38].

11.2.3.6 Adenoviruses

Ads are non-enveloped viruses with a linear double-stranded DNA genome. There are 50 distinct serotypes in humans. They are associated with the common cold and they can cause respiratory, intestinal and eye infections in humans [32, 39]. Ads have been widely used for gene transfer since they have a broad host range and can infect both proliferating and non-dividing cells. However, the Ad genome does not integrate into the host cell genome, so infection with Ad vectors will only lead to transient gene expression.

Based on the time of transcription of each gene after infection, the Ad genome is divided into the early (E) and late (L) regions, with an ITR at either end [40]. Ad enters the host cell via specific cell surface receptors; *in vitro*, these include the coxsackievirus and Ad receptor (CAR [41]). *In vivo*, the receptor usage remains controversial [42, 43]. Ad is internalized rapidly via receptor-mediated endocytosis, facilitated via receptors including integrins $\alpha v\beta 3$ and $\alpha v\beta 5$ [44].

A number of Ad vectors have been developed, including replication-competent and replication-defective vectors, mostly based on serotype 2 (Ad2) or serotype 5 (Ad5). The first generation of replication-defective Ad vectors was E1- and E3-deleted. The second generation includes E1-, E3- and E4- or E2-deleted vectors based on Ad5. E1-deleted Ad5 vectors can be grown in specific cell lines such as human embryonic kidney 293 cells [45] transformed with Ad E1, to supply E1 in *trans*. The left-hand ITR and packaging signals from the left-hand 300 bp of the genome are required for replication in 293 cells [46].

First-generation Ad vectors were strongly immunogenic, with immunosuppressive drugs needed to extend transgene expression [47, 48]. However, second-generation Ad vectors have overcome Ad immunogenicity to some extent by introduction of a mutation in the Ad E2a gene [49], or deletion in E4 [50] with only modest success [51].

To reduce immunogenicity, a gutted (or gutless) Ad vector was developed containing only the ITR required for replication and 5′-*cis*-acting Ad encapsulation signals necessary for packaging [52]. However, this vector is difficult to produce, requiring the use of helper virus to provide all the viral proteins in *trans* [32].

Outlined below is a protocol for co-transfection of FG293 cells with Ad5 shuttle vector pDC516 and pBHGfrt genomic plasmid (see Protocol 11.1), followed by a protocol to purify these Ad5 plaques by end-point dilution (see Protocol 11.2). Protocols for bulk preparation of crude lysates of high titre Ad5 stock (see Protocol 11.3) and preparation of CsCl$_2$ purified recombinant Ad5 (see Protocol 11.4) follow.

PROTOCOL 11.1   Co-Transfection of FG293 Cells with Ad5 Shuttle Vector pDC516 and pBHGfrt Genomic Plasmid

Equipment and reagents

- FG293 human embryonic kidney cells (Microbix Systems Inc., http://microbix.com)

- Cell culture medium I: minimal essential medium, glutamine, 10% fetal calf serum (FCS) (all Sigma)

- Cell culture medium II: minimal essential medium, glutamine, penicillin/streptomycin, 10% newborn calf serum (NBS) (all Sigma)

- T25 tissue culture flasks (Falcon; Beckton Dickinson)

- pBHGfrt genomic adenovirus plasmid (Microbix Systems Inc.)

- pDC516 containing cDNA insert of interest (Microbix Systems Inc.)

- HEPES-buffered saline (HeBS) (5 g 4-(2-hydroxyethyl)-1-piperazineethanesulfonic acid (HEPES)-free acid, 8 g NaCl, 0.37 KCl, 1 g glucose, make up to 1 l and pH to 7.1) (see note a)

- 2.5 M $CaCl_2{}^a$

- Phosphate-buffered saline (PBS)++ (PBS supplemented with 0.01% $CaCl_2$ and 0.01% $MgCl_2$)

- 100% glycerol sterilized by autoclaving.

Method

1 Day 0: set up six T25 flasks of FG293 cells from growing stocks.

2 Day 1: change the medium to cell culture medium I and incubate cells for 4 h at 37 °C.

3 Set up 5 vol of DNA precipitate:

1 vol:	5 µg pBHGfrt (genomic adenovirus plasmid)
	3 µg plasmid containing insert (e.g. pDC516-ICAM-1 [53])
	500 µl HeBS
	50 µl 2.5 M $CaCl_2$ (added dropwise)

Leave to precipitate for 20–30 min at room temperature.

4 Add 500 µl of precipitate per flask in five flasks (without removing medium) and leave for 24 h at 37 °C (or up to 48 h). One flask will remain untransfected as a control.

5 Day 2: remove medium and replace with fresh cell culture medium I.

6 Thereafter replace medium twice per week with cell culture medium II and observe cells every 2 days.

7 When cytopathic effect is observed (i.e. large 'holes' in monolayer with lots of rounded-up cells, this can take up to 3 weeks[b] and looks different to overgrown monolayer on control flask; see Figure 11.2). Do not replace medium and leave for a further 48 h before harvesting.

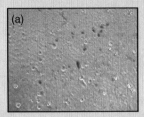 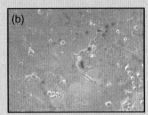 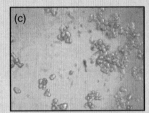

Figure 11.2 Monitoring of infection of 293 cells with Ad5 vectors. Sub-confluent 293 cells were infected with Ad vectors and observed after 36 h. (a) Non-infected 293 cells. (b) Infection of 293 with low titer Ad5. (c) Infection of 293 with high titer Ad5. Note the 'bunch of grapes' morphology of the infected cells as the virus's lytic life cycle has a cytopathic effect on the mammalian cells.

8 To harvest, remove medium from transfected flasks and pool in 50 ml Falcon tube, add 5 ml 1× PBS++ to each flask and scrape remaining cells from flasks.

9 Pool with the spent medium and centrifuge at 1000g for 15 min.

10 Discard the supernatant and resuspend the cell pellet in 150–200 µl PBS++ supplemented with 10% glycerol per T25 harvested flask.

11 Freeze at −70 °C and thaw at 37 °C three times prior to titration.

12 Aliquot and store at −70 °C.

Notes

[a]Sterilize by filtration or autoclaving; store at 4 °C.

[b]If cytopathic effect is observed rapidly (death of all cells within a few days) then collect the supernatant and debris. Follow steps 9–11 above and then reinfect new T25 flasks of FG293 cells, using increasing dilutions of the crude lysate prepared in step 11 (e.g. 10 µl lysate; 5 µl lysate; 1 µl lysate).

PROTOCOL 11.2 Plaque Purification by End-Point Dilution and Titration of Crude Virus Stock

Equipment and reagents

- FG293 human embryonic kidney cells (Microbix Systems Inc, http://microbix.com)
- Cell culture media I: minimal essential medium, glutamine, 10% FCS (all Sigma)
- 96-well tissue culture plates

- Phase microscope
- 80 °C freezer.

Method

1 Day 0: seed FG293 cells onto 96-well plate from growing stocks, to achieve 50–60% confluency the following day (approximately 2×10^4 cells per well).

2 Day 1: prepare serial dilutions of crude adenovirus lysate (see Table 11.2 in cell culture medium I.

Table 11.2 Dilution curve for titration of recombinant adenovirus.

Row number	Final virus dilution	Virus volume (μl)	Media volume (μl)
–	10^{-2}	50 stock	4950
A	10^{-4}	50 of 10^{-2} dil	4950
B	10^{-6}	50 of 10^{-4} dil	4950
C	10^{-7}	500 of 10^{-6} dil	4500
D	10^{-8}	500 of 10^{-7} dil	4500
E	10^{-9}	500 of 10^{-8} dil	4500
F	10^{-10}	500 of 10^{-9} dil	4500
G	10^{-11}	500 of 10^{-10} dil	4500
H	0	0	5000

3 Remove the medium from all wells and replace with diluted virus stock (200 μl/well). Use medium alone for control wells.

4 Incubate the plate overnight at 37 °C.

5 Day 2: remove medium and replace with 200 μl fresh cell culture medium I.

6 Replace medium every 2–3 days for 8 days.[c] Before removing the medium, observe each well; once the cytopathic effect is apparent in a well, mark it and stop changing the medium in that well.

7 Day 7: prepare a fresh 96-well plate of FG293 cells for a further round of plaque purification.

8 Day 8: collect the cells and medium from three wells at the highest dilution (lowest starting concentration of lysate) for which the cytopathic effect is apparent.

9 Store plaques in separate sterile eppendorfs. These plaques will have arisen from a single adenovirus virion. Freeze two and extract the third by three freeze–thaw cycles.

10 Repeat the plaque purification assay with the plate prepared on day 7 using the plaque isolated and lysed by freeze–thaw in step 9 above.

11 Repeat titration again (i.e. three times in total).

12 The plaques at the greatest dilution for which the cytopathic effect is apparent can be considered 'plaque pure.'

Notes

[c]We expect that the cytopathic effect will be observed by day 3 in the wells infected with the most concentrated lysate, but that no effect will be observed over the full 8 days in wells infected at the highest dilution. If a cytopathic effect is observed in wells infected at the highest dilution after 3 days, then collect media and debris as in step 8 and repeat titration but start at higher dilution (e.g. use lysate diluted 10^{-11} as starting dilution).

PROTOCOL 11.3   Preparation of Crude Lysates of High Titer Adenovirus Stock from Bulk Cultures of FG293 Cells

Equipment and reagents

- FG293 cells infected with titrated adenovirus stock (infection of between 5 × T75 flasks to 15 × T175 flasks)

- 1× PBS (Sigma)

- Cell scrapers for T75 or T175 flasks as appropriate (Falcon, Becton Dickinson)

- 50 ml conical tubes (Falcon)

- Bench-top serological centrifuge capable of taking 50 ml Falcon tubes

- PBS++ with 10% glycerol

- 5% sodium deoxycholate solution

- DNase I (Sigma): dilute to 1 mg/ml in 10 mM Tris-HCl, pH 7.4; 50 mM NaCl; 0.1 mg/ml bovine serum albumin; 1 mM dithiothreitol; 50% glycerol (see note d)

- 1 M $MgCl_2$

- 0.45 µm low-protein syringe filters (Sartorius)

- 20 ml syringes.

Method

1 Harvest Ad5 transduced FG293 cells by removing medium and debris from flasks and pooling in a 50 ml Falcon tube.

2 Add 1× PBS to each flask and scrape any remaining cells from flasks, and pool with the spent medium.

3 Centrifuge the pooled media and cells at 1000g for 15 min.

4 Discard the supernatant and resuspend cell pellet in 2 ml PBS++ plus 10% glycerol per T175 flask harvested.

5 Freeze at $-70\,^{\circ}$C until lysis is required.

6 Thaw the cellular material and lyse with a 5% solution of sodium deoxycholate (add 1.5 ml solution/15 ml lysate).

7 Add 75 µl of 1 mg/ml DNase I and 300 µl 1 M MgCl$_2$ for every 15 ml lysate, and incubate at 37 $^{\circ}$C for 1 h.[d]

8 Centrifuge at 1000g 15 min.

9 Discard the pellet and filter sterilize the supernatant through 0.45 µm low-protein syringe filters.

10 Confirm there is adenovirus in the crude lysate by FG293 cell plaque assay (see Protocol 11.2) and a suitable functional assay to check that the protein of interest is expressed.

11 Aliquot and store at $-70\,^{\circ}$C.

Notes

[d]You will need 75 µl DNase I solution for every 15 ml lysate.

PROTOCOL 11.4 Preparation of CsCl$_2$ Purified Recombinant Adenovirus

Equipment and reagents

- Class II safety cabinet

- Ultracentrifuge and fixed-angle rotor

- Ultracentrifuge tubes and sealing assembly

- 70% ethanol

- Sterile distilled H$_2$O

- Glass Pasteur pipettes

- PBS++ (PBS supplemented with 0.01% CaCl$_2$ and 0.01% MgCl$_2$)

- CsCl$_2$ solution, $\rho = 1.34$: 51.2 g CsCl$_2$ dissolved in 100 ml buffer DG (750 mM NaCl; 50 mM KCl; 250 mM Tris-HCl, pH 7.4)

- CsCl$_2$ solution, $\rho = 1.40$: 62 g CsCl$_2$ in 100 ml DG

- Sterile 21-gage needles

- Sterile 2 ml syringes

- Slide-a-lyse dialysis cassette (Pierce; Perbio Science UK Ltd)

- 2 l dialysis buffer (10 mM Tris-HCl, pH 8.0; 135 mM NaCl; 1 mM $MgCl_2$)
- 100% glycerol
- Magnetic stirrer and sterile stir bar
- Cryotubes for aliquoting purified recombinant adenovirus.

Method

1 Add 3 ml low-density $CsCl_2$ solution ($\rho = 1.34$) to each of two ultracentrifuge tubes.[e]

2 Use a glass Pasteur pipette to underlay with 1.6 ml high-density $CsCl_2$ ($\rho = 1.4$).

3 Carefully add the crude adenovirus lysate on top of the gradient.

4 If necessary, fill the tubes to the top with PBS++ by underlaying the top of the virus layer to avoid air bubbles.

5 Ensure tubes are balanced (same mass) and then seal tubes as appropriate.

6 Centrifuge using an ultracentrifuge at 9000g in a fixed-angle rotor for 2 h at 18 °C.[f,g]

7 Pierce the top of each tube with a 21-gage needle then using a second sterile 21-gage needle and syringe carefully remove the white virus layer by piercing the tube underneath the virus. Avoid taking up surrounding $CsCl_2$ or perturbing the virus layer as it is removed.

8 Prepare a second set of $CsCl_2$ gradient tubes.

9 Dilute purified virus 1 : 2 with PBS and overlay onto new $CsCl_2$ gradients.

10 Balance and seal tubes as in step 5

11 Centrifuge at 100 000g for 18 h at 18 °C without brake.

12 Remove discrete band of adenovirus as before in the smallest possible volume.

13 Dialyze against 1 l sterile dialysis buffer for 2 h at room temperature in a laminar safety cabinet.

14 Dialyze virus for a further 18 h against 1 l fresh dialysis buffer containing 10% glycerol.

15 Aliquot adenoviral stocks and store at −80 °C.

16 Carry out titration of stock as in Protocol 11.2.

Notes

[e]Prepare ultracentrifuge tubes (two per 15 ml of crude lysate) in a safety cabinet by soaking in 70% ethanol for 10 min, followed by rinsing with copious amounts of sterilized water. Allow to dry in the cabinet.

[f]Ensure brake is switched off.

[g]The virus should form a discrete white layer between the two CsCl2 layers.

11.2.4 Non-viral DNA vectors

Research into alternative non-viral gene delivery methods has been gathering pace in recent years, following mounting concerns into the use of viral vectors. Despite their high efficiency, possible insertional activation of oncogenes or adverse immune responses pose very real threats to the success of viral gene therapy treatments. There have been several incidents of serious illness and death following treatment with viral vectors. These include the death of a teenager from adverse immunological reactions after adenoviral treatment for a liver enzyme deficiency [54] and reports of 4 out of 11 boys developing leukemia after being treated in France with retroviruses for SCID [55]. The use of non-viral vectors offers many safety advantages over viruses (see Table 11.1) but, to date, remains less efficient.

Non-viral vector delivery methods fall broadly into two categories: physical and chemical methods.

1 **Physical methods**: Simple direct injection of plasmid DNA, without the inclusion of any vector, has shown remarkable success in a number of tissues, including muscle, liver, skin and solid tumors (reviewed in Herweijer and Wolff [56]). Direct injection into skeletal muscle shows the most promise for DNA-based immunization procedures, although gene therapy applications are also under development.

   Untargeted intravascular delivery of plasmid DNA into whole animals, either systemic or regional, does result in low-level widespread transient transgene expression [57]. However, increasing hydrodynamic pressure by rapid delivery of large volumes of DNA substantially increases efficiency [58], localizing largely to the liver. Transgene expression can be further improved by localized hydrodynamic delivery to specific organs, including the liver [59] and the kidney [60].

   The addition of physical methods to temporarily disrupt membranes and enhance DNA transfer have been explored, including particle bombardment by gene gun [61], ultrasound [62] and electroporation [63]. Whilst all of these methods have shown successful gene delivery, their harsh and almost always unphysiological nature lead to unwanted cytotoxicity. Therefore, clinical application is realistically highly limited.

2 **Chemical methods**: Chemical methods almost exclusively contain a polycationic component, enabling DNA binding and condensation, as well as electrostatic interaction with cell surface molecules.

11.2.4.1 Cationic liposomes

Liposomal gene delivery, first pioneered by Felgner and colleagues in 1987 [64], uses cationic lipids that readily bind the negative phosphate backbone of DNA, spontaneously condensing the DNA into small particles and protecting it from intracellular degradation [65]. Internalization of the DNA is thought to occur via both coated pit and non-coated endocytosis pathways, depending on the charge and size of the liposomal complexes.

A cationic lipid is fundamentally composed of four functional domains: a positively charged head group (usually a single or multiple amine-derived group), a spacer of varying lengths, a linker bond and a hydrophobic anchor (see Figure 11.3). The relationship between structure and efficiency is an area of intense research [66]. Although some cationic lipids are effective at DNA delivery alone, they are often formulated with a non-charged 'helper' phospholipid or cholesterol to improve stability and transfection efficiency [67]. It is thought

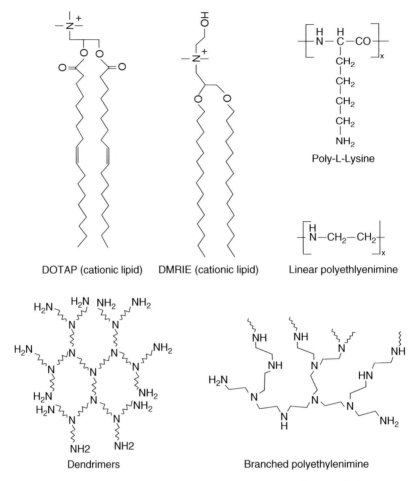

DOTAP (cationic lipid)   DMRIE (cationic lipid)   Linear polyethlyenimine

Poly-L-Lysine

Dendrimers   Branched polyethylenimine

Figure 11.3 The structures of some commonly used chemical non-viral vectors. DMRIE: 1,2-dimyriotloxypropyl-3-dimethyl-hydroxy ethyl ammonium bromide; DOTAP: dioleoyltrimethy-lamino propane.

that the neutral lipid helps to facilitate endosomal membrane disruption, enabling better access of the DNA to the nucleus.

Highly efficient lipid-mediated delivery of DNA and RNA has been attained both *in vitro* and *in vivo*, providing transient and stable transfectants to a wide range of tissues and organs in many animal species (reviewed in Karmali and Chaudhuri [68]). *In vivo* efficiency is, however, diminished due to interaction of the positively charged lipid/DNA particles with the negatively charged blood components, such as serum, forming large aggregates that are unable to reach their intended cellular target. By incorporating a hydrophilic polymer, polyethylene glycol (PEG), which shields the cationic charge, an increase in stability and transfection efficiency has been achieved [69]. Furthermore, natural targeting ligands, such as transferrin [70], folate [71] or various antibodies [72], can be added to increase tissue specificity. Additional enhancements to lipid-mediated delivery systems have been explored by adding polylysines [73] and membrane permeabilizing agents [74].

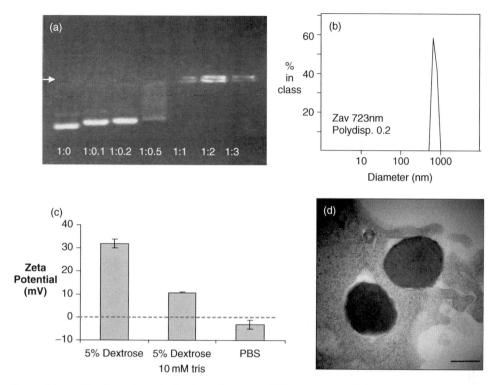

Figure 11.4 Physical characterization of vector–DNA complexes. (a) *DNA retardation assay.* The (Lys)$_{16}$-molossin peptide [91] was added to pGL3 plasmid DNA, containing the luciferase reporter gene, in PBS at the various w/w ratios indicated and loaded on to an 0.8% agarose gel after 30 min (loaded at arrow). Electrophoretic mobility is reduced as the vector concentration increases. Charge neutralization occurs at a peptide/DNA w/w ratio of ∼1:1. (b) *Size analysis of peptide–DNA particles by DLS.* (Lys)$_{16}$-molossin peptide mixed with pGL3 plasmid DNA at a ratio of 3:1(w/w), at 10 μg/ml DNA, in PBS, analyzed at 30 min on a zetasizer 3000HS (Malvern Instruments, Malvern, UK). Average diameter (Zav) of the particles is 723 nm with a polydispersity of 0.2 (range 0–1) indicating a relatively homogeneous population. (c) *Zeta potential analysis of peptide–DNA particles by DLS.* Peptide–DNA complexes prepared as for (b) but in PBS, 5% dextrose or 5% dextrose 10 mM Tris solution. Net surface charge measured on the zetasizer 3000HS at 30 min. In the presence of salt the charge is substantially reduced. (d) *Macropinocytosis of peptide–DNA complexes viewed by transmission electron microscopy.* (Lys)$_{16}$-molossin–DNA particles prepared as in (b) but using biotin-labeled peptide and conjugated to 10 nm streptavidin gold, were transfected into HUH7 hepatoma cells. After a 4 h incubation the cells were fixed and processed. Two particles are visualized, both with a halo of gold beads. One inside the cell is surrounded by a vesicle and the other is entering through the cell membrane. The bar is 500 nm.

Liposomes themselves have also been used together with other non-viral vectors, such as the arginine–glycine–aspartate (RGD)-containing peptides [75–77] and some viruses, including adenovirus [78] and the hemaglutinating virus of Japan [79], to improve DNA delivery.

Some cell toxicity and acute inflammation reactions have been reported in both cell and animal models [80], as well as following clinical trials [81], which currently pose limitations on the success of lipid-mediated gene therapy in the clinic.

11.2.4.2 Cationic polymers

Cationic polymers for gene delivery may be synthetic or naturally occurring. The most effective and extensively researched is the biodegradable peptide poly(L-lysine), first reported as an effective gene delivery agent by Wu and Wu [82] in 1988. Following its success there have been numerous adaptations and manipulations of the polylysine peptide, as well as a huge number of different cationic polymers in linear or branched configurations explored as DNA delivery vehicles. These include cationic proteins (e.g. histones, protamines, spermine) and peptides, polyethyleneimine (PEI), dendrimers, polyaminoesters, cationic dextran and chitosans.

Polylysine has not only been shown to be a good DNA condensing agent, shielding it from degradation, but also has been suggested to possess some nuclear trafficking properties. The length and type of positively charged amino acid chain influences the stability and efficacy of vector/DNA particles [83].

Alone, it is capable of gene delivery; however, polylysine chains of varying length are most effective when linked with a targeting ligand, forming a receptor-mediated gene delivery approach. A wide range of peptide vectors have been explored [84]. The ligand takes advantage of the ability of receptors on a cell's surface to bind and internalize recognized molecules, adding specificity to the DNA delivery system. Targeting ligands include naturally occurring proteins (such as transferrin [85], insulin [86] or asialorosomucoid [82, 87]), structural motifs from natural receptor binding ligands (e.g. sugar residues [88] or synthetic peptides [89–91]) or antibodies against an epitope on the extracellular portion of the receptor (e.g. polymeric immunoglobulin receptor [92]).

There is some evidence that polylysine can induce an inflammatory response following direct administration *in vivo* [93]; however, the same is not seen with DNA–ligand–polylysine complexes [94].

One drawback to polylysine-based vectors is the requirement for assistance with endosomolysis. These delivery systems lack any buffering capacity and, therefore, the DNA is readily degraded following the drop in pH within the endosomal–lysosomal pathway. The simple addition of the organic lysosomotropic agent chloroquine reduces enzyme degradation and allows escape of the DNA [95]. Its widely reported toxicity *in vitro*, however, limits its clinical applications.

A more sophisticated method of preventing endosomal–lysosomal degradation has been adapted from the many destabilizing proteins found in viruses. The best studied is a 20-amino-acid peptide from the hemagglutinin protein of the influenza virus [96]. The peptide gains fusogenic activity once in the acidic environment of the lysosome, resulting in disruption of the membrane, releasing the DNA. This has been successfully incorporated into receptor-mediated gene delivery systems [97].

PEI is a widely used polymer in the manufacturing industry, but has more recently been exploited as a vehicle for gene delivery [98]. PEI is an efficient and economic gene transfer agent available in both linear and branched forms in a variety of molecular weights (see Figure 11.3, reviewed in Kircheis *et al.* [99]). Its highly positive charge enables effective DNA condensation to form nuclease-protected stable particles, allowing electrostatic binding to the cell surface and subsequent endocytic internalization. Once inside the cell, the high number of protonable nitrogen groups results in a high buffering capacity, called the 'proton sponge' hypothesis. This buffers the endosomal compartment, leading to rupture of the membrane, allowing the DNA increased access to the nucleus, without the addition of any membrane-disruption agents. PEI-mediated gene delivery has been achieved in numerous *in vitro* and *in vivo* models [100].

A disadvantage of the use of PEI as a transfection agent is its nonbiodegradable nature. Toxicity has also been associated with the excessive positive charges on the polymer [98]. Shielding the PEI/DNA complexes with PEG, similar to liposomes, does significantly reduce toxicity [101]. Further targeting of the system can be achieved by the addition of ligands either with or without the PEG shield [102, 103].

Polyamidoamine cascade polymers, or Starburst dendrimers, are another highly positively charged polymer. They are spherical and highly branched (see Figure 11.3), with varying degrees of branching forming the different generations of dendrimer that are commercially available. Like PEI, their high amine density enables neutralization of the acid pH within the endosomal vesicle, allowing DNA to escape degradation. They have been shown to be effective as a transfection agent, without the need for additional endosomolytic agents [104].

11.2.5 Assessing the physical properties of a non-viral vector

Physical characterization of the non-viral vector is an important initial step in the development of non-viral gene delivery systems. The size and surface charge of the vector/DNA particle are both crucial parameters, influencing the passage of the DNA through the cell and its ultimate expression. Preliminary experiments to explore these characteristics are relatively easy and highly worthwhile in order to enhance future understanding and aid optimization of the delivery system.

Simple mixing of the polycationic delivery agent and the DNA results in electrostatic binding, condensation and charge neutralization of the DNA to form particles (see Protocols 11.5 and 11.6). Visualization of neutralization can be demonstrated through a DNA retardation assay (Figure 11.4a). Inhibition of electrophoretic mobility indicates when sufficient vector has been added to the DNA to neutralize charges. Particle formation and subsequent size and charge can be more readily assessed using a technique called dynamic light scattering (DLS), alternatively called photon correlation spectroscopy. This measures the scattering of laser light upon a population of particles, resulting in values of average particle diameter (Z-average) and average net surface charge of a particle (zeta potential). It also establishes some level of homogeneity of the particle population (polydispersity). It is not, however, quantitative for particle numbers or relative numbers within a heterogeneous population. Figure 11.4b shows the size profile for a peptide–DNA complex, giving its average diameter (Z-average) and an indication of the level of homogeneity within the sample of particles (polydispersity, range 0–1).

The cation-to-anion ratio (also calculated in terms of the N : P ratio, i.e. amine to phosphate) within the complex is an important consideration, with a slightly positive net charge being the optimal for delivery to cell lines *in vitro*. DLS is capable of measuring the net surface charge of a complex, which unquestionably plays a key role in initial binding to the cell surface. The zeta potential reading is a value in millivolts depicting the movement of the particles within an electric field, giving a positive or negative value depending on its speed and direction of movement. Figure 11.4c shows the net surface charge of the same peptide–DNA complex prepared in different solutions. The surface charge has been found to be highly influenced by the presence of salts within the buffer.

Many factors have been found to influence the size and charge of vector–DNA complexes. Factors include buffer composition (presence of salts), nature of the cationic vector, DNA concentration and time. These can easily be investigated using the DLS technique [77, 97, 105].

To date, by far the most visually informative data has come from transmission electron microscopy of the particles within transfected cells (unpublished Collins and Fabre) [106, 107]. Pre-coating of the formed particles with gold beads using a biotin–streptavidin link where possible ensures easy and unambiguous identification following cell entry. Figure 11.4d shows two peptide–DNA particles entering a cell by macropinocytosis; the inner particle is enclosed within a vesicle. The particles are clearly labeled by a halo of gold beads. Electron microscopy gives us not only information regarding shape and size of the complexes, but also reveals valuable information about the movement of the particle through the cell.

Other ultrastructural techniques, such as atomic force microscopy and scanning electron microscopy, have shown a variety of different shapes of vector–DNA complexes [83, 108]. However, one has to be wary of the accuracy of these findings, since preparation of the samples for these procedures is highly unphysiological, including drying and/or high salt concentrations, all of which greatly influence the particle structure.

11.2.6 Optimizing *in vitro* gene delivery

Prior to physical characterization of a new non-viral vector system, work must begin on establishing its ability to delivery DNA *in vitro*. Choose a familiar cell line that is easy to grow and manipulate, and if possible relevant to your intended end target tissue.

Outlined below are the protocols for preparation of DNA/Lipid complexes (see Protocol 11.5), preparation of DNA/peptide complexes (see Protocol 11.6) followed by the protocol for a 'typical' transfection (see Protocol 11.7). Vector/DNA preparation and suggested incubation times do not usually vary greatly between the different chemical vectors and as explained, it is vital that some optimisation is carried out to ensure that the vector is used at its full potential. For commercially bought non-viral vectors it is however best to begin by following the manufacturer's guidelines but nevertheless, optimisation is still important to achieve the highest efficiency for your application.

PROTOCOL 11.5   Preparation of Lipid–DNA Complexes

Equipment and reagents

- Lipofectamine 2000 at 1 mg/ml (Invitrogen)
- Endotoxin-free plasmid DNA at 1 mg/ml in water[h]
- Culture medium without supplements.[i]

Method

1 Dilute 15 μl DNA in 135 μl culture medium (without supplements).[j]

2 In a second tube, dilute 30 μl Lipofectamine 2000 in 120 μl culture medium (without supplements).

3 Mix diluted DNA and diluted lipid gently together and incubate for 5 min at room temperature.[k]

4 Incubate for a further 20 min at room temperature, before adding 1.2 ml culture medium (no supplements). Complexes are ready for assessing (as described in Section 11.2.5) or adding to cells.

Notes

[h]Lipofectamine 2000 can be used to deliver any DNA expression plasmid. Begin by using a plasmid containing a reporter gene (see Section 11.2.8).

[i]Prepare the complexes in the same culture medium used to grow the target cells, but without any added supplements (i.e. no serum).

[j]Volume of complexes prepared for samples in triplicate (i.e. 1.5 ml total, 0.5 ml per well). DNA used at 10 μg/ml with a weight : weight ratio of 2 : 1 for lipid–DNA complexes.

[k]Do not vortex.

PROTOCOL 11.6　Preparation of Peptide–DNA Complexes

Equipment and reagents

- Peptide at 1 mg/ml in sterile PBS[l]
- Endotoxin-free plasmid DNA at 1 mg/ml in water[m]
- Culture medium without supplements[n]
- Vortex mixer.

Method

1 Dilute 15 μl DNA with 1455 μl culture medium (no supplements).[o]

2 Add 30 μl peptide dropwise whilst vortexing the dilute DNA solution.

3 Vortex solution for a further 10 s.

4 Incubate the complexes for 30 min before assessing (as described in Section 11.2.5) or adding to cells.

Notes

[l]This method is applicable to any synthetic peptide to be tested as a non-viral vector.

[m]Any DNA expression plasmid can be used. Begin by using a plasmid containing a reporter gene (see Section 11.2.8).

[n]Prepare the complexes in the same culture medium used to grow the target cells, but without any added supplements (i.e. no serum).

[o]Volume of complexes prepared for samples in triplicate (i.e. 1.5 ml total, 0.5 ml per well). DNA used at 10 μg/ml with a weight : weight ratio of 2 : 1 for peptide–DNA complexes.

PROTOCOL 11.7 Transfection

Equipment and reagents

- Cells to be transfected[p]
- Complete culture medium[q]
- Culture medium without supplements[r]
- Culture medium with double concentration of supplements[s]
- Transgene detection assay.[t]

Method

Day 1

1 Seed cells into 24-well plates at 2×10^5 per well in complete culture medium without antibiotics.[u]

2 Incubate overnight at 37 °C, 95% air/5% CO_2.

Day 2

1 Prepare complexes as described above in Protocols 11.5 and 11.6.

2 Remove medium and add 1 ml per well culture medium (without supplements) to wash cells.

3 Remove culture medium and add 0.5 ml vector–DNA complexes per well.

4 Incubate at 37 °C in 95% air/5% CO_2 for 4 h.

5 Add 0.5 ml per well culture medium containing twice the concentration of supplements and place cells back into the incubator.[v]

Day 3

1 Harvest cells for transgene expression, according to detection assay.

Notes

[p]When commencing transfection experiments, choose a relevant target cell line that is easy to grow. Cell lines and stocks of DNA and vectors should be prepared and cultured under mycoplasma-free conditions. The presence of mycoplasma can reduce transfection levels.

[q]Culture medium specific to the target cells containing all the necessary growth supplements, including serum, l-glutamine, and so on. Concentrations of supplements are cell-type dependent.

[r]The same culture medium as in note q, but containing no added supplements.

[s]Complete culture medium, as in note q, containing double the concentration of added supplements.

[t]Assay to measure the level of transgene expression. Reporter gene assays are discussed in Section 11.2.8.

[u]Cells need to be 80–90% confluent on the day of the transfection.

[v]If harvesting at any time point >48 h, the medium must be replaced at 24 h with complete culture medium.

11.2.7 Optimization strategies

Optimal gene delivery conditions vary with every new vector and cell type. It is recommended, therefore, to begin with a basic transfection as described above, including all appropriate controls (see Section 11.3). Subsequent experiments can then be centered around the conditions producing the best results, looking at variables listed below to attempt to improve gene expression:

1 **Ratios of vector–DNA**: For both lipid-mediated and peptide-mediated delivery, it has been shown that there needs to be a slight excess of the cationic moiety, resulting in vector–DNA particles with a net positive charge. This is thought to aid electrostatic binding to the cell surface. Transfection data can be supported by DNA retardation assays and zeta potential values.

2 **Concentration of DNA**: *In vitro* low concentrations of DNA are generally used to minimize concentrations of vector needed (usually 5–10 µg/ml DNA). It is advisable to try increasing DNA concentration to find the point at which expression levels plateau. *In vivo*, however, higher concentrations may be required. Be aware that, at high concentrations, solubility of the vector in the buffer may become a problem.

3 **Time course**: Look at both the length of incubation of the vector–DNA complexes with the cells and the time point of harvest. Some cells are particularly sensitive to the absence of serum, and viability may be compromised if incubation time is too long. This can be established by including a toxicity assay (see Section 11.2.9) in parallel with the transfection. Likewise, the presence of the vector–DNA complexes in the overnight incubation may be harmful to the cell, in which case they should be removed after the initial 4 h incubation and replaced with complete medium. The standard transfection method harvests the cells on day 1, 24 h after complexes are added. It is worthwhile looking at the expression levels over a number of days.

4 **Influence of serum**: The presence of serum *in vivo* is an unavoidable issue. It is important, therefore, to establish whether gene delivery is affected by serum. If possible, it is highly beneficial to the cells to keep it in the medium throughout the procedure.

5 **Buffer**: Try preparing the complexes in different solutions. The presence of salt greatly increases the size of many of the vector–DNA particles. Other solutions that can be used include PBS, 5% dextrose, 5% dextrose + 10 mM Tris, pH 7.4. There may be increased cytotoxicity with non-buffered solutions; therefore, a toxicity assay should be performed (see Section 11.2.9).

6 **Components of vector system**: Some non-viral vector systems, such as PEI, contain intrinsic abilities for endosomal disruption. Other systems, such as peptide vectors, require added membrane disruption agents, such as chloroquine, fusogenic peptide or Lipofectamine (see earlier). Concentrations and different combinations of the above can be varied to establish optimal conditions. Some components, such as chloroquine, can cause substantial toxicity to the cells at high concentrations.

11.2.8 Reporter genes and assays

The preliminary optimization experiments using a new non-viral vector are best carried out using a reporter gene with a straightforward assay for detection of gene expression.

Many commercial plasmids are available which contain reporter genes, including the following:

Reporter gene	Assay for detection
Green fluorescent protein (GFP); for example, phGFP-S65T (Clontech)	Fluorescence visualization and flow cytometry.
LacZ (β-galactosidase); for example, pCMV-β (Clontech)	Colorimetric and luminescence assays. Antibody localization.
Luciferase; for example, pGL3-control (Promega)	Luminescence assay. Antibody localization.
Secreted placental alkaline phosphatase (SEAP) (Clontech)	Luminescence and colorimetric assays.

Quantitative assays are often carried out in combination with a further assay to measure protein concentration in the sample, the Bradford assay [109]. This standardizes the values in each sample to give units per milligram of protein.

11.2.9 Cytotoxicity assays

It is important to assess the toxicity of the transfection agent on the cells. Looking at the protein levels per well gives some indication of cell death, but measures only the concentration of remaining cells and assumes that cell loss has occurred through death rather than overgrowth. A more accurate test of toxicity is advisable. There are many cytotoxicity assays, including: the traditional trypan blue exclusion test [110] and fluorescence viability staining with ethidium bromide and acridine orange [111]; 3-(4,5-dimethylethiazol-2-yl)-2,5-diphenyl tetrazolium bromide assay and 3-(4,5-dimethylethiazol-2-yl)-5-(3-carboxymethoxyphenyl-2(4-sulfonphenyl)-2H-tetrazolium, inner salt assays, both measuring mitochondrial activity [112, 113]; or the lactate dehydrogenase release assay [114].

11.2.10 Future steps for non-viral vector development

From cell lines, the next stage is to look at primary cells, such as hepatocytes, neurons or endothelial cells. It is possible also to culture some tissues *ex vivo* in the laboratory, including corneas [97] and aorta [115], and these can be transfected with the non-viral vector system and reporter gene plasmid.

Testing the system *in vivo*, of course, is the intended and desirable next step. Depending on the final target tissue or organ, there are many different possible routes of administration. Be aware, though, that conditions that proved successful *in vitro* may not give optimal results *in vivo*, as *in vivo* conditions are much less controlled and there are a large number of potential obstacles that every vector must overcome to deliver the DNA to the nucleus of the cells.

11.3 Troubleshooting

11.3.1 General points

11.3.1.1 Poor transfection or transduction efficiency

- Ensure that plasmid DNA is of good quality and is of the desired concentration. In particular, ensure that all possible contaminants have been removed. If ethidium bromide/cesium chloride (EtBr/CsCl) ultracentrifugation has been used for purification of DNA (this method is not widely used, however), ensure that all EtBr and CsCl has been removed. If resins/columns have been used, ensure that there has been no leaching of the matrix into the final preparation.

- Ensure that cells are of the correct confluency and are morphologically as you would expect at the start of any transfection procedure. Cells that are too confluent often do not take up DNA at the desired efficiency. Likewise, if they have not adhered and spread onto dishes sufficiently they will take up DNA less efficiently.

11.3.1.2 Viral vectors

- No viral plaques observed. This is probably due to poor transfection efficiency.

  —Ensure that FG293 cells are not of too high passage number, since according to the laboratory where these cells were first transformed they become less efficient at taking up DNA/producing infective virions after approximately passage 40 (ref. http://www.microbix.com/Public/Default.aspx?I=64&n=FAQ).

  —Ensure that transfection is commenced 18–30 h after plating FG293 cells, otherwise they will become too confluent for efficient uptake of DNA.

  —Ensure that DNA is of sufficient quality; for example, by UV spectrophotometry.

  —Gel electrophoresis after restriction digest – there should be no smearing on the gel which could indicate contamination of genomic DNA or RNA.

  —Ensure that DNA is of the expected concentration – use UV spectrophotometry.

    ◦ Read the absorbance of plasmid DNA at A_{260} and A_{320} against a suitable blank and calculate the concentration using the equation:

$$\text{Concentration}(\mu g/ml) = (A_{260} \text{ reading} - A_{320} \text{ reading}) \times 50 (\text{Beer'slaw})$$

  —Ensure that stocks of $CaCl_2$ and HeBS are not contaminated and that salts have not come out of solution. These should be stored away from natural light and have a shelf life of approximately 6 months.

- Cell death occurs very rapidly (within 24 h) after transfection.

  —Ensure that stocks of DNA, $CaCl_2$ and HeBS are not contaminated and that concentrations are correct.

- Cell death occurs rapidly after infection with virus containing cell lysate.

  — Dilute the virus stock further before addition to cells; it may be of a very high titer (see note c in Protocol 11.2).

  — Ensure that virus is not contaminated with fungus or bacteria. It is possible to filter sterilize the original cell lysate using a 45 μm syringe filter; however, the titer may be significantly reduced.

11.3.1.3 Non-viral vectors

- **Poor transfection efficiency**. Try optimization strategies suggested in Section 11.2.7 to improve transfection efficiency. Look at cytotoxicity (Section 11.2.9).

  With polycationic vectors, transfection efficiency is often reduced in the presence of serum and other proteins. Therefore, they should be avoided in the culture medium, along with other polyanions or cations such as HEPES which have also shown to disrupt gene delivery.

  Likewise, antibiotics should be avoided in all cell culture media. Antibiotics mask problems with contamination in the cell culture, which could affect transfection efficiency. It has also been found to interfere with some non-viral vector systems (e.g. Lipofectamine, Invitrogen).

- **Variable or irreproducible results**. Standardize all components of the delivery system. Prepare a large batch (5–10 mg) of endotoxin-free DNA (i.e. using the Gigaprep assay from Qiagen), resuspending the final pellet in sterile filtered water and storing in small aliquots at −35 °C. Do not refreeze DNA once thawed.

  Where possible, similar concentrations of non-viral vectors should be prepared in the appropriate sterile filtered buffers and stored in small aliquots at −35 °C. This can be easily carried out for peptides and PEI. Lipofectamine products, however, must usually be stored at 4 °C; follow manufacturer's guidelines for shelf life.

  There is often variability between different experiments. It is essential to include controls in each experiment, since precise levels of transfection cannot be effectively compared between two separate experiments. This is due to assay sensitivity with some reporter genes, in particular the luciferase assay, which is highly sensitive to temperature and light. Controls should include cells alone, DNA alone, a positive control and a control vector where possible. For some assays, internal standards can be included by aliquoting and storing a known positive sample which can be included in each assay.

- **Toxicity**. If toxicity is observed, try different optimization strategies (see Section 11.2.7), including reducing the time of cell exposure to transfection agents, reducing the concentration of vector/DNA complexes and addition of serum within the culture medium.

References

1. Neschadim, A., McCart, J.A., Keating, A. and Medin, J.A. (2007) A roadmap to safe, efficient, and stable lentivirus-mediated gene therapy with hematopoietic cell transplantation. *Biology of Blood and Marrow Transplantation*, **13** (12), 1407–1416.

2. Gordillo, G.M., Xia, D., Mullins, A.N. *et al.* (1999) Gene therapy in transplantation: pathological consequences of unavoidable plasmid contamination with lipopolysaccharide. *Transplant Immunology*, **7** (2), 83–94.

3. Riddell, S.R., Elliott, M., Lewinsohn, D.A. *et al.* (1996) T-cell mediated rejection of gene-modified HIV-specific cytotoxic T lymphocytes in HIV-infected patients. *Nature Medicine*, **2** (2), 216–223.

4. Apparailly, F., Millet, V., Noel, D. *et al.* (2002) Tetracycline-inducible interleukin-10 gene transfer mediated by an adeno-associated virus: application to experimental arthritis. *Human Gene Therapy*, **13** (10), 1179–1188.

5. Modlich, U., Pugh, C.W. and Bicknell, R. (2000) Increasing endothelial cell specific expression by the use of heterologous hypoxic and cytokine-inducible enhancers. *Gene Therapy*, **7** (10), 896–902.

6. Morgenstern, J.P. and Land, H. (1990) Advanced mammalian gene transfer: high titre retroviral vectors with multiple drug selection markers and a complementary helper-free packaging cell line. *Nucleic Acids Research*, **18** (12), 3587–3596. Describes the pBABE retroviral vectors and eukaryotic packaging cell lines.

7. Chen, P., Tian, J., Kovesdi, I. and Bruder, J.T. (2007) Promoters influence the kinetics of transgene expression following adenovector gene delivery. *Journal of Gene Medicine*, **10** (2), 123–131.

8. Chuah, M.K., Schiedner, G., Thorrez, L. *et al.* (2003) Therapeutic factor VIII levels and negligible toxicity in mouse and dog models of hemophilia A following gene therapy with high-capacity adenoviral vectors. *Blood*, **101** (5), 1734–1743.

9. Sadeghi, H. and Hitt, M.M. (2005) Transcriptionally targeted adenovirus vectors. *Current Gene Therapy*, **5** (4), 411–427.

10. Gruh, I., Wunderlich, S., Winkler, M. *et al.* (2008) Human CMV immediate-early enhancer: a useful tool to enhance cell-type-specific expression from lentiviral vectors. *Journal of Gene Medicine*, **10** (1), 21–32.

11. Lewandoski, M. (2001) Conditional control of gene expression in the mouse. *Nature Reviews Genetics*, **2** (10), 743–755.

12. Gossen, M. and Bujard, H. (1992) Tight control of gene expression in mammalian cells by tetracycline-responsive promoters. *Proceedings of the National Academy of Sciences of the United States of America*, **89** (12), 5547–5551.

13. Miyazaki, S., Miyazaki, T., Tashiro, F. *et al.* (2005) Development of a single-cassette system for spatiotemporal gene regulation in mice. *Biochemical and Biophysical Research Communications*, **338** (2), 1083–1088.

14. Pluta, K., Diehl, W., Zhang, X.Y., *et al.* (2007) Lentiviral vectors encoding tetracycline-dependent repressors and transactivators for reversible knockdown of gene expression: a comparative study. *BMC Biotechnology*, **7**, 41.

15. Semple-Rowland, S.L., Eccles, K.S. and Humberstone, E.J. (2007) Targeted expression of two proteins in neural retina using self-inactivating, insulated lentiviral vectors carrying two internal independent promoters. *Molecular Vision*, **13**, 2001–2011.

16. Gonzalez-Nicolini, V., Sanchez-Bustamante, C.D., Hartenbach, S. and Fussenegger, M. (2006) Adenoviral vector platform for transduction of constitutive and regulated tricistronic or triple-transcript transgene expression in mammalian cells and microtissues. *Journal of Gene Medicine*, **8** (10), 1208–1222. Describes the use of IRES to drive transcription of more than one gene from a viral promoter.

17. Lawson, C. (2006) Strategies for gene transfer to solid organs: viral vectors. *Methods in Molecular Biology* **333**, 175–200.

18. Kay, M.A., Glorioso, J.C. and Naldini, L. (2001) Viral vectors for gene therapy: the art of turning infectious agents into vehicles of therapeutics. *Nature Medicine*, **7** (1), 33–40.

19. Hu, W.S. and Pathak, V.K. (2000) Design of retroviral vectors and helper cells for gene therapy. *Pharmacological Reviews*, **52** (4), 493–511.

20. Coffin, J.M., Hughes, S.H. and Varmus, H.E. (1997) *Retroviruses*, Cold Spring Harbor Laboratory Press, New York.

21. Rivière, I., Brose, K. and Mulligan, R.C. (1995) Effects of retroviral vector design on expression of human adenosine deaminase in murine bone marrow transplant recipients engrafted with genetically modified cells. *Proceedings of the National Academy of Sciences of the United States of America*, **92** (15), 6733–6737. Describes the pMFG retroviral vector.

22. Markowitz, D., Goff, S. and Bank, A. (1988) A safe packaging line for gene transfer: separating viral genes on two different plasmids. *Journal of Virology*, **62** (4), 1120–1124.

23. Lewis, P.F. and Emerman, M. (1994) Passage through mitosis is required for oncoretroviruses but not for the human immunodeficiency virus. *Journal of Virology*, **68** (1), 510–516.

24. Sanders, D.A. (2002) No false start for novel pseudotyped vectors. *Current Opinion in Biotechnology*, **13** (5), 437–442.

25. Kafri, T., van Praag, H., Ouyang, L. *et al.* (1999) A packaging cell line for lentivirus vectors. *Journal of Virology*, **73** (1), 576–584.

26. Mitta, B., Rimann, M., Ehrengruber, M.U. *et al.* (2002) Advanced modular self-inactivating lentiviral expression vectors for multigene interventions in mammalian cells and *in vivo* transduction. *Nucleic Acids Research*, **30** (21), e113.

27. Buchschacher, G.L. Jr. and Wong-Staal, F. (2001) Approaches to gene therapy for human immunodeficiency virus infection. *Human Gene Therapy*, **12** (9), 1013–1019.

28. Linial, M. (2000) Why aren't foamy viruses pathogenic? *Trends in Microbiology*, **8** (6), 284–289.

29. Trobridge, G., Josephson, N., Vassilopoulos, G. *et al.* (2002) Improved foamy virus vectors with minimal viral sequences. *Molecular Therapy*, **6** (3), 321–328.

30. Bauer, T.R. Jr., Allen, J.M., Hai, M. *et al.* (2008) Successful treatment of canine leukocyte adhesion deficiency by foamy virus vectors. *Nature Medicine*, **14** (1), 93–97.

31. Berns, K.I. and Giraud, C. (1996) Biology of adeno-associated virus, in Adeno-Associated Virus (AAV) Vectors in Gene Therapy (eds K.I. Berns and C. Giraud), Springer-Verlag, Berlin, pp. 1–24. An overview of the biology of AAV.

32. Lai, C.M., Lai, Y.K. and Rakoczy, P.E. (2002) Adenovirus and adeno-associated virus vectors. *DNA and Cell Biology*, **21** (12), 895–913.

33. Dong, J.Y., Fan, P.D. and Frizzell, R.A. (1996) Quantitative analysis of the packaging capacity of recombinant adeno-associated virus. *Human Gene Therapy*, **7** (17), 2101–2112.

34. Zaiss, A.K., Liu, Q., Bowen, G.P. *et al.* (2002) Differential activation of innate immune responses by adenovirus and adeno-associated virus vectors. *Journal of Virology*, **76** (9), 4580–4590.

35. Xiao, W., Chirmule, N., Berta, S.C. *et al.* (1999) Gene therapy vectors based on adeno-associated virus type 1. *Journal of Virology*, **73** (5), 3994–4003.

36. Chirmule, N., Xiao, W., Truneh, A. *et al.* (2000) Humoral immunity to adeno-associated virus type 2 vectors following administration to murine and nonhuman primate muscle. *Journal of Virology*, **74** (5), 2420–2425.

37. Chirmule, N., Propert, K., Magosin, S. *et al.* (1999) Immune responses to adenovirus and adeno-associated virus in humans. *Gene Therapy*, **6** (9), 1574–1583.

38. Büning, H., Perabo, L., Coutelle, O. *et al.* (2008) Recent developments in adeno-associated virus vector technology. *Journal of Gene Medicine*, **10** (7), 717–733.

39. Rowe, W.P., Huebner, R.J., Gilmore, L.K. *et al.* (1953) Isolation of a cytopathogenic agent from human adenoids undergoing spontaneous degeneration in tissue culture. *Proceedings of the Society for Experimental Biology and Medicine*, **84** (3), 570–573.

40. Ginsberg, H.S. (1984) The Adenoviruses, Plenum Press, New York.

41. Bergelson, J.M., Cunningham, J.A., Droguett, G. *et al.* (1997) Isolation of a common receptor for coxsackie B viruses and adenoviruses 2 and 5. *Science*, **275** (5304), 1320–1323.

42. Parker, A.L., Waddington, S.N., Nicol, C.G. *et al.* (2006) Multiple vitamin K-dependent coagulation zymogens promote adenovirus-mediated gene delivery to hepatocytes. *Blood*, **108** (8), 2554–2561.

43. Waddington, S.N., Parker, A.L., Havenga, M. *et al.* (2007) Targeting of adenovirus serotype 5 (Ad5) and 5/47 pseudotyped vectors *in vivo*: fundamental involvement of coagulation factors and redundancy of CAR binding by Ad5. *Journal of Virology*, **81** (17), 9568–9571.

44. Wickham, T.J., Mathias, P., Cheresh, D.A. and Nemerow, G.R. (1993) Integrins $\alpha v \beta 3$ and $\alpha v \beta 5$ promote adenovirus internalization but not virus attachment. *Cell*, **73** (2), 309–319.

45. Graham, F.L., Smiley, J., Russell, W.C. and Nairn, R. (1977) Characteristics of a human cell line transformed by DNA from human adenovirus type 5. *Journal of General Virology*, **36** (1), 59–74. Ad5 vectors and suitable packaging cells.

46. Hearing, P., Samulski, R.J., Wishart, W.L. and Shenk, T. (1987) Identification of a repeated sequence element required for efficient encapsidation of the adenovirus type 5 chromosome. *Journal of Virology*, **61** (8), 2555–2558.

47. Shen, W.Y., Lai, M.C., Beilby, J. *et al.* (2001) Combined effect of cyclosporine and sirolimus on improving the longevity of recombinant adenovirus-mediated transgene expression in the retina. *Archives of Ophthalmology*, **119** (7), 1033–1043.

48. Yap, J., O'Brien, T., Tazelaar, H.D. and McGregor, C.G. (1997) Immunosuppression prolongs adenoviral mediated transgene expression in cardiac allograft transplantation. *Cardiovascular Research*, **35** (3), 529–535.

49. Engelhardt, J.F., Ye, X., Doranz, B. and Wilson, J.M. (1994) Ablation of E2A in recombinant adenoviruses improves transgene persistence and decreases inflammatory response in mouse liver. *Proceedings of the National Academy of Sciences of the United States of America*, **91** (13), 6196–6200.

50. Qian, H.S., Channon, K., Neplioueva, V. *et al.* (2001) Improved adenoviral vector for vascular gene therapy : beneficial effects on vascular function and inflammation. *Circulation Research*, **88** (9), 911–917.

51. Wen, S., Schneider, D.B., Driscoll, R.M. *et al.* (2000) Second-generation adenoviral vectors do not prevent rapid loss of transgene expression and vector DNA from the arterial wall. *Arteriosclerosis, Thrombosis, and Vascular Biology*, **20** (6), 1452–1458.

52. Kumar-Singh, R., Yamashita, C.K., Tran, K. and Farber, D.B. (2000) Construction of encapsidated (gutted) adenovirus minichromosomes and their application to rescue of photoreceptor degeneration. *Methods in Enzymology*, **316**, 724–743.

53. Lawson, C., Holder, A.L., Stanford, R.E. *et al.* (2005) Anti-intercellular adhesion molecule-1 anti-bodies in sera of heart transplant recipients: a role in endothelial cell activation. *Transplantation*, **80** (2), 264–271.

54. Marshall, E. (1999) Gene therapy death prompts review of adenovirus vector. *Science*, **286** (5448), 2244–2245.

55. Cavazzana-Calvo, M. and Fischer, A. (2007) Gene therapy for severe combined immunodeficiency: are we there yet? *Journal of Clinical Investigation*, **117** (6), 1456–1465.

56. Herweijer, H. and Wolff, J.A. (2003) Progress and prospects: naked DNA gene transfer and therapy. *Gene Therapy*, **10** (6), 453–458. Review of physical DNA vectors.

57. Kawabata, K., Takakura, Y. and Hashida, M. (1995) The fate of plasmid DNA after intravenous injection in mice: involvement of scavenger receptors in its hepatic uptake. *Pharmaceutical Research*, **12** (6), 825–830.

58. Liu, F., Song, Y. and Liu, D. (1999) Hydrodynamics-based transfection in animals by systemic administration of plasmid DNA. *Gene Therapy*, **6** (7), 1258–1266.

59. Zhang, X., Dong, X., Sawyer, G.J. *et al.* (2004) Regional hydrodynamic gene delivery to the rat liver with physiological volumes of DNA solution. *Journal of Gene Medicine*, **6** (6), 693–703.

60. Maruyama, H., Higuchi, N., Nishikawa, Y. *et al.* (2002) Kidney-targeted naked DNA transfer by retrograde renal vein injection in rats. *Human Gene Therapy*, **13** (3), 455–468.

61. Yang, N.S., Burkholder, J., Roberts, B. *et al.* (1990) *In vivo* and *in vitro* gene transfer to mammalian somatic cells by particle bombardment. *Proceedings of the National Academy of Sciences of the United States of America*, **87** (24), 9568–9572.

62. Taniyama, Y., Tachibana, K., Hiraoka, K. *et al.* (2002) Development of safe and efficient novel nonviral gene transfer using ultrasound: enhancement of transfection efficiency of naked plasmid DNA in skeletal muscle. *Gene Therapy*, **9** (6), 372–380.

63. Weaver, J.C. (1993) Electroporation: a general phenomenon for manipulating cells and tissues. *Journal of Cellular Biochemistry*, **51** (4), 426–435.

64. Felgner, P.L., Gadek, T.R., Holm, M. *et al.* (1987) Lipofection: a highly efficient, lipid-mediated DNA-transfection procedure. *Proceedings of the National Academy of Sciences of the United States of America*, **84** (21), 7413–7417. First use of liposomes as a transfection agent.

65. Felgner, P.L. and Ringold, G.M. (1989) Cationic liposome-mediated transfection. *Nature*, **337** (6205), 387–388.

66. Niculescu-Duvaz, D., Heyes, J. and Springer, C.J. (2003) Structure–activity relationship in cationic lipid mediated gene transfection. *Current Medicinal Chemistry*, **10** (14), 1233–1261.

67. Farhood, H., Serbina, N. and Huang, L. (1995) The role of dioleoyl phosphatidylethanolamine in cationic liposome mediated gene transfer. *Biochimica et Biophysica Acta*, **1235** (2), 289–295.

68. Karmali, P.P. and Chaudhuri, A. (2007) Cationic liposomes as non-viral carriers of gene medicines: resolved issues, open questions, and future promises. *Medicinal Research Reviews*, **27** (5), 696–722.

69. Kim, J.K., Choi, S.H., Kim, C.O. *et al.* (2003) Enhancement of polyethylene glycol (PEG)-modified cationic liposome-mediated gene deliveries: effects on serum stability and transfection efficiency. *Journal of Pharmacy and Pharmacology*, **55** (4), 453–460. Review of liposomes as DNA vectors.

70. Cheng, P.W. (1996) Receptor ligand-facilitated gene transfer: enhancement of liposome-mediated gene transfer and expression by transferrin. *Human Gene Therapy*, **7** (3), 275–282.

71. Hofland, H.E., Masson, C., Iginla, S. *et al.* (2002) Folate-targeted gene transfer *in vivo*. *Molecular Therapy*, **5** (6), 739–744.

72. Tan, P.H., Manunta, M., Ardjomand, N. *et al.* (2003) Antibody targeted gene transfer to endothelium. *Journal of Gene Medicine*, **5** (4), 311–323.

73. Vitiello, L., Chonn, A., Wasserman, J.D. *et al.* (1996) Condensation of plasmid DNA with polylysine improves liposome-mediated gene transfer into established and primary muscle cells. *Gene Therapy*, **3** (5), 396–404.

74. Legendre, J.Y. and Szoka, F.C. Jr. (1993) Cyclic amphipathic peptide-DNA complexes mediate high-efficiency transfection of adherent mammalian cells. *Proceedings of the National Academy of Sciences of the United States of America*, **90** (3), 893–897.

75. Hart, S.L., Arancibia-Carcamo, C.V., Wolfert, M.A. *et al.* (1998) Lipid-mediated enhancement of transfection by a nonviral integrin-targeting vector. *Human Gene Therapy*, **9** (4), 575–585.

76. Li, J.M., Collins, L.,Zhang, X. *et al.* (2000) Efficient gene delivery to vascular smooth muscle cells using a nontoxic, synthetic peptide vector system targeted to membrane integrins: a first step toward the gene therapy of chronic rejection. *Transplantation*, **70** (11), 1616–1624.

77. Zhang, X., Collins, L. and Fabre, J.W. (2001) A powerful cooperative interaction between a fusogenic peptide and lipofectamine for the enhancement of receptor-targeted, non-viral gene delivery via integrin receptors. *Journal of Gene Medicine*, **3** (6), 560–568.

78. Dodds, E., Piper, T.A., Murphy, S.J. and Dickson, G. (1999) Cationic lipids and polymers are able to enhance adenoviral infection of cultured mouse myotubes. *Journal of Neurochemistry*, **72** (5), 2105–2112.

79. Kaneda, Y. (2001) Improvements in gene therapy technologies. *Molecular Urology*, **5** (2), 85–89.

80. Filion, M.C. and Phillips, N.C. (1997) Toxicity and immunomodulatory activity of liposomal vectors formulated with cationic lipids toward immune effector cells. *Biochimica et Biophysica Acta*, **1329** (2), 345–356.

81. Ruiz, F.E., Clancy, J.P., Perricone, M.A. *et al.* (2001) A clinical inflammatory syndrome attributable to aerosolized lipid–DNA administration in cystic fibrosis. *Human Gene Therapy*, **12** (7), 751–761.

82. Wu, G.Y. and Wu, C.H. (1988) Receptor-mediated gene delivery and expression *in vivo*. *Journal of Biological Chemistry*, **263** (29), 14621–14624.

83. Wolfert, M.A. and Seymour, L.W. (1996) Atomic force microscopic analysis of the influence of the molecular weight of poly(L)lysine on the size of polyelectrolyte complexes formed with DNA. *Gene Therapy*, **3** (3), 269–273.

84. Fabre, J.W. and Collins, L. (2006) Synthetic peptides as non-viral DNA vectors. *Current Gene Therapy*, **6** (4), 459–480. Review of the use of peptides as DNA vectors.

85. Wagner, E., Zenke, M., Cotten, M. *et al.* (1990) Transferrin–polycation conjugates as carriers for DNA uptake into cells. *Proceedings of the National Academy of Sciences of the United States of America*, **87** (9), 3410–3414.

86. Huckett, B., Ariatti, M. and Hawtrey, A.O. (1990) Evidence for targeted gene transfer by receptor-mediated endocytosis. Stable expression following insulin-directed entry of NEO into HepG2 cells. *Biochemical Pharmacology*, **40** (2), 253–263.

87. Wu, G.Y. and Wu, C.H. (1988) Evidence for targeted gene delivery to Hep G2 hepatoma cells *in vitro*. *Biochemistry*, **27** (3), 887–892.

88. Nishikawa, M., Yamauchi, M., Morimoto, K. *et al.* (2000) Hepatocyte-targeted *in vivo* gene expression by intravenous injection of plasmid DNA complexed with synthetic multi-functional gene delivery system. *Gene Therapy*, **7** (7), 548–555.

89. Collins, L., Sawyer, G.J., Zhang, X.H. *et al.* (2000) *In vitro* investigation of factors important for the delivery of an integrin-targeted nonviral DNA vector in organ transplantation. *Transplantation*, **69** (6), 1168–1176.

90. Patel, S., Zhang, X., Collins, L. and Fabre, J.W. (2001) A small, synthetic peptide for gene delivery via the serpin–enzyme complex receptor. *Journal of Gene Medicine*, **3** (3), 271–279.

91. Shewring, L., Collins, L., Lightman, S.L. *et al.* (1997) A nonviral vector system for efficient gene transfer to corneal endothelial cells via membrane integrins. *Transplantation*, **64** (5), 763–769.

92. Schachtschabel, U., Pavlinkova, G., Lou, D. and Kohler, H. (1996) Antibody-mediated gene delivery for B-cell lymphoma *in vitro*. *Cancer Gene Therapy*, **3** (6), 365–372.

93. Gill, T.J. III, Papermaster, D.S., Kunz, H.W. and Marfey, P.S. (1968) Studies on synthetic polypeptide antigens. XIX. Immunogenicity and antigenic site structure of intramolecularly cross-linked polypeptides. *Journal of Biological Chemistry*, **243** (2), 287–300.

94. Wilson, J.M., Grossman, M., Wu, C.H. *et al.* (1992) Hepatocyte-directed gene transfer *in vivo* leads to transient improvement of hypercholesterolemia in low density lipoprotein receptor-deficient rabbits. *Journal of Biological Chemistry*, **267** (2), 963–967.

95. Tietz, P.S., Yamazaki, K. and LaRusso, N.F. (1990) Time-dependent effects of chloroquine on pH of hepatocyte lysosomes. *Biochemical Pharmacology*, **40** (6), 1419–1421.

96. Carr, C.M. and Kim, P.S. (1993) A spring-loaded mechanism for the conformational change of influenza hemagglutinin. *Cell*, **73** (4), 823–832.

97. Collins, L. and Fabre, J.W. (2004) A synthetic peptide vector system for optimal gene delivery to corneal endothelium. *Journal of Gene Medicine*, **6** (2), 185–194.

98. Boussif, O., Lezoualc'h, F., Zanta, M.A. *et al.* (1995) A versatile vector for gene and oligonucleotide transfer into cells in culture and *in vivo*: polyethylenimine. *Proceedings of the National Academy of Sciences of the United States of America*, **92** (16), 7297–7301.

99. Kircheis, R., Wightman, L. and Wagner, E. (2001) Design and gene delivery activity of modified polyethylenimines. *Advanced Drug Delivery Reviews*, **53** (3), 341–358. Review of polyethyleneimine as a DNA vector.

100. Lungwitz, U., Breunig, M., Blunk, T. and Gopferich, A. (2005) Polyethylenimine-based non-viral gene delivery systems. *European Journal of Pharmaceutics and Biopharmaceutics*, **60** (2), 247–266.

101. Ogris, M., Brunner, S., Schuller, S. *et al.* (1999) PEGylated DNA/transferring–PEI complexes: reduced interaction with blood components, extended circulation in blood and potential for systemic gene delivery. *Gene Therapy*, **6** (4), 595–605.

102. Erbacher, P., Remy, J.S. and Behr, J.P. (1999) Gene transfer with synthetic virus-like particles via the integrin-mediated endocytosis pathway. *Gene Therapy*, **6** (1), 138–145.

103. Kircheis, R., Kichler, A., Wallner, G. *et al.* (1997) Coupling of cell-binding ligands to polyethylenimine for targeted gene delivery. *Gene Therapy*, **4** (5), 409–418.

104. Tang, M.X., Redemann, C.T. and Szoka, F.C. Jr. (1996) *In vitro* gene delivery by degraded polyamidoamine dendrimers. *Bioconjugate Chemistry*, **7** (6), 703–714.

105. Collins, L., Kaszuba, M. and Fabre, J.W. (2004) Imaging in solution of $(Lys)_{16}$-containing bifunctional synthetic peptide/DNA nanoparticles for gene delivery. *Biochimica et Biophysica Acta*, **1672** (1), 12–20.

106. Labat-Moleur, F., Steffan, A.M., Brisson, C. *et al.* (1996) An electron microscopy study into the mechanism of gene transfer with lipopolyamines. *Gene Therapy*, **3** (11), 1010–1017.

107. Grosse, S., Aron, Y., Thevenot, G. *et al.* (2005) Potocytosis and cellular exit of complexes as cellular pathways for gene delivery by polycations. *Journal of Gene Medicine*, **7** (10), 1275–1286.

108. Budker, V.G., Slattum, P.M., Monahan, S.D. and Wolff, J.A. (2002) Entrapment and condensation of DNA in neutral reverse micelles. *Biophysical Journal*, **82** (3), 1570–1579.

109. Bradford, M.M. (1976) A rapid and sensitive method for the quantitation of microgram quantities of protein utilizing the principle of protein–dye binding. *Analytical Biochemistry* **72**, 248–254.

110. Freshney, R. (1987) *Culture of Animal Cells. A Manual of Basic Techniques*, 2nd edn, Wiley-Liss.

111. Parks, D.R., Bryan, V.M., Oi, V.T. and Herzenberg, L.A. (1979) Antigen-specific identification and cloning of hybridomas with a fluorescence-activated cell sorter. *Proceedings of the National Academy of Sciences of the United States of America*, **76** (4), 1962–1966.

112. Mosmann, T. (1983) Rapid colorimetric assay for cellular growth and survival: application to proliferation and cytotoxicity assays. *Journal of Immunological Methods*, **65** (1–2), 55–63.

113. Cory, A.H., Owen, T.C., Barltrop, J.A. and Cory, J.G. (1991) Use of an aqueous soluble tetrazolium/formazan assay for cell growth assays in culture. *Cancer Communications*, **3** (7), 207–212.

114. Korzeniewski, C. and Callewaert, D.M. (1983) An enzyme-release assay for natural cytotoxicity. *Journal of Immunological Methods*, **64** (3), 313–320.

115. Merrick, A.F., Shewring, L.D., Sawyer, G.J. *et al.* (1996) Comparison of adenovirus gene transfer to vascular endothelial cells in cell culture, organ culture, and *in vivo*. *Transplantation*, **62** (8), 1085–1089.

12
Gene Therapy Strategies: Constructing an AAV Trojan Horse

M. Ian Phillips[1], Edilamar M. de Oliveira[2], Leping Shen[1], Yao Liang Tang[1] and Keping Qian[1]
[1] *Keck Graduate Institute, Claremont University Colleges, Claremont, California, USA*
[2] *Laboratory of Biochemistry, School of Physical Education and Sport, Sao Paulo University, Sao Paulo, Brazil*

12.1 Introduction

There are many gene therapy strategies for a variety of diseases that could be treatable or even curable by gene therapy. However, success has been limited. This reflects the state of the field which is at various stages, and the limitation of tools. In theory, gene therapy is very obvious: if a disease is monogenetic and due to a missing gene, then it should be possible, if not simple, to provide the missing gene by inserting it into the chromosomal position to which it belongs. The problem has been how to deliver the gene safely and effectively.

Even if the disease is multigenetic, we know from drug therapy that manipulating a specific gene or blocking a gene product can control the disease. For example, hypertension undoubtedly involves multiple genes. But there are successful drugs to control high blood pressure that inhibit specific gene products such angiotensin type 1 receptors, angiotensin converting enzyme or renin. Therefore, even in cases of multigenetic diseases, inhibition of a single gene delivered to cells could be an effective gene therapy.

To deliver DNA to replace genes or inhibitors of gene expression, researchers have favored the Trojan horse approach with viral vectors. In the ancient story of Homer, the Greeks, having besieged the city of Troy for 10 years, apparently gave up and sailed away.

Genomics: Essential Methods   Edited by Mike Starkey and Ramnath Elaswarapu
© 2011 John Wiley & Sons, Ltd.

They left behind a gift for the unconquered Trojans – a huge wooden horse. The Trojans of Troy willingly pulled the horse into their impregnable city to celebrate. During the night, from inside the deceptively silent horse, Greek soldiers emerged, opened the gates for the rest of their army to enter and destroyed the city from within. The idea of entering a cell with the help of the cell and then attacking it from the inside is even more ancient than this story. It is the process used by viruses, plasmids, bacteria and parasites. By using viruses and plasmids as the Trojan horse, rendered harmless, we can modify genes or introduce new ones to make the cell die or survive longer, secrete proteins or switch off genes, differentiate or not differentiate.

In this chapter we discuss some of the strategies for gene delivery and give detailed protocols on adeno-associated virus (AAV) as a gene delivery method which is safe, reliable and long lasting.

12.1.1 General strategies for gene therapy: Basic methods

12.1.1.1 Transgenics

Studies using transgenics have revealed which genes are important or vital. Transgenic mice with genes knocked out, or genes 'knocked in' to increase the number of copies of genes [1], are the fodder of gene studies in living animals. They have been very useful and practical for studying the role of specific genes producing specific human proteins. The method involves harvesting embryonic stem cells from the inner cell mass of the blastocyst. A desired gene is made, using recombinant DNA (rDNA), and inserted in a vector together with promoter sequences to regulate the gene expression. A neo^r gene, which is resistant to lethal effects of neomycin, is added to the DNA cassette. A thymidine kinase gene (tk) is also added. Those cells that fail to take the vector inside their walls can be killed by neomycin. A few of the remaining cells allow the vector in, but the gene is inserted randomly. Cells with random insertion of the gene are killed by gangcyclovir. The presence of Tk phosphorylates gangcyclovir, which protects the gene. That leaves only those cells in which homologous recombination has occurred. The normal gene has been knocked out and a new, specified gene knocked in. These cells are then injected into a blastocyst and implanted in the uterus to produce offspring that can be bred for more generations of animals lacking a gene or possessing a new gene. If the new gene is nonfunctional (i.e. a null allele), the function of the former gene may be revealed through breeding the mice with the knockout gene to homozygosity. If the gene knocked out is absolutely essential then it may be embryonically lethal.

Ideally, the function of the missing gene should be as obvious as if a limb had been cut off. In actuality, several things can happen. The knocked out gene may prevent the embryo from developing (it is embryonically lethal), or the missing gene is fully compensated by other genes or subtle changes occur in development or in different organs also that the effect is not obvious. Nevertheless, the technique has had a huge influence on revealing functional effects of proteins, especially where antibodies have not been developed. The opposite of knocking in copies of a gene has been used to reveal mechanisms of diseases caused by overexpression of a protein [1]. The transgenic animal approach requires going through embryonic development. This limits the technique when a knocked out gene is embryonically lethal. However, a method first used by Gu *et al.* [2] is able to induce the same mutation and avoid lethality.

12.1.1.2 Cre/Lox P System

To knockout a target gene in specific cell groups or tissue, in adult animals, the Cre/lox P system is a suitable technique. It is based on the viral bacteria phage P1, which produces Cre, a recombinase enzyme. Cre cuts its viral DNA into packages. Cre cuts all the DNA out between two separate lox P sites. The DNA ends, which each have a half lox P site, are then ligated by the recombinase. Gu *et al.* [2] used this principle with a strategy of a conventional transgenic mouse, in which the Cre transgene plus a promoter was inserted by homologous recombination in a cell-specific type. This mouse was crossed with a second mouse strain that had a target gene flanked by two lox P sites. In the offspring the target gene was only deleted in those specific cells that contained Cre. The target gene remained functional in all the other cells and the animals survived development, so the function of the targeted gene in specific cells could be studied.

More recent developments have made the technique less laborious to use [3, 4]. An example is a study by Sinnayah *et al.* [4], who made transgenic mice with lox P insertions flanking the gene for angiotensinogen. Angiotensinogen is a substrate for the enzyme renin and is one of the critical components for the synthesis of the peptide angiotensin. Instead of making a separate strain of Cre mice and proceeding with breeding, they simply injected Cre into the floxed mice. This had the advantage of not only being time saving, but also of opening up a new way to study genes with site-directed gene ablation in specific cells. As they were working on the brain they were able to pinpoint anatomically a very small brain structure. Their study addressed a long simmering debate of whether the brain makes its own angiotensin [5, 6] or whether the angiotensin that has been found in the brain is taken up from angiotensin in the blood. By injecting Cre into a brain structure they showed that angiotensin synthesis could be blocked and, therefore, is made in the brain.

12.1.1.3 Antisense mRNA

To inhibit synthesis of proteins by inhibiting gene translation there are two methods: antisense mRNA and RNA interference (RNAi). Antisense was discovered in 1977 [7], but it was not until 1993 that its potential for inhibition by *in vivo* delivery was clearly demonstrated. Antisense to neuropeptide Y Y-1 receptor was injected into the rat brain and the injected animals showed a decrease of behavioral anxiety [8]. Antisense to angiotensinogen, angiotensin-converting enzymes and angiotensin type 1 receptor genes was injected into the brains of spontaneously hypertensive rats and they showed decreases in high blood pressure [9]. Antisense is based on the fact that mRNA is in the 'sense' direction from 5' to 3'. Antisense is a limited sequence of DNA in the antisense direction 3' to 5' designed from knowing the sequence of a target gene. Antisense oligodeoxynucleotides (AS-ODN) are usually built around the initiation codon of a gene (the AUG start site) but may be shorter than the full-length gene. This is because the AS-ODN binds to part of the appropriate mRNA sequences and prevents the mRNA from translating the protein it would otherwise produce.

For gene modification within a cell with antisense a viral vector can be fitted with DNA in the antisense direction. We have designed these in the adenoassociated virus and shown them to have long-lasting inhibitory effects on designated cell protein synthesis [10].

Antisense inhibition, although widely used in research and approved for some clinical treatments [11], is not perfect. When antisense is put into a cell it is competing with the cells' own mRNA copying machinery. The presence of AS-ODN may actually increase the number of cell-produced mRNA copies, thereby overcoming the endogenously administered AS-ODN.

Because of this, antisense as a treatment has not proven to be a killer of cells and, therefore, not a revolutionary anti-cancer agent, as it was originally hoped. However, it has played a pivotal role in leading to the next advance in cellular gene inhibition – small interfering RNA.

12.1.1.4 siRNA

Fire and Mello [12] were using antisense to study behavioral effects on the primitive worm *Caenorhabditis elegans*. They tested sense RNA and antisense RNA on the worms but there was no effect. They then tested a combination of sense and antisense RNA. The effect was dramatic. The worms started to twitch spontaneously. The gene that was holding back the twitching had been silenced. Fire and Mello had discovered gene silencing by double-stranded (ds) RNA which acted as small interfering RNA (siRNA). They were rewarded for the discovery with the Nobel Prize in 2006. RNAi has become widely recognized as a biological mechanism for the regulation of gene expression and used for intracellular inhibition. dsRNA is produced in the nucleus. In the cytoplasm it binds to an enzyme, Dicer. Dicer literally dices up the dsRNA into short strands (15–20 nucleotides).

One of the strands is loaded into a protein complex, RNA-induced silencing complex (RISC). The RISC now has the single strand of short RNA as a binding site to bind to a complementary sequence on the cell's mRNA. This binding leads to cleavage of mRNA, degrading the message and stopping it from translating a specific protein. And hence it is silenced.

RNAi is a fundamental cellular process of gene regulation in the cells of animals and plants. Since both animals and plants are subject to diseases induced by viruses, RNAi may have evolved to protect cells from invasion by viruses. The genome of retroviruses is in double strands of RNA. A retrovirus, lacking cellular mechanisms and DNA, injects its genomic dsRNA into a cell to reproduce itself using the DNA of the invaded cell. RNAi protects the cell by destroying the viral RNA through the RISC mechanism.

siRNA is more powerful than antisense in silencing genes, but it has its difficulties. It is not long lasting, it may silence off-target sites and it has been not been easy to inject systemically as a therapy. We have directly compared siRNA with AS-ODN to inhibit the Beta-1 adrenergic receptor gene [13]. The effect was measured on blood pressure in hypertensive rats and on measures of heart performance, because beta blockers have long been used for hypertension and heart failure treatments. The siRNA and AS-ODN were injected systemically in a Lipofectamine vehicle. The result was a significantly better effect on lowering blood pressure and improving heart performance with the siRNA than with the AS-ODN. Both approaches lasted about 1 week with a single injection [13].

12.1.1.5 MicroRNA

MicroRNA (miRNA) offers completely new possibilities for gene modification, cell therapy and drug development. It is involved in almost every biological process regulated by genes and its absence or mutations could be the cause of many disease states from birth defects to cancer.

Although miRNAs were discovered over 20 years ago in *C. elegans* [14] and later found in mammals, we are still in an early stage of discovering how many there are, what they do and how they do it. Over 500 miRRNAs have been found in the human genome. A recent review in *Nature Reviews* suggests that miRNAs regulate one-third of human genes [15]. MiRNAs have become recognized as a new class of gene regulators and, therefore, important for gene modification of cells. MiRNAs are small non-coding RNAs that modify gene expression by post-transcriptional inhibition of targeted mRNA. In the nucleus, miRNA is formed from introns and exons as 'primary' or 'pri-miRNA.' But it is not a messenger RNA – it does not specify or generate a protein. The pri-RNA, a folded-back structure of 60–70 nucleotides, is processed in the nucleus by the enzymes Drosha and Pasha. Drosha cuts out the stem–loop structure which is the 'pre-miRNA.' The pre-miRNA is exported out of the nucleus by exportin and into the cytoplasm where it is diced up by the enzyme Dicer RNase III, mentioned above in the siRNA process. The same effect occurs when Dicer cuts the stem–loop into short-length (19–25 nucleotides) inverted 'mature miRNA.' As with siRNA, one strand of the mature miRNA becomes part of the RISC and targets mRNA by binding to antisense complementary regions and cleaving or degrading the targeted mRNA. Multiple roles for miRNAs in gene regulation have been revealed by gene expression analysis polymerase chain reaction (PCR), and by transgenic mice with knockouts of specific miRNA. Expression arrays are revealing specific miRs in different tissues and cells from invertebrates to humans. Many miRNAs (miR-1, miR-34, miR-60, miR-87, miR-124a) are highly conserved between vertebrates and invertebrates [16], including the small temporal (st)RNAs discovered in *C. elegans* (e.g. let-7 RNA, lin-4) that are similar to miRNAs in humans. As these stRNAs are critical for cell differentiation and timing of neural connections, the conservation may indicate functional evolution.

A survey of mouse tissues with northern blotting [16] showed miR-1 is dominant in the heart (45%). In the liver, miR-122 was 72% of all miRNAs tested and miR-124a was profound in the mouse brain.

Although the mechanism of miRNA action is principally inhibitory on targeted mRNA, which is essential for normal growth and differentiation in cell and tissue development, miRNAs can be involved in cancer. They can be depleted or suppressed, allowing oncogenes to be overproduced. Kumar *et al.* [17] recently showed that global suppression of miRNAs in various cancer cell lines increased cancer cell transformation and enhanced tumorogenesis in mice. To suppress miRNA they targeted Drosha and Dicer with siRNA. Non-cancerous cells did not become cancerous, but did not grow. This suggests that increasing miRNAs could be a new gene therapy approach to treating cancer by suppressing oncogenes.

12.1.2 Gene therapy strategies: Delivering genes to cells

12.1.2.1 Non-viral delivery

Non-viral delivery for gene transfer refers to formulations of liposomes, naked DNA and dendrimers. The main advantage is that they are non-pathogenic and have low toxicity, they can be easily purified and most significant they can carry a transgene of almost any size. The disadvantages are low efficiency and transient transgene expression. Naked DNA injected into heart or muscle cells can be expressed for 2 weeks to several months [18].

12.1.2.2 Electroporation

Electroporation (EP) is used for *in vitro* transfection of DNA into cells. It was first used by Neumann *et al.* in 1982 [19] to transfer genes into mouse lyoma cells. Later, in 1987,

Okino *et al.* [20] used EP for *in vivo* delivery of chemotherapeutic agents to solid tumors. Since then EP has been used in humans and animals for the delivery of drugs and genes. EP is now frequently used for plasmid DNA gene transfer as an alternative to delivery with viral vectors or lipid-soluble products. Heller and Heller [21] have listed the numerous reports of using EP to deliver many types of transgenes genes into multiple cell types and organs to test their efficacy in reducing cancer and immune reactions. EP is an efficient method of increasing gene expression several fold over direct injection of naked DNA (plasmids).

EP is not a Trojan horse but a bombardment to breach the cell wall. The mechanism of EP delivery is straightforward. Electrodes use voltage or current in pulses to make the membrane of cells semi-permeable by briefly opening pores in the lipid membrane of the cell without doing permanent damage. Molecules that are usually excluded from entering the cell (hydrophilic, non-lipid-soluble or anionic molecules) can physically pass through the pores. The challenge is to find the right parameters for different genes, cell types and purposes. The voltage (volts), current (amps), pulse width (short or long), pulse interval (seconds), pulse shape (square, bipolar, exponential), pulse number (single, multiple, trains) and frequency of stimulation can all affect the outcome. Long low-voltage pulses yield high expression in muscle but not in cancer cells. To avoid direct current damage, high field strength pulses (>700 V/cm) are usually mixed with short pulses (microseconds) and low field strength (<700 V/cm) by long pulses (9 ms). Several systems are sold commercially. If in doubt, refer to the numerous papers which have reported enhanced DNA delivery each with their own set of parameters. For a review, see Heller and Heller [21].

12.1.2.3 Liposomes

DNA is negatively charged. Cationic liposomes bind to DNA and form a liposome–DNA complex, which fuses with lipid membranes of the cell and allows DNA to enter. Many purified cationic liposomes are commercially available and are widely used (e.g. Lipofectamine, DOPE, DOTAP). They are best for *in vitro* delivery into cells, but are used for *in vivo* delivery, although there has to be a check for complement activation or toxicity.

12.1.2.4 Dendrimers

Dendrimers have the potential for highly efficient gene delivery [22]. They are a class of synthetic spherical polymers which are highly branched. They can be made to almost any size and are stable and soluble in water. They have a positive charge on the surface to bind to lipid cell membranes and carry negatively charged DNA on the inside. They enhance the delivery of DNA and RNA, single- and double-stranded. They have been shown to enhance the delivery of plasmids-mediated gene transfer in the heart [23]. A DNA : dendrimer ratio of $1:20$ was used to show the uptake and expression of β-galactosidase in mouse hearts [23].

12.1.3 Viral delivery

Gene modification for cell transplantation has mainly focused on transferring genes or recombinant genes in cassettes by viral delivery for high transfection efficiency and high expression level that can be long lasting. Several reviews and books have adequately covered the pros and cons of different viruses as vectors [24] and so these will only be briefly summarized here. The main issues to consider are: (i) whether the cells to be transfected are dividing or non-dividing; (ii) the carrying capacity of the virus – that is, whether the vector can hold and deliver a large transgene or gene regulatory cassette; (iii) whether the vector is stable or transient; (iv) whether it is safe and free from host immune reaction; (v) whether the transfection is very efficient; and (vi) whether high titers canbe produced.

12.1.3.1 Retroviruses

The original retrovirus used as a vector was the Moloney murine leukemia virus (MMLV). Retroviruses are RNA viruses that work in dividing cells but not in non-dividing cells. The carrying capacity of retroviruses is about 10 kb when fully deleted. Retroviruses are easily produced in large amounts by being secreted from packaging cells and can be concentrated. The retrovirus integrates into a host chromosome and can produce very long-term stable transgene expression. Over time, however, gene silencing does occur. One of the biggest drawbacks is random insertion, which could lead to mutagenesis. For experimental use retroviruses have been widely employed, but for human use they are risky.

12.1.3.2 Lentivirus

Lentiviruses are retroviruses derived from human immundodeficiency virus (HIV) or feline immundodeficiency virus (FIV), simian immundodeficiency virus (SIV) or equine infectious anemia virus (EIAV). Like other retroviruses, they integrate into the host chromosome and are actively taken up by dividing cells and provide long-lasting transgenic expression. They have a large carrying capacity of about 10 kb, and can also enter non-dividing cells. The transduction efficiency is 20–50%. Therefore, lentivirus vectors are suitable for gene delivery in cancer cells and cancer stem cells where the goal is to stop cell division. They have advantages for delivering transgenes that can induce differentiation or apotosis. For producing lentivirus, three components are used: (i) the expressing plasmid, which contains the transgene or shRNA; (ii) a packaging plasmid; and (iii) an envelope plasmid. 293T/293FT cells are normally used for packaging lentivirus via transient transfection. In the development of lentivirus vectors, the first-generation lentiviral vector had three plasmids: (i) the expressing plasmid with *cis*-acting elements of HIV and the gene of interest; (ii) packaging plasmid with all HIV viral genes, except the envelope gene; and (iii) envelope plasmid vesicular stomatitis virus (VSV G). Although the envelope gene is separated from helper plasmid, first-generation lentivirus contains HIV accessory genes in its packaging plasmid and has serious safety concerns. The second-generation lentiviral vector is improved by removing accessory genes from the packaging plasmid, which increases the safety. The common packaging plasmid contains both psPAX2 and pCMV-dR8.2 dvpr with components of gag, rev, cPPT and RRE. The third-generation lentiviral vector is self-inactivating (SIN) vector, which is removed from the enhancer region of 3' U3 of long terminal repeat (LTR), leading to a transcriptionally inactive vector, but offering maximal biosafety. To decrease the chance of retroviral recombination, the packaging plasmid is split into two separate plasmids, one containing rev and the other containing gag-pol. Since it needs the transfection of four different plasmids in 293T/293FT cells, the packaging efficiency is lower than the first- and second-generation systems.

12.1.3.3 Herpes simplex virus (HSV)

HSV is a dsDNA virus with a very large capacity (20–30 kb). Since HSV normally infects neurons its potential use has been mostly studied in neuronal cells. It infects non-dividing cells and does not integrate into the genome. The carrying capacity is attractive for the insertion of large transgenes or complex gene cassettes. However, the safety of HSVs is unknown.

12.1.3.4 Adenovirus

Adenovirus is easy to produce in high titers ($10^{10} - 10^{12}$ pfu/ml), can infect non-dividing cells and has a capacity of 7–8 kb. It is transient because it does not integrate into the genome. It has been a preferred vector for numerous experiments, but it was a disastrous failure in a human trial. The adenovirus is not safe because it produces proteins that cause a graft–host reaction. The immune and inflammatory responses to adenovirus and the proteins it generates was predictable and its use in humans was not only tragic for the loss of a subject's life (17-year-old Jeremy Gelsinger), but also a serious setback to other gene therapies as a novel therapeutic approach.

12.1.3.5 Adeno-associated virus

AAV is not related to adenovirus, despite the name. AAV is a human parvovirus that can infect non-dividing cells. It is actually very common in humans and apparently harmless. AAV is long lasting and stable. The wild type integrates into chromosome 19. Recombinant adeno-associated virus (rAAV) also integrates [25], although each new rAAV needs to be tested for which chromosome it integrates into. When stripped of its gag and pol genes down to the bare inverted terminal repeats, an AAV vector appears to be non-toxic and does not induce host immune responses because it does not produce foreign proteins. Uptake efficiency depends on multiple of infection (MOI, i.e. number of viruses, or DNA particles, per cell) (see Figure 12.1).The main disadvantage for AAV is its low carrying capacity (4.7 kb) .But it can be produced in very high titers (more than 10^{12} pfu/ml). It is being used in clinical trials and has several advantages for gene modification and cell transplantation. Recently, the serotype of AAV has become recognized as an important consideration in gene transfer into particular tissue. Cheng *et al.* [26] have shown that serotype 8 is more efficient for transfecting the pancreas than the standard serotype 2. Others have shown preferential tissue transfection for serotype 5 in the liver [27] and lung [28].

12.1.3.6 Reporter genes

To label cells with an internal marker so that the cells can be identified after transfection with AAV, a reporter gene can be inserted into the AAV. To label cells with an internal marker so that cells transfected with AAV can report activity of a viable cell, fluorescent genes [29–31] can be inserted into the AAV; for example, intracellular green fluorescent protein (GFP) delivered by AAV, as shown in Figure 12.1. Similar effects are seen with Luciferase (Luc) or β-galactosidase (Lac Z). The gene sequence of these reporters can be inserted into any of the vectors described above and mark the uptake and expression of invisible proteins expressed by the transgene delivered by the vector. Each cell marker has its own advantage or disadvantage. An advantage of GFP is that it is visible using a highly sensitive fluoroscope so that the cells can be located, even under the skin in tissues and tumors. Luciferase has the advantage that it is quantifiable using luminometers, dual luciferase assays or relative luciferase gene expression. Lac Z is quantifiable but not visible *in vivo*. At the next level of sophistication, a cell- or tissue-specific promoter is spliced with

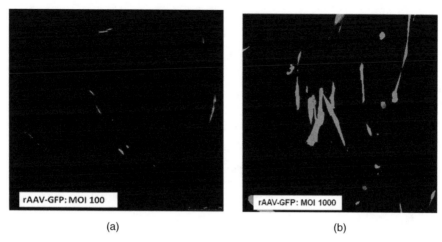

(a)   (b)

Figure 12.1 AAV-GFP transduction. The green fluorescent cells indicate the *gfp* expression ana-lyzed by confocal microscopy ($\times 250$) (Leica Microsystems-TCS SP5). (a) Transduction efficiency was 10% for MOI 100. (b) Transduction efficiency was 40% for MOI 1000. A vector **pUF^{11}** (an AAV-based plasmid vector) consisting GFP (535 bp) was transduced in human fetal fibroblasts (IMR-90). IMR-90 cells were plated in six-well plates (5×10^4 cells/well) and were cultured in minimum essential medium (MEM Engle) supplemented with 10% heat-inactivated fetal bovine serum (FBS), 0.1 mM non-essential amino acids and 1.0 mM sodium pyruvate. The cells were cultured for 24 h at 37 °C, 5% CO_2. When cells were approximately 70–80% confluent, the cells were washed in PBS and transduced with AAV-GFP vector. (See Plate 12.1.)

the selected cell marker transgene so that the transgene can be observed to be expressed in one type of cell.

12.1.4 Production, purification and titration of recombinant adeno-associated virus (rAAV)

The parvovirus AAV has a single-stranded genome of approximately 4.7 kb. rAAV in which the two open reading frames of AAV, designated rep and cap, have been replaced by a gene of interest has become an important tool for gene delivery [32].

The production method described in this chapter was developed by Muzyscka and co-workers [33]. The protocol used is the rAAV plasmid (pTR-UF^3) [33] as an example to describe the steps. The rAAV plasmid (pTR-UF^3) contains the GFP gene under control of cytomegalovirus (CMV) promoter through the polio virus type 1 internal ribosomal entry site (IRES) (Figure 12.2) and the helper plasmid pDG [34] which contains both the AAV genes (rep and cap) for AAV propagation.

Figure 12.2 Plasmid (pTR-UF3) containing a CMV promoter with a GFP reporter gene. The plasmid was developed at the University of Florida. ITR: AAV inverted terminal repeats; AMP^R: Ampicillin as antibiotic; other abbreviations in text.

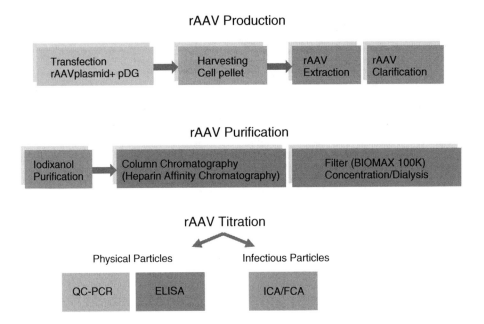

Figure 12.3   Steps for rAAV production, purification and titration.

After harvesting the virus from the HEK 293 cells, the virus is extracted by freezing and thawing the cell and clarified by low-speed centrifugation.

rAAV was purified by an efficient method based on density gradient centrifugation using iodixanol. To purify rAAV further, column chromatography (such as heparin affinity chromatography) is used as a second step in purification following the iodixanol gradient (see below).

An important index of virus quality is the ratio of physical particles (PPs) to infectious particles (IPs) in a given preparation. Quantitative competitive polymerase chain reaction (QC-PCR) and enzyme-linked immunosorbent assay (ELISA) were used for PPs. Infectious titers were determined by infectious center assay (ICA)/fluorescence center assay (FCA). An overview of the procedures for rAAV production, purification and titration is shown in Figure 12.3. The protocols (see Protocols 12.1–12.7) are described in detail individually.

12.2 Methods and approaches

PROTOCOL 12.1   Transfection by $Ca_3(PO_4)_2$ (Calcium Phosphate) for 10 of 15 cm Dishes

Equipment and reagents

- HEK 293 cells[a] (ATCC)

- Dulbecco's modified Eagle's medium (DMEM) (Cellgro)

- FBS

- Antibiotic–antimycotic (100×) (Invitrogen)

- Trypsin–EDTA buffer (Invitrogen)

- Phosphate-buffered saline (PBS)

- 2× HBS[b] (Hepes-buffered saline) 1000 ml: 16.4 g NaCl, 0.74 g KCl, 0.21 g Na_2HPO_4, 2.0 g dextrose, 10.0 g Hepes, titrate to pH 7.05 with 5 M NaOH; filter sterilize through a 0.45 μm nitrocellulose filter and store at −20 °C in 13.5 ml aliquots

- rAAV plasmid (such as pTR-UF3)

- Helper plasmid: pDG

- 2.5 M $CaCl_2$ (1.4 ml aliquot at −20°C)

- 100 × 20 mm TC dishes (WVR)

- 150 × 25 mm TC Dishes (Fisher)

- 50 ml conical tubes

- Pipettes (5, 10 and 25 ml)

- Biological safety cabinet

- CO_2 incubator.

Method

1 Split HEK 293 cells by trypsin–EDTA into 1 : 3 the day before the experiment; check cell for confluency.[c]

2 Thaw one aliquot of 2× HBS and keep at 37 °C until ready to use.

3 Pre-warm 210 ml of complete media (DMEM/5% FBS/1× antibiotic–antimycotic).

4 Prepare transfection mixture:

   (a) Calculate how much input DNA (X ml) to add. Input DNA: 150 μg of rAAV plasmid and 450 μg of pDG in 10 of 15 cm dishes

   (b) In a 50 ml conical tube, add following ingredients in order:

      (11.25 − X) ml of H_2O

      1.25 ml of 2.5 M $CaCl_2$

      X ml of input DNA (rAAV plasmid and pDG)

      12.5 ml of 2× HBS[d]

      Gently mix the transfection mixture well.

5 Discard old media from each dish before adding 2× HBS in the step 4b.

6 Let the mixture (from step 4b) incubate 2–3 min[e](less than 5 min) to form the $Ca_3(PO_4)_2$ precipitate.

7 Transfer the mixture into pre-warmed complete media (210 ml) drop by drop.

8 Dispense 22 ml of media from step 7 into each 15 cm dish immediately.

9 Equalize by forth and back motion.

10 Incubate at 37 °C and 5% CO_2 in an incubator for about 48 h.

Notes

[a] Using low passage number (<P50) of HEK 293 cells.

[b] An exact pH of 2× HBS is extremely important for efficient transfection. The optimal pH range is 7.05–7.12.

[c] About 70–80% of confluency is ideal.

[d] Discard old media before adding 2× HBS.

[e] This is the key step for transfection. Check Ca3(PO4)2 precipitate when you prepare new 2.5 M CaCl2 and 2× HBS.

PROTOCOL 12.2 Harvesting Transfected Cells for 10 of 15 cm Dishes

Equipment and reagents

- Cell scraper
- DMEM (Cellgro)
- FBS
- Antibiotic–antimycotic (100×) (Invitrogen)
- 50 ml conical tubes
- Pipettes (5, 10 and 25 ml)
- Centrifuge
- Biological safety cabinet
- Lysis buffer: 150 mM NaCl and 50 mM Tris (pH 8.4).

Method

1 Take the dishes from the CO_2 incubator and scrape them with cell scrapers to dislodge all cells with media.

2 Collect cells and media in six 50 ml conical tubes.

3 Add 30 ml of fresh media to the dishes (one by one) to rinse remaining calls and distribute the media into the above six 50 ml conical tubes.

4 Centrifuge the conical tubes at 1000 rpm, 4 °C for 10–15 min to pellet the cells.

5 Discard supernatant.

6 Add 2.0 ml of lysis buffer into each 50 ml conical tube, up and down four to five times to re-suspend pellets by 5 ml pipette, combine all of pellets in to one of 50 ml conical tube

7 Transfer 2 ml more lysis buffer to five of the 50 ml conical tubes (one by one to resuspend the remaining pellets and pour all pellets suspension (about 15 ml) in one of the 50 ml conical tubes.

8 Store cell pellets suspension in $-20\,°C$ freezer for extraction and clarification of virus.

PROTOCOL 12.3 Purification of AAV: Iodixanol Purification

Equipment and reagents

- Dry ice
- Ethanol
- Benzonase (Sigma)
- Oak Ridge tubes
- Quick Seal 50 ml ultracentrifuge tubes (Beckman Coulter)
- 12 cm^3 syringes
- 19-GA and 16-GA needles
- 100 µl micro-capillary pipettes (Kimble)
- Parafilm
- Iodixanol 60% (w/v) (Accurate Chemical & Scientific Corp.)
- 5 M NaCl
- 1× PBS-MK: 1× PBS, 1 mM $MgCl_2$ and 2.5 mM KCl
- Phenol red (Sigma)
- Lysis buffer: 150 mM NaCl and 50 mM Tris (pH 8.4)
- Heat sealer
- 70 Ti/70.1 Ti rotor (Beckman Coulter)
- Optima L-90K ultracentrifuge (Beckman Coulter).

Method

1 Extract virus from the cells by freezing (dry ice–ethanol) and thawing (at 37°C) at least three times.[f]

2 Add benzonase into about 15 ml cell lysate from harvesting (50 units/ml) and incubate at 37°C for 30 min.

3 Transfer the 15 ml lysates to one Oak Ringe tube and centrifuge at 400 rpm (2000g) and 4°C for 20 min.

4 Prepare one 50 ml ultracentrifuge tube per 15 ml lysates for iodixanol gradients through a 12 cm^3 syringe connected to a 19-GA needle and 100 μl micro-capillary pipettes sealed by parafilm as follows:

(a) Pour the supernatant from step 3 into the 50 ml ultracentrifuge tube.

(b) Begin building the iodixanol gradient in the following order:[g]

~15 ml cell lysate from step (a)

7.5 ml of of 15% iodoxanol

7.5 ml of 40% iodixanol

5.0 ml of 60% iodixanol

	15%[h]	25%	40%	60%
Iodixanol 60% (ml)	12.50	20.84	33.25	50
1× PBS-MK (ml)	27.50	29.16	16.75	0
5 M NaCl (ml)	10.0	0	0	
Phenol red (μl)	0	100	0	125

5 Clean top and remove liquid in the neck of the ultracentrifuge tube by Kimwipes.

6 Add lysis buffer into the top of the ultracentrifuge tube to balance the tubes.[i]

7 Seal the tubes with heat sealer.[j]

8 Centrifuge at 18 °C, 64 000g (69 000 rpm)in Optima L-90K ultracentrifuge with 70Ti rotor.

9 Using a 16-GA needle connected to a 12 cm^3 syringe, carefully remove the virus-containing fractions (about 4 ml of 40% iodixanol and 3 ml of 60% iodixanol)[k] into a 15 ml tube and store at −20 °C for further purification by column chromatography.

Notes

[f] Each time: freezing for 10 min and thawing for 15 min, shake vigorously.

[g] Mix the iodixanol fractions thoroughly before adding.

[h] The 15% iodixanol fraction contains 1 M NaCl to destabilize ionic interaction between macromolecules.

[i] Add lysis buffer slowly and carefully; do not interfere with the gradients.

[j] To make sure the tubes are sealed completely without leaking.

[k] Do not take any of inter-phase between 40% iodixanol and 25% iodixanol.

PROTOCOL 12.4   Column Chromatography (Heparin Column Chromatography)

Equipment and reagents

- Econo-Pac chromatography columns (Bio-Rad)

- Heparin–agarose, type I (Sigma)

- 5 M NaCl

- 1× PBS-MK: 1× PBS, 1 mM $MgCl_2$ and 2.5 mM KCl

- Pipettes (10 and 25 ml)

- Ultrafree centrifugal filter devices (Millipore: BIOMAX 100 K NMWL membrane 15 ml vol)

- PBS

- Lactated Ringer's solution

- Centrifuge.

Method

1 Take a new 20 ml Econo-Pac chromatography columns, set up stand and place the column in a clamp.

2 Add 5 ml slurry[l] of heparin–agarose, type I, into the column to drop the liquid by gravity flow.

3 When the last drop is left in the columns, place a filter above the heparin,[m] forming about 2.5 ml bed of heparin column.

4 Pre-equilibrate the heparin column with 20 ml of 1× PBS-MK by gravity flow.

5 Pre-elute the heparin column with 10 ml of 1× PBS-MK/1 M NaCl by gravity flow.

6 Equilibrate the heparin column with 25 ml of 1× PBS-MK under gravity twice.

7 Load the virus-containing fractions (around 7 ml up to 28 ml) from purification by the iodixanol step gradient into the heparin column.

8 Allow the virus-containing fraction through the heparin column under gravity; discard it because AAV is reabsorbed onto column.

9 Wash the heparin column with 25 ml of 1× PBS-MK under gravity twice.

10 Put an Ultrafree centrifugal filter device (BIOMAX 100 K) under the heparin column.

11 Elute the virus into the BIOMAX 100 K filter with 7 ml of 1× PBS-MK/1 M NaCl.

12 Centrifuge the eluted virus at maximum 2000g at 10 °C until the volume is reduced to 300–500 μl.

13 Add 5 ml of 1× PBS (or 5 ml of Lactated Ringer's solution) into the filter, centrifuge at 2000g at 10 °C until the volume is around 300–500 μl.

14 Repeat step 13 two more times and transfer the concentrated virus (300–500 μl) to a 1.5 ml tube.

15 Store the concentrated virus at 4 °C for characterization of the purified rAAV.[n]

Notes

[l]Shake heparin–agarose vigorously and add it into the column immediately.

[m]Use flat end of a 25 ml sterile pipette to push the filter into the column until touching the heparin–agarose without any air bubbles.

[n]Store the concentrated virus at 4 °C for short periods and store at −80 °C in small aliquots for the long term.

PROTOCOL 12.5   Titration of AAV: Determining rAAV Physical Particles by QC-PCR[o]

Equipment and reagents

- 10× DNase buffer: 500 mM Tris pH 7.5 and 100 mM $MgCl_2$

- DNase I

- 10× proteinase K buffer: 100 mM Tris pH 8.0, 100 mM EDTA and 10% SDS

- Proteinase K (20 mg/ml)

- 3 M NaAC pH 7.0

- Ethanol (absolute – 200 proof)

- Blue dextran (Sigma): 20 mg/ml

- Internal standard control (ISC)[p]

- PCR reagents: 10× PCR buffer, 50 mM $MgCl_2$, 10 mM dNTPs, 10 μM template forward primer, 10 μM template reverse primer and Taq polymerase

- Nuclease-free water

- Water bath

- Thermal cycler (iCycler, Bio-Bad)

- Epi Chemi II darkroom (UVP).

Method

1 DNase I digestion: 5 μl of purified rAAV virus, 10 μl of 10× DNase buffer with 10 U of DNAse I in 100 μl of reaction mixture, incubate at 37 °C for 1 h.

2 At the end of the incubation, add 11.2 μl of 10× proteinase K buffer and 1 μl of proteinase K (20 mg/ml).

3 Incubate the mixture at 42 °C for 1 h.

4 In the mixture, add 13 μl of 3 M NaAC (pH 7.0), 280 μl of ethanol (from −20 °C) and 0.5 μl of blue dextran^q (20 mg/ml).

5 Incubate at −20 °C at least for 30 min.

6 Centrifuge at >12 000 rpm, 4 °C for 20 min; discard the supernatant.

7 Wash the pellet with 70% ethanol (from −20 °C); discard the supernatant.

8 Dry the pellet in air.

9 Dissolve the pellet in 10 μl of nuclease-free water.

10 Make a serial dilution of the pellet (from purified rAAV) (mostly dilute by 1 : 10 and 1 : 100) for subsequent QC-PCR.

11 Each test sample (purified virus) normally to set four or eight reactions^r as follows:

	Internal standard control	Purified virus (unknown PP)
(1)	1 μl (200 pg/μl)	1 μl (1 : 100 dilution)
(2)	1 μl (100 pg/μl)	1 μl (1 : 100 dilution)
(3)	1 μl (20 pg/μl)	1 μl (1 : 100 dilution)
(4)	1 μl (10 pg/μl)	1 μl (1 : 100 dilution)
(5)	1 μl (200 pg/μl)	1 μl (1 : 10 dilution)
(6)	1 μl (100 pg/μl)	1 μl (1 : 10 dilution)
(7)	1 μl (20 pg/μl)	1 μl (1 : 10 dilution)
(8)	1 μl (10 pg/μl)	1 μl (1 : 10 dilution)

12 Add three controls: one positive control of full-length gene, a second positive control of ISC and one negative control.

13 PCR reactions mixture: a 25 μl reaction volume contains 1× PCR buffer, 1.5 mM MgCl$_2$, 0.2 μM dNTPs, 0.2 μM of template forward primer, 0.2 μM template reverse primer and 1 unit of Taq polymerase.

14 Amplification: 4 min at 94 °C; 25 cycles of 30 s at 94 °C, 30 s at 53 °C (annealing), 1 min at 72 °C and a final extension period of 10 min at 72 °C in a thermal cycler.

15 Electrophoresis: 10 μl of amplification products (with 2 μl of 6× loading buffer) were analyzed on 1.5% agarose stained with ethidium bromide.

16 The image of the stained gel was taken using an Epi Chemi II darkroom.

17 The densities of the target and competitor bands in each lane were measured using VisionWorks®LS Analysis Software.

Notes

[o]PCR is a very popular experimental method in laboratories now. QC-PCR requires almost the same reagents and equipment. However, the disadvantage of QC-PCR is the need to construct the ISC as a competitor individually according to the different genes (or promoter) in the purified rAAA virus.

[p]The appendix A will briefly describe how to construct an ISC for the rAAV containing GFP gene.

[q]Using 10 µg of blue dextran as a carrier, which does not interfere in PCR reactions.

[r]You could perform the dilution of 1 : 100 first to save time and reagents.

PROTOCOL 12.6   Determining rAAV Physical Particles by ELISA

This is an alternative method for determining PPs to avoid the disadvantage of QC-PCR.

Equipment and reagents

- AAV 2 Titration ELISA Kit (American Research Products, Inc)
- Precision sterile pipettes and tips
- 12-well channel pipette
- Distilled water
- Vials
- Incubator for 37°C
- Microtiter plate spetrophotometer (450 nm).

Method

The titration of rAAV titers is performed using the kit according to the manufacturer's instruction. Following is a list of key steps in the protocol for convenience.

1 Preparation of reagents

  (a) Sample buffer: dilute 20× sample buffer with distilled water to appropriate volume of 1× sample buffer.

  (b) Standard
      samples: reconstitute one vial of the kit control (empty AAV 2 capsids containing defined amount of particles/ml in the label) with 500 µl of distilled water, serial dilutions of 1 : 2, 1 : 4, 1 : 8, 1 : 16, 1 : 32 and 1 : 64 with 1× sample buffer.[s]

  (c) Tested samples: dilute the tested samples with 1× sample buffer to appropriate (estimated) titers located in optimal measurement range.

(d) Wash buffer: dilute 20× wash buffer with distilled water to appropriate volume of 1× wash buffer.

(e) Biotin conjugate (anti-AAV 2): reconstitute the lyophilized biotin conjugate with 750 μl of distilled water, forming 20× biotin conjugate at first time, then dilute 20× biotin conjugate with 1× wash buffer[t] to appropriate volume of 1× biotin conjugate: B*.

(f) Streptavidin peroxidase conjugate: dilute 20× streptavidin peroxidase conjugate with 1× wash buffer[u] to appropriate volume of 1× streptavidin peroxidase conjugate: C*.

(g) Substrate: dilute 20× substrate with distilled water to appropriate volume of 1× substrate: S*.

2 Take appropriate numbers of eight-well microtiter strips coated with mouse monoclonal antibody and place on the microtiter plate.

3 Pipette 100 μl per well of 1× sample buffer (as blank), serial dilution of standard samples and tested samples into the wells of the microtiter strips, seal strips with adhesion foil and incubate at 37 °C for 1 h.

4 Wash the wells with 200 μl of 1× wash buffer/per well (incubate for 5–10 s) twice.

5 Pipette 100 μl per well of 1× biotin conjugate (B*), seal strips with adhesion foil and incubate at 37 °C for 1 h.

6 Repeat washing step in 4.

7 Pipette 100 μl per well of 1× streptavidin peroxidase conjugate (C*), seal strips with adhesion foil and incubate at 37 °C for 1 h.

8 Repeat washing step in 4.

9 Pipette 100 μl per well of 1× substrate (S*) and incubate at room temperature for 10–15 min.

10 Add 100 μl per well of stop solution to stop color reaction.

11 Measure intensity of color reaction in plate reader at 450 nm within 30 min.

12 Calculate the titers of rAAV virus according to the intensities and the standard curve.

Notes

[s]Prepare serial dilutions of standard and diluted tested rAAV samples with 1× sample buffer.

[t]Dilute 20× biotin conjugate into 1× biotin conjugate with 1× wash buffer.

[u]Dilute 20× streptavidin peroxidase conjugate into 1× streptavidin peroxidase conjugate with 1× wash buffer also.

PROTOCOL 12.7 Determining rAAV Infectious Particles by Infectious Center Assay (ICA)/Fluorescent Cell Assay (FCA)

Equipment and reagents

- C-12 cell line (ATCC)
- DMEM (Cellgro)
- FBS
- Geneticin 50 mg/ml (Invitrogen)
- Gentamicin 50 mg/ml (Invitrogen)
- 0.025% trypsin–EDTA
- PBS
- Purified rAAV virus (pTR-UF³ as an example)
- Adenovirus serotype 5 (Ad5)
- T-75 flask
- 96-well plate
- Sterile pipettes and tips
- Sterile 50 ml tubes
- Biological safety cabinet
- CO$_2$ incubator
- Fluorescence microscope (Nikon).

Method

Day 1: Seed C-12 Cell into a 96-Well Plate

1 Take a T-75 flask of C-12 cells; pour off old media.[v]

2 Wash cells with 10 ml of PBS and decant.

3 Add 2 ml of 0.025% trypsin–EDTA to cells; incubate about 2 min.

4 Knock on the T-75 flask until the cells are detached completely.

5 Add 8 ml of fresh media (DMEM/200 µg/ml of geneticin/200 µg/ml of gentamicin/5% FBS) to the trypsinized cells; resuspend the cells by repeatedly pipetting up and down, forming cell suspension.

6 Transfer 0.75 ml of the cell suspension to 10 ml of fresh media in a 50 ml tube.[w]

7 Add 100 µl per well of the cell suspension into a 96-well plate (or into appropriate number of wells of a 96-well plate); incubate at 37 °C, 5% CO$_2$ overnight.

Day 2: Infect Cells

1 Make serial dilutions of purified rAAV virus (pTR-UF³ as an example):

(a) Take a fresh 96-well plate and put 250 µl[x] of fresh media in the first well (A1).

(b) Put 225 µl of fresh media into the wells from A2 to A10.

(c) Add 2.5 µl of purified rAAV virus of pTR-UF^3 into A1 and mix well by pipetting.

(d) Make serial dilutions (from 10^{-2} to10^{-11}) by transferring 25 µl of well-mixed solution beginning from A1 to the next well (A2) and continually from A9 to A10, changing pipette tip after each transfer.

2 In a new tube, add 25 µl of Ad5^y into 975 µl of fresh media; mix well.

3 Take day 1 seed 96-well plate from incubator; decant old media.

4 Transfer 100 µl of serial dilutions (normally select from 10^{-8} to 10^{-11}) to per well (duplicate or triple).

5 Pipette appropriate volume^z of Ad5 (MOI 20) to each well.

6 Pipette 100 µl of fresh media to per well cells (duplicate or triple) and pipette the same volume as in step 5 of Ad5 (MOI 20) to each well.^{aa}.

7 Pipette 100 µl of fresh media to per well (duplicate or triple) without adding Ad5.^{bb}

8 Incubate at 37 °C, 5% CO$_2$ for 48 h.^{cc}

9 Score the cells infected with rAAV-UF3 visually using a fluorescence microscope.

10 Calculate the IP/ml according the score and dilution.^{dd}

Notes

^{v}C-12 cell line contains integrated wild-type AAV rep and cap genes for both the ICA and fluorescent cell assay.

^{w}Transfer 1.5 ml of the cell suspension into a new T-75 flask containing 15 ml of fresh media; incubate at 37 °C, 5% CO2 for further use.

^{x}The accured volume should be put in 247.5 µl fresh media.

^{y}Ad5 is co-infected C-12 along with rAAV. It is titrated using the same C-12 cell line in a serial dilution cytopathic effect (CPE).

^{z}The volume (µl) depends on the titer of Ad5 and the confluency of C-12 cells for MOI 20.

^{aa}A negative control (Ad5).

^{bb}A negative control (medium).

^{cc}The amount of Ad producing well-developed CPE in 48 h on C-12 was used to provide a helper function for both the ICA and FCA.

^{dd}The PP/ml and the IP/ml differ by a factor 2 or less in most cases.

12.3 Troubleshooting

- **The quality of plasmids**: Impurities in the plasmid preparation can be deleterious to transfection efficiency. The plasmids (rAAV plasmids and helper plasmid of pDG), which are purified by CsCl–ethidium bromide equilibrium centrifugation, are good quality to be transfected. The quality of the plasmids, which are purified by QIAGEN Plasmid Midi and Maxi Kits, are acceptable.

- **Low passages (<50) of HEK 293 cells**: The higher titer of AAV would be produced in low passage of HEK 293 cells. It has recently been found [35] that the tumorigenicity of the HEK 293 cell line reached 100% when the passage exceeded 65, whereas using low-passage (<52) HEK 293 cell line no tumor could be induced under the same condition. Thus, more attention should be paid to the passage level (<50) of the HEK 293 cell line.

- **Appropriate pH of 2× HBS**: The optimum pH range for high transfection efficiency is extremely narrow (between 7.05 and 7.12). The pH meter used to measure the pH of 2× HBS should be calibrated several times until very accurate. The pH of 2× HBS could change during thawing and storage. The solution should be stored at −20 °C in small aliquots. The solution should be made fresh if transfection has not worked well.

- **Precipitates of Ca$_3$(PO4)$_2$–DNA complexes**: Highly effective precipitates for transfection purposes can be generated only in a very narrow range of physico-chemical conditions that control the initiation and growth of the Ca$_3$(PO$_4$)$_2$–DNA complexes. The pH value and concentrations of phosphate (HBS solution) and calcium (CaCl$_2$ solution) are the main factors influencing the characteristics of the Ca$_3$(PO$_4$)$_2$–DNA complexes. It is necessary to check Ca$_3$(PO$_4$)$_2$ precipitates when you prepare new 2.5 M CaCl$_2$ and 2× HBS. Also, the reaction time of the precipitate complex is very important. Theoretically, the standing time can be optimized for each set of parameters. It has been found in most cases that almost all of the soluble DNA in the reaction mix can be bound into an insoluble complex with Ca$_3$(PO$_4$)$_2$ in 1–3 min. Extending the reaction time to more than 5 min results in reducing the level of expression.

- **Discontinuous iodixanol gradients**: It is important the first iodixanol step (15%) contains 1 M NaCl to destabilize ionic interactions between macromolecules. Without 1 M NaCl in 15% iodixanol the gradients would distribute the vector along the whole length of the gradient, making separation of virus unsuccessful. High salt needs to be excluded from the rest of the gradient in order to band the virus under iso-osmotic conditions, allowing direct loading in subsequent chromatographic steps.

References

1. Smithies, O. (2005) Many little things: one geneticist's view of complex diseases. *Nature Reviews Genetics*, **6**, 419–425. Oliver Smithies Nobel Prize 2007.

2. Gu, H., Marth, J.D., Orban, P.C. *et al.* (1994) Deletion of a DNA polymerase beta gene segment in T cells using cell type-specific gene targeting. *Science*, **265**, 103–106.

3. Sakai, K., Agassandian, K., Morimoto, S. *et al.* (2007) Local production of angiotensin II in the subfornical organ causes elevated drinking. *Journal of Clinical Investigation*, **117**, 1088–1095.

4. Sinnayah, P., Lindley, T.E., Staber, P.D. *et al.* (2004) Targeted viral delivery of Cre recombinase induces conditional gene deletion in cardiovascular circuits of the mouse brain. *Physiological Genomics*, **18**, 25–32.

5. Phillips, M.I. and Sumners, C. (1998) Angiotensin II in central nervous system physiology. *Regulatory Peptides*, **78**, 1–11.

6. Phillips, M.I. (2004) A Cre–loxP solution for defining the brain rennin–angiotensin system. Focus on 'Targeted viral delivery of Cre recombinase induces conditional gene deletion in cardiovascular circuits of the mouse brain'. *Physiological Genomics*, **18**, 1–3.

7. Zamecnik, P.C. and Stephenson, M.L. (1978) Inhibition of Rous sarcoma virus replication and cell transformation by a specific oligodeoxynucleotide. *Proceedings of the National Academy of Sciences of the United States of America*, **75**, 280–284.

8. Wahlestedt, C., Pich, E.M., Koob, G.F. *et al.* (1993) Modulation of anxiety and neuropeptide Y–Y1 receptors by antisense oligodeoxynucleotides. *Science*, **259**, 528–531.

9. Gyurko, R., Wielbo, D. and Phillips, M.I. (1993) Antisense inhibition of AT1 receptor mRNA and angiotensinogen mRNA in the brain of spontaneously hypertensive rats reduces hypertension of neurogenic origin. *Regulatory Peptides*, **49**, 167–174.

10. Kimura, B., Mohuczy, D., Tang, X. and Phillips, M.I. (2001) Attenuation of hypertension and heart hypertrophy by adeno-associated virus delivering angiotensinogen antisense. *Hypertension*, **37**, 376–380.

11. Crooke, S.T. (2004) Progress in antisense technology. *Annual Review of Medicine*, **55**, 61–95.

12. Fire, A., Xu, S., Montgomery, M.K. *et al.* (1998) Potent and specific genetic interference by double-stranded RNA in *Caenorhabditis elegans*. *Nature*, **391**, 806–811. Andrew Fire and Craig Mello Nobel prize 2006.

13. Arnold, A.S., Tang, Y.L., Qian, K. *et al.* (2007) Specific β_1-adrenergic receptor silencing with small interfering RNA lowers high blood pressure and improves cardiac function in myocardial ischemia. *Journal of Hypertension*, **25**, 197–205.

14. Lee, R.C., Feinbaum, R.L. and Ambros, V. (1993) The *C. elegans* heterochronic gene lin-4 encodes small RNAs with antisense complementarity to lin-14. *Cell*, **75**, 843–854.

15. Esquela-Kerscher, A. and Slack, F.J. (2006) Oncomirs – microRNAs with a role in cancer. *Nature Reviews Cancer*, **6**, 259–269.

16. Lagos-Quintana, M., Rauhut, R.,Yalcin, A. *et al.* (2002) Identification of tissue-specific microRNAs from mouse. *Current Biology*, **12**, 735–739.

17. Kumar, M.S., Lu, J., Mercer, K.L. *et al.* (2007) Impaired microRNA processing enhances cellular transformation and tumorigenesis. *Nature Genetics*, **39**, 673–677.

18. Wolff, J.A. and Budker, V. (2005) The mechanism of naked DNA uptake and expression. *Advances in Genetics*, **54**, 3–20.

19. Neumann, E., Schaefer-Ridder, M., Wang, Y. and Hof-Schneider, P.H. (1982) Gene transfer into mouse lyoma cells by electroporation in high electric fields. *EMBO Journal*, **1**, 841–845.

20. Okino, M., Marumoto, M., Kanesada, H. *et al.* (1987) Electrical impulse chemotherapy for rat solid tumors. *Japanese Journal of Cancer Research*, **46**, 420.

21. Heller, L.C. and Heller, R. (2006) *In vivo* electroporation for gene therapy. *Human Gene Therapy*, **17** 890–897.

22. Bromberg, J.S., Boros, P., Ding, Y. *et al.* (2002) Gene transfer methods for transplantation. *Methods in Enzymology*, **346**, 199–224.

23. Wang, Y., Boros, P., Liu, J. *et al.* (2000) DNA/dendrimer complexes mediate gene transfer into murine cardiac transplants *ex vivo*. *Molecular Therapy*, **2**, 602–607.

24. Phillips, M.I. (ed.) (2002) *Gene Therapy Methods*, vol. **346**, Academic Press, New York, pp. 1–728. Useful collection of gene therapy methods.

25. Wu, P., Phillips, M.I., Bui, J. and Terwilliger, E.F. (1998) Adeno-associated virus vector-mediated transgene integration into neurons and other nondividing cell targets. *Journal of Virology*, **72**, 5919–5926. Addresses the question of integration of AAV.

26. Cheng, H., Wolfe, S.H., Valencia, V. *et al.* (2007) Efficient and persistent transduction of exocrine and endocrine pancreas by adeno-associated virus type 8. *Journal of Biomedical Science*, **14** (5), 585–594.

27. Mingozzi, F., Schuttrumpf, J., Arrunda, V.R. *et al.* (2002) Improved hepatic gene transfer by using an adeno-associated virus serotype 5 vector. *Journal of Virology*, **76**, 10497–10502.

28. Zabner, J., Seiler, M., Walters, R. *et al.* (2000) Adeno-associated virus type 5 (AAV5) but not AAV2 binds to the apical surfaces of airway epithelia and facilitates gene transfer. *Journal of Virology*, **74**, 3852–3858.

29. Shimomura, O. (1991) Preparation and handling of aequorin solutions for the measurement of cellular Ca^{2+}. *Cell Calcium*, **12** (9), 635–643. Preparation and handling of aequorin solutions for the measurement of cellular Ca^{2+}, Osamu Shimomura, Nobel Prize 2008.

30. Chalfie, M., Tu, Y., Euskirchen, G. *et al.* (1994) Green fluorescent protein as a marker for gene expression. *Science*, **263**, 802–805. Martin Chalfie Nobel Prize 2008 (but unfortunately, not the deserving DC Prasher).

31. Heim, R., Prasher, D.C. and Tsien, R.Y. (1994) Wavelength mutations and posttranslational autoxidation of green fluorescent protein. *Proceedings of the National Academy of Sciences of the United States of America*, **91**, 12501–12504. Roger Y. Tsien Nobel Prize 2008.

32. Wu, Z., Asokan, A. and Samulski, R.J. (2006) Adeno-associated virus serotypes: vector toolkit for human gene therapy. *Molecular Therapy*, **14**, 316–327.

33. Zolotukhin, S., Byrne, B.J., Mason, E. *et al.* (1999) Recombinant adeno-associated virus purification using novel methods improves infectious titer and yield. *Gene Therapy*, **6**, 973–985. Originators of AAV vector production method.

34. Grimm, D., Kern, A., Rittner, K., and Kleinschmidt, J. (1998). Novel tools for production and purification of recombinant adenoassociated virus vectors. *Human Gene Therapy*, **9**, 2745–2760.

35. Shen, C., Gu, M., Song, C. *et al.* (2008) The tumorigenicity diversification in human embryonic kidney 293 cell line cultured *in vitro*. *Biologicals*, **4**, 213–276.

13

An Introduction to Proteomics Technologies for the Genomics Scientist

David B. Friedman
Vanderbilt University School of Medicine, Nashville, Tennessee, USA

13.1 Introduction

Several fields of study, most notably mass spectrometry and protein/peptide separations, contribute to what is collectively referred to as proteomics today. A vital aspect that ties both of these technology-oriented approaches together is the richness of protein sequence databases that are available from a wide variety of organisms. Importantly, these protein sequence databases are inferred from genomic sequence databases, which flourished in the 1990s as a direct result of the human genome project in the mid 1980s. This chapter is designed to introduce the wide spectrum of proteomics technologies to researchers experienced in genomics technologies, with a major focus on two-dimensional (2D) gel-based ('difference gel electrophoresis' (DIGE)), liquid chromatography coupled with tandem mass spectrometry (LC/MS/MS)-based ('shotgun') and mass spectrometry (MS)-based (matrix-assisted laser desorption/ionization (MALDI)-imaging) technologies.

The term 'proteomics' was perhaps first coined at the Lorne protein chemistry meeting in Australia in 1994 and started to appear in publications circa 1995 (e.g. see Ref. [1]). Proteomics in modern practice utilizes proteomics technologies in a variety of ways that are dictated by the experimental design. For example, many experiments are focused at mining proteomes or sub-proteomes (sub-cellular fractionation) to produce a detailed list of protein constituents. Many other proteomic strategies are designed to analyze dynamic protein expression in response to experimental perturbation. This includes changes in protein

---

Genomics: Essential Methods   Edited by Mike Starkey and Ramnath Elaswarapu
© 2011 John Wiley & Sons, Ltd.

expression levels and in post-translational alterations, including biologically significant pro-teolysis and modification (via phosphorylation, acetylation and glycosylation, among many others) that are completely hidden in the static DNA code.

Many proteomic strategies contain protein identification as a major component, and this is routinely performed using MS followed by a statistical comparison of the mass spectral data with theoretical data generated from protein sequence databases that themselves are mostly generated from the genomic sequences. In some cases, quantification of protein abundance changes can also be made using MS (described in more detail in Sections 13.2.2 and 13.2.3). For identification purposes, MS is used to produce data characteristic of individual proteins, usually at the level of amino acid sequence or peptides that are generated after digestion with a site-specific protease, and sometimes include accurate measurements of the intact protein mass. Powerful bioinformatics algorithms can then be applied to search ever-expanding databases for proteins that match these experimentally derived mass spectral signatures [2–9].

The implementation of new ionization sources for MS in the mid 1980s allowed for dramatic increases in sensitivity, resolution and mass accuracy for peptide and protein MS. These ionization sources are electrospray ionization (ESI) and MALDI, both of which can be coupled to a wide variety of mass analyzers. Many instrument configurations are avail-able that have complementary and overlapping capabilities, and it is easy for a novice to get lost in the technical jargon. Typical configurations for proteomics experiments include ESI sources coupled with a variety of medium–high-resolution mass analyzers, including linear quadrupoles, time-of-flight (TOF), quadrupolar ion traps (three-dimensional or linear ion traps) and ultra-high resolution orbitrap and Fourier-transform ion cyclotron resonance (FTICR) mass analyzers. MALDI sources can also be coupled to this variety of mass analyzers, but are more typically in line with TOF mass analyzers.

Strategies for protein characterization using MS come in two basic forms, colloqui-ally termed a 'top-down' and a 'bottom-up' approach. Top-down strategies are named so because they begin with the acquisition of the intact mass of the protein, typically using a high-resolution instrument such as a MALDI-TOF or ESI-FTICR mass spectrometer, fol-lowed by subsequent fragmentation analysis. This approach is considerably difficult to adapt to a global-scale discovery-phase proteomics experiment, but can be extremely effective for targeted studies. The bottom-up approach begins by digesting the target protein(s) or pro-teome into much smaller fragments. This approach works because the peptides are very easy to acquire mass spectral information on, and the resulting peptide masses and fragmentation patterns can be extremely predictive for protein identification using software search tools.

In a typical bottom-up protein identification strategy, individual proteins are first digested with a site-specific protease (such as trypsin) to produce a discrete set of peptides that can be collectively mass analyzed. This is easily performed on biochemically- or gel-resolved proteins. The collection of peptide ion masses, which can be measured under conditions of high mass accuracy (less than 10 ppm using MALDI-TOF MS), can be used to interrogate protein databases directly for statistically significant candidate protein matches, a process referred to as peptide mass mapping or fingerprinting [6–9]. In many cases, peptide mass mapping is more than sufficient to generate unambiguous matches.

Additional mass spectral data can be generated from individual peptide ions from the mix-ture to reveal information about the amino acid sequence. In such a tandem MS experiment, a fragmentation step is incorporated where individual ions, selected in a first-stage mass analyzer, are now fragmented about the amide bonds (between the amino acids), and the fragment ions are subsequently mass analyzed. As ions are fragmented about once or twice per molecule, the overall fragmentation pattern is indicative of the amino acid sequence of the selected peptide ion, and this pattern can be correlated with predicted fragmentation patterns from theoretical peptide digests from selected protein databases to produce statistically significant candidate protein identifications. Database search algorithms, such as Sequest and Mascot, are designed to interrogate databases with either peptide mass map and/or fragmentation data ([10], and www.matrixscience.com), and most modern ESI and MALDI instruments are capable of performing tandem MS experiments.

The peptide mass mapping strategy for resolved proteins is quick and reliable (especially fast using MALDI-TOF), but the power of this approach falls off when the collection of peptides is derived from three or more proteins. For more complex protein mixtures and more global-scale analyses, tandem MS for peptide fragmentation analysis proves to be the most sensitive detection analysis for MS, and is the method of choice for acquiring mass spectral data from sub-proteomes, immuno-affinity captured-complexes and even from unfractionated proteomes. This approach, colloquially termed 'shotgun' proteomic analysis, is described further in Section 13.2.3.

Since protein identification by MS is reliant on pattern matching to protein sequences present in databases, both the peptide mass mapping and the peptide fragmentation approach fail when the protein sequences are not present in the selected database. For these and other situations where identification is ambiguous, fragmentation data from tandem MS can sometimes be used to identify homologous regions in other proteins or to identify sequences that are only represented in expressed sequence tag (EST) databases. In extreme cases, fragmentation data can be interpreted *de novo* and used to direct the design of degenerate oligonucleotides for cloning purposes.

13.2 Methods and approaches

13.2.1 Gel-based strategies

Sodium dodecyl sulfate polyacrylamide gel electrophoresis (SDS-PAGE), first described by Laemmli [11], has long been the method of choice for resolving intact proteins (based on apparent molecular mass, commonly referred to as molecular weight, MW) for a variety of biochemical analyses. For more complex mixtures, 2D gel electrophoresis (2DE) is typically the method of choice to resolve intact protein species using two orthoganol separations, the first based on charge (isoelectric point, pI) and the second by apparent molecular mass. The 2D gel experiments are particularly powerful for visualizing protein isoforms that result from charged post-translational modification, such as phosphorylation and sulfation (which add charge) or acetylation (which neutralizes charge). They are also useful in detecting splice variants and proteolytic cleavages that result in protein species with altered MW and pI.

First introduced by O'Farrell and coworkers in 1975 [12], modern 2D gel technology makes use of first-dimension isoelectric focusing through highly reproducible immobilized pH gradient strips (IPGs) that are commercially available from a number of vendors [13, 14]. Proteins resolved by SDS-PAGE are often directly amenable to bottom-up protein identification strategies after gel-excision and digestion of the target protein into peptides directly within the gel slice. Although the 2D gel approach has historically been used for cataloguing experiments, more recently it has been employed mostly for differential expression studies on a global scale, where proteomes are compared between multiple experimental conditions. For these comparative approaches, replicate gels are required to ensure that changes are biologically significant and not due to experimental/sample preparation variation and analytical gel-to-gel variation.

However, until recently, these strategies lacked the ability to easily quantify abundance changes due mainly to the inability to directly correlate migration patterns and protein staining between gel separations (analytical gel-to-gel variation). Stable isotopes have been used in gel-based proteomics, whereby different proteomes have been separately labeled with different stable isotopes (e.g. growing cells using ^{14}N- versus ^{15}N-labeled medium) prior to mixing and running together through the same 2DE separation [15]. In this case, abundance changes are monitored during the MS stage, which must be performed on each resolved protein, and are typically limited to comparisons made within a single gel.

DIGE technology has more recently been used for direct quantification of abundance changes on a global scale without interference from gel-to-gel variation. This is done using spectrally resolvable MW and charge-matched fluorescent dyes (Cy2, Cy3 and Cy5) to pre-label protein samples which are then multiplexed onto 2D gels. These dyes offer sub-nanogram detection limits and a linear dynamic range of circa four orders of magnitude [16, 17], and when used with an internal standard methodology, enable facile comparison of global expression changes between multiple experimental conditions [18–22], with each condition represented by independent replicates to provide high statistical power [23].

The DIGE approach is most beneficial when a Cy2-labeled internal standard (comprised of a mixture of all samples in the experiment) is co-resolved on a series of gels that each contains individual samples labeled with Cy3 or Cy5. Because these individual samples are multiplexed with an equal aliquot of the same Cy2-standard mixture, each resolved feature can be directly related to the cognate feature in the Cy2 standard mixture within that gel. These intra-gel ratios can then be normalized to all other ratios for that feature from the other samples across the other gels in the experiment with extremely low technical (analytical) noise and high statistical power [23–25] (Figure 13.1). This approach is also directly amenable to multivariate statistical analyses such as principle component analysis and hierarchical clustering, which can be extremely beneficial in visualizing the variation within a set of experimental samples, to help determine if the major source of variation is describing the biology or indicating unanticipated variation between samples (or introduced during sample preparation), in addition to pinpointing subsets of proteins that respond collectively to an experimental stimulus or classification [19–22, 26].

13.2.1.1 Gel-based strategies: Procedure and applications

- Resolution of intact protein species based on size (MW) and charge (pI).

- Resolves intact protein species, including charged isoforms, splice-variants and proteolytic products.

- Proteins identified directly from gel via in-gel digestion and MS.

- DIGE for quantitative studies on intact proteins with low technical noise and high statistical power.

- DIGE is directly amenable to independent replicates from multiple experimental conditions, as well as multivariate statistical analysis.

- High-resolution (24 cm × 20 cm) separations commercially available, using either low-resolution (pH 3–11), medium-resolution (pH 4–7, 7–11) or high-resolution (e.g. pH 5–6) IPGs for the first-dimension isoelectric focusing.

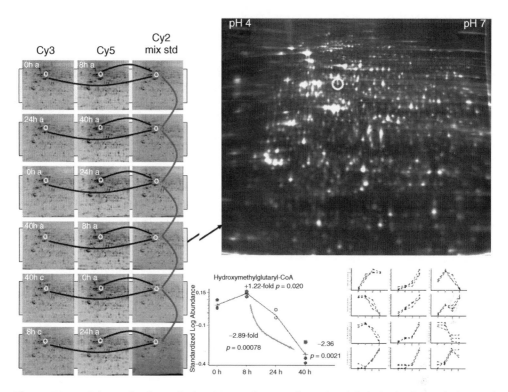

Figure 13.1 Schematic of a typical DIGE experiment using a Cy2-labeled mixed-sample internal standard to coordinate independent replicate samples from multiple conditions into a single analysis. Tri-partite images from six gels are shown at left, with lines indicating Cy3 : Cy2 and Cy5 : Cy2 ratios made within each gel followed by the normalization of ratios between gels as described in the text. The enlarged gel indicates the typical resolution of intact protein forms with isoelectric points between pH 4 and 7. Shown below are examples of protein expression profiles. Adapted from Friedman *et al.* [20].

The strengths of 2D gel proteome analysis lie mainly in the resolution and quantification of intact protein species, including charged post-translational modifications as well as proteolytic forms, and facile protein identification directly from the resolved intact protein by gel excision, in-gel digestion and MS. For quantitative studies, this platform provides the highest statistical power when using the DIGE technology with the mixed-sample internal standard method. Several studies have now been published that address the validation of the DIGE technique with respect to quantitative assessment [27], as well as same/same experiments that demonstrate low technical noise, obviating the need for technical replicates and enabling a statistically powered experiment using a minimal number of independent biological replicates (e.g. twofold changes with $N = 4$ independent samples [23]).

The limitations of 2D gel proteome analysis are almost exclusively tied to the isoelectric focusing step of the 2D gel separation technology, which make it difficult to monitor proteins of extreme pI ($3 < $ pI < 11) or MW ($10 < M_r < 150$ kDa), lower abundance proteins (typically below 20 fmol), and hydrophobic integral membrane proteins. In addition, the overall sensitivity of 2DE is limited by the amount of material accommodated by the 2DE gel and how those proteins are distributed in abundance throughout the proteome. Total protein fluorescent stain detection systems can detect levels as low as 1 ng (e.g. 20 fmol of a 50 kDa protein), which is nearing the lower limit of detection in most mass spectrometers after in-gel digestion [25].

Although some of these limitations can now be addressed using a series of overlapping 2D gel analyses each using a narrow pH range with increased protein loads [13, 14, 20], more sensitive mining of a given proteome or sub-proteome can be afforded by using an LC/MS/MS approach on digested proteins, as discussed in Section 13.2.2.

13.2.2 LC/MS strategies

Liquid chromatography coupled with mass spectrometry (LC/MS)-based approaches offer greater sensitivity than is typically afforded by the protein-staining detection limits from gel-based strategies. With LC/MS-based approaches, protein identification is performed at the level of peptide fragmentation patterns acquired during tandem MS (LC/MS/MS), and which are indicative of the amino acid sequence. In what is commonly referred to as 'shotgun' analysis, proteins from a complex mixture are collectively digested with a site-specific protease (e.g. trypsin). The resulting complex mixture of peptides is separated by high-performance liquid chromatography (HPLC) and introduced directly into a sensitive (low femtomole) tandem mass spectrometer capable of isolating and fragmenting individual peptides 'on the fly' (e.g. see Ref. [2]).

In a typical shotgun LC/MS/MS experiment, individual fragmentation spectra are generated, often in large volumes that can be used to collectively interrogate protein databases to generate lists of candidate protein identifications. This turns out to be a very powerful approach because each component of the starting protein complex can be identified using mass spectral data from only a handful of 'surrogate' peptides derived from that protein [10]. These experiments incorporate reverse-phase high performance liquid chromatography (RP-HPLC) directly upstream, but in-line to a tandem mass spectrometer to separate the individual peptides in time, thereby allowing for tandem MS on an optimal number of individual peptide ions. The pattern of fragment ions (m/z distribution) from each spectrum is then individually compared with theoretical fragmentation patterns generated by database search algorithms to produce statistically significant candidate protein identifications (Figure 13.2).

LC/MS/MS experiments typically utilize ESI sources to generate multiply charged ions for tandem MS, although LC-MALDI/TOF/TOF workflows are also possible. For more complex protein mixtures, an additional ion-exchange HPLC separation is often employed to further pre-fractionate the peptides in what is often termed multidimensional HPLC (MDLC)/tandem MS (LC/LC/MS/MS; MDLC/MS/MS; multidimensional protein identification technology (MudPIT) [2]; DALPC [3]). This approach is particularly powerful at analyzing/identifying defined protein subsets, such as components of organelles [28], multi-protein complexes [3] or co-immunoprecipitations [29, 30], and it has even been applied to whole proteomes [31, 32].

The LC/MS/MS strategy can also be performed quantitatively. This is traditionally done using stable isotope labeling, where two or more samples can be individually labeled and multiplexed into the same analytical run. This is similar to what was explained in Section 13.2.1 for DIGE using 2D gels to remove analytical variation from a given run (but does not currently include an internal standard for normalization). The first commercially available technology for this was termed isotope-coded affinity tag (ICAT) [33], where proteins were labeled on cysteine sulfhydryls using different stable isotope reagents (and included a

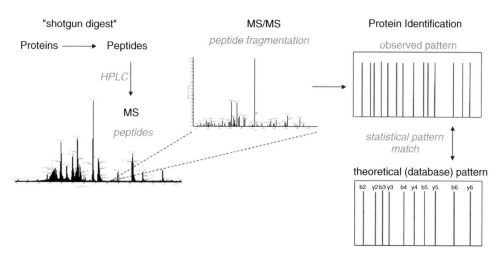

Figure 13.2 Schematic of a typical LC/MS/MS 'shotgun' experiment. Proteins are digested into peptides using a site-specific protease and the resulting peptides are resolved by HPLC directly in line with tandem MS. Individual fragmentation patterns from tandem MS spectra are then searched against theoretical patterns generated by database search algorithms to produce statistically significant candidate protein identifications.

cysteine-containing peptide enrichment step using a biotin−streptavidin interaction). With this approach, quantification is performed on the basis of the ion intensity of the intact peptides, which can be obscured by co-eluting peptides and affected by low signal noise.

Another quantitative strategy involves the use of similar stable isotope tags that contain isobaric reporter ions that are indistinguishable in the mass spectrometer until a selected peptide is fragmented during tandem MS. Termed isobaric tag for relative and absolute quantitation (iTRAQ) [34] or tandem mass tags [35], differentially labeled samples can be co-resolved in a single LC/MS/MS analysis, and the relative quantification can be derived from the ion intensities of the reporter fragmentation ions (since they are indistinguishable in the intact peptide, they are all selected for simultaneous fragmentation during MS/MS). More recently, the LC/MS/MS field has seen an increase in the use of so-called 'label-free' approaches, where individual samples are analyzed by LC/MS/MS separately, and relative protein abundance is related between samples using the peptide MS/MS fragmentation spectra that give rise to statistically significant protein identifications (termed 'spectral counting') [36, 37]. Whereas these approaches are suitable for global discovery-phase experiments, selected single-reaction monitoring/multiple-reaction monitoring (SRM/MRM) is another label-free LC/MS/MS technique, whereby a collection of previously selected peptides can be targeted for subsequent (or validation) quantitative analyses [38, 39].

13.2.2.1 Multidimensional LC/MS/MS strategies: Procedure and applications

- Resolution of peptides derived from proteolytic cleavage.

- Based on peptide-fragmentation patterns from MS data followed by database searching.

- Involves RP-HPLC separations directly in-line with tandem MS of peptides derived from proteins via controlled proteolytic cleavage.

- Can be coupled with a second chromatographic separation (e.g. strong cation exchange) for MudPIT for very complex mixtures with high sensitivity and dynamic range.

- Isotope-labeled protein/peptide tags (e.g. ICAT, iTRAQ) or label-free spectral counting techniques for quantitative analysis.

As LC/MS-based strategies are performed on peptides derived from intact proteins, this technology generally delivers the most sensitive survey of proteins present in a sample owing to the fact that peptides are generally homogeneous in their physicochemical properties and can be resolved in time using HPLC, and often only a handful of peptides may be necessary to identify a protein unambiguously. Furthermore, the additional resolving power of multiple HPLC separations during MudPIT experiments affords excellent sensitivity and dynamic range [2, 3].

The limitations of LC/MS-based technologies are in the ability to easily distinguish between post-translationally modified isoforms and/or proteolytic cleavage products, as well as lower statistical power for global/discovery quantitative studies. This is mainly due to the fact that the analysis is performed on the peptide products of the starting mixture *en masse* without any measurement of the intact proteins, such that isoforms remain unresolved and proteolytic products are masked by the intact species if both are present. The lower statistical power stems mainly from the fact that the mass spectrometer is sampling ions for tandem MS based on relative signal intensities that can vary from run to run [40, 41], requiring many technical replicates for each independent sample to attain acceptable levels of statistical power.

13.2.3 MALDI imaging and profiling

A main feature of MALDI imaging mass spectrometry (IMS) is that it is able to map the spatial distribution of proteins within intact tissue specimens and produce molecular images of specific protein species. This is in contrast to the 2D-gel based and LC/MS/MS-based technologies that are performed on protein homogenates, where this spatial information is not retained. With MALDI-IMS, samples such as thin sections cut from fresh-frozen tissue biopsies [42–44], cells procured via laser capture microdissection [45] and fluids [46] can be analyzed directly in the mass spectrometer (Figure 13.3). As is the case for the 2D-gel technologies, MALDI-IMS also quantifies expression at the level of intact proteins (where modified and proteolytic forms are resolved), and is directly amenable to complex experimental designs including independent replicates from multiple experimental conditions, as well as multivariate statistical analysis. This process can resolve and analyze intact molecular signals from a single MS acquisition within a tissue section (e.g. as small as 25–50 μm laser diameter in the MALDI source) [47].

In most cases, cryostatically derived tissue sections are mounted onto a MALDI target plate, prepared for MS on-target and introduced directly into the ion source (reviewed in Refs [48, 49]). This type of experiment can be performed in low-resolution 'profiling' mode where the tissue of interest is interrogated in multiple discrete locations (often directed by histology [50]), or in high-resolution 'imaging' mode whereby MALDI-TOF spectra are acquired at every position along a rastered $X-Y$ array across the tissue. In such an imaging experiment, individual mass/charge (m/z) values (each representing a discrete intact molecular species) can be tracked throughout a tissue section where relative ion signal is proportional to protein expression. This generates an ion density map for each resolved m/z species in the analysis. The resulting molecular profiles and high-resolution images can be compared within and

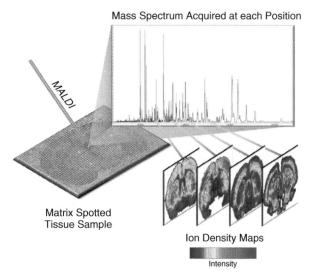

Figure 13.3 Schematic of a MALDI-IMS experiment. Intact tissue (cryostat thin-section) is mounted on a MALDI target and overlaid with matrix. Individual MALDI-MS spectra are acquired at individual positions across the tissue at high resolution. The enlarged MS spectrum contains singly charged molecular ions that can be mapped across the tissue by intensity. The intensity distribution for four individual species is shown below. Image courtesy of R.M. Caprioli and R. Groseclose.

between tissue sections, and statistically-significant signatures indicative of disease can be identified [51, 52]. This approach has been particularly useful to investigate prognostic and diagnostic markers for a variety of cancers [51, 53–58]. Proteins that give rise to the molecular signals of interest can be biochemically isolated from the intact tissue, enabling identification using the standard peptide-based protein identification strategies described above.

Another commercially available MALDI-based profiling platform, surface-enhanced laser desorption ionization (SELDI), is used for serum and other biofluid analysis [59–61]. The main feature of SELDI is several different proprietary sample preparation methods that simplify/enrich the protein mixture based on different physicochemical parameters (i.e. hydrophobicity, ionic interaction, metal affinity [59–61]). This approach continues to be employed in many research studies, with several technical studies devoted to issues regarding experimental design and reproducibility [62–64].

13.2.3.1 Imaging: Procedure and applications

- Resolution of molecular ions of intact protein species directly from intact samples (e.g. tissue thin-sections) in a mass spectrometer.

- Retains spatial information across intact tissue.

- Can be run in both high-resolution mode (imaging) and low-resolution mode (profiling).

- Can be directly correlated with histology on intact tissue.

- Is directly amenable to independent replicates from multiple experimental conditions, as well as multivariate statistical analysis.

The major strengths of MALDI imaging are in the quantification of intact protein species directly from intact tissue sections where the spatial/topological distribution is retained. As with the DIGE technology described in Section 13.2.1, large datasets can be easily normalized and analyzed using multivariate statistical analyses that enable the visualization of variation between samples as well as identify prognostic and diagnostic indicators of phenotype. This technique generally consumes small sample amounts relative to the other technologies (e.g. see Ref. [65]), although this will vary depending on the experiment. The major limitation of MALDI imaging is that only those proteins that are soluble under matrix application conditions that retain the *in vivo* spatial protein distribution are analyzed. In addition, the identification of proteins occurs in a separate experiment, usually after biochemical enrichment/fractionation, although methods to identify proteins directly from tissue specimens are progressing [66].

13.3 Troubleshooting

No single proteomics technology platform is capable of accomplishing a true global analysis of the entire proteome. Each of these major technology platforms has strengths and limitations, and in many cases these complement each other across platforms. Ultimately, all analytical platforms have fundamental limits in total protein amount before resolution is affected, and this is constrained in some sample types by the presence of a few proteins comprising the bulk of the sample (this is most problematic in serum/plasma studies [67]). The numbers provided in the following comparisons are approximate and reflect typical ranges, but of course will be dependent on the nature/quality of the samples being analyzed.

13.3.1 Number of resolved features and modifications

All three of the major technology platforms discussed here have the ability to provide for resolution on hundreds to thousands of features. The majority of intact signals tractable by the MALDI-IMS technology typically fall below the 30 000 MW range, and this technology retains the spatial distribution within intact tissues. The intact features that can be resolved in sample homogenates by 2D gels for the DIGE technology typically fall within the MW range of circa 10–200 kDa, and the isoelectric focusing range of pH 3–11. Both DIGE/MS and MALDI-IMS analyze and quantify intact protein species, and both are amenable to the analysis of independent (biological) replicates across multiple experimental conditions to enable facile multivariate statistical analysis. In some cases (e.g. phosphorylated proteins), modified isoforms can be directly visualized by the 2D gel pattern (a charge-train of proteins in the first dimension) and easily verified by protein excision, in-gel digestion and MS (even without MS information on the modified peptide(s) per se). Such isoforms would be apparent in the MALDI-IMS experiment as a series of m/z peaks (e.g. offset by 80 Da for phosphorylation). Unless these modified peptides were readily detected (or specifically targeted) in an LC/MS/MS experiment, they may be difficult to capture/analyze in a complex mixture (especially in a global-scale discovery study) from a bottom-up shotgun analysis.

Proteins that are 'difficult' to resolve by intact-protein methods are often readily identified using the LC/MS/MS shotgun approach because sufficient surrogate peptides can be obtained and mass analyzed (assuming that the proteins remain soluble during the various isolation, extraction and digestion steps). The shotgun approach also provides for greater sensitivity and dynamic range, especially via MudPIT analysis. However, when using peptide-based approaches it is more challenging to analyze modifications and/or perform quantitatively in a non-targeted discovery analysis, especially for experiments that require independent replicates across multiple experimental conditions.

13.3.2 Sample consumption, protein identification and depth of coverage

In comparison with the 2D-gel/DIGE approach that can consume up to hundreds of micrograms of sample per gel, both the MALDI-IMS and shotgun LC/MS/MS techniques require a relatively small amount of material, often as low as high femtomole–low picomole range, or as few as several thousand cells procured by laser-capture microdissection. MALDI-IMS consumes perhaps the smallest amount of material (e.g. see Ref. [65]); and when coupled with unsupervised multivariate statistical clustering algorithms, the patterns produced by MALDI-IMS can provide robust prognostic and diagnostic disease markers [51, 52] even though a fraction of the proteome is covered. DIGE technology can similarly produce quantitative data from large cohorts [23]. As the quantifiable features are resolved on a 2D gel, proteins of interest can be readily identified using standard 'bottom-up' MS-based approaches directly from the DIGE gels, whereas protein identification from a MALDI-IMS analysis usually occurs in a separate experiment (but see Groseclose *et al.* [66]).

For LC/MS/MS shotgun analyses, proteins are identified and quantified directly from the MS data based on surrogate peptides that were generated by protease digestion of the starting protein sample. This approach generally provides a greater depth of coverage in a given proteome because peptides are relatively homogeneous in their physicochemical properties and can be resolved over multiple chromatographic phases. Analysis of surrogate peptides also enables the identification of proteins that are otherwise refractory to 2D gels and MALDI-IMS due to issues such as solubility, hydrophobicity and extreme MW/pI of the intact proteins. However, as mentioned above, important post-translational modifications and processed forms would only be detected if the important peptides are captured or directly targeted with this technology (because information on the intact protein is not retained), and it also becomes increasingly challenging to perform complex experimental comparisons that are relatively straightforward by DIGE or MALDI-IMS [23, 40, 41]. Whether this becomes a strength or a limitation depends on the experimental goals; for a global, discovery-based experiment these intact protein attributes can go undetected, but in a targeted approach, especially using MRM-directed MS approaches [39], specific modifications can be quantified under conditions where they might go undetected in the 2D gel or MALDI approaches.

13.3.3 Statistical power

Technical and biological replicates are important for ensuring the accuracy of quantitative measurements. However, biological replicates are vital to assess whether or not changes in protein abundance/modification are descriptive of the biology rather than arising from unanticipated sources of experimental variation (e.g. inter-subject variation, sample preparation variation, analytical variation of the instrument).

'Statistical power' is the ability to visualize an X-fold magnitude change (effect size) at a $Y\%$ confidence interval (e.g. $p < 0.05$) and be 'correct' (usually expressed as power of 0.8 or 80% of the time it is 'correct'). Statistical power depends heavily on the analytical variation of the instrumentation making the measurement, as well as the number of independent (biological) replicates. For example, the experimental noise of DIGE is extremely low owing to the internal standard experimental design, enabling statistically powered experiments with very few biological replicates [23], whereas LC/MS experiments have a relatively high degree of analytical variation with little to no internal standard methodologies being employed, often resulting in underpowered experiments unless sufficient biological and technical replicates are employed [40, 41].

Pooling independent (biological) replicates either to produce sufficient material or to minimize costs (the 'economics of proteomics') should be done with extreme caution with respect to the statistical power of the resulting data. Pooling samples can be effective if the technical variation between samples is very low (e.g. sample preparation, analytical platforms with low noise) and disastrous if it is high (e.g. biological variation unrelated to experiment, analytical platforms with high noise). Even with sufficient technical replicates of the same samples to account for high technical variability of the sample preparation and analysis, performing the $N = 1$ experiment on pooled samples assumes that the averaging of populations is reflective of biological signal and that the technical noise is low. Technical replicates provide confidence in the result from the tested samples, but do not provide any confidence to the biological relevance. In some cases it may be valid to create sub-pools from a larger experiment, but in these cases it is still essential to maintain some degree of individualization of samples to retain statistical power [68].

13.3.4 Conclusions

Each of the three major technology platforms described here provides complementary strengths to a proteomics experiment. Although it is difficult to hold any one of these strengths paramount, perhaps most often the decision can come down to what you can see and how well you can see it. This trade-off between sensitivity (depth of coverage) and statistical power (is it biologically significant?) is quickly becoming a major focus of investigation in quantitative proteomics studies (e.g. see Refs [23, 40, 41, 68]). It is often the case that sample fractionation and enrichment strategies are necessary to provide greater depth of coverage. However, this increased sensitivity comes at a price, as it introduces technical variation/noise into the system, thereby lowering the statistical power unless a sufficient number of technical replicates are analyzed in addition to the requisite number of biological replicates to provide biological significance to the experiment. Analyzing unfractionated samples introduces the least amount of technical variation/noise, enabling a much higher statistical power, but at the cost of overall sensitivity/depth of coverage. Ultimately, the decision of which approach to utilize is best directed by the nature of the experimental question being asked, and also by the experimental instrumentation/expertise available. In the best-case scenarios, a proteomics project will not be limited by the utilization of only one technology platform.

References

1. Wasinger, V.C., Cordwell, S.J., Cerpa-Poljak, A. *et al.* (1995) Progress with gene-product mapping of the Mollicutes: *Mycoplasma genitalium*. *Electrophoresis*, **16** (7), 1090–1094.

2. Wolters, D.A., Washburn, M.P. and Yates, J.R. III (2001) An automated multidimensional protein identification technology for shotgun proteomics. *Analytical Chemistry*, **73** (23), 5683–5690. Multidimensional protein identification technology (MudPIT).

3. Link, A.J., Eng, J., Schieltz, D.M. *et al.* (1999) Direct analysis of protein complexes using mass spectrometry. *Nature Biotechnology*, **17** (7), 676–682. Direct analysis of large protein complexes (DALPC).

4. Ducret, A., Van Oostveen, I., Eng, J.K. *et al.* (1998) High throughput protein characterization by automated reverse-phase chromatography/electrospray tandem mass spectrometry. *Protein Science*, **7** (3), 706–719.

5. Eng, J.K., McCormack, A.L. and Yates, J.R. (1994) An approach to correlate tandem mass spectral data of peptides with amino acid sequences in a protein database. *Journal of the American Society for Mass Spectrometry*, **5** (11), 976–989. Database searching with tandem MS data.

6. Yates, J.R. III, Speicher, S., Griffin, P.R. and Hunkapiller, T. (1993) Peptide mass maps: a highly informative approach to protein identification. *Analytical Biochemistry*, **214** (2), 397–408.

7. Mann, M., Hojrup, P. and Roepstorff, P. (1993) Use of mass spectrometric molecular weight information to identify proteins in sequence databases. *Biological Mass Spectrometry*, **22** (6), 338–345.

8. James, P., Quadroni, M., Carafoli, E. and Gonnet, G. (1993) Protein identification by mass profile fingerprinting. *Biochemical and Biophysical Research Communications*, **195** (1), 58–64.

9. Henzel, W.J., Billeci, T.M., Stults, J.T. *et al.* (1993) Identifying proteins from two-dimensional gels by molecular mass searching of peptide fragments in protein sequence databases. *Proceedings of the National Academy of Sciences of the United States of America*, **90** (11), 5011–5015.

10. Yates, J.R. III, Eng, J.K., McCormack, A.L. and Schieltz, D. (1995) Method to correlate tandem mass spectra of modified peptides to amino acid sequences in the protein database. *Analytical Chemistry*, **67** (8), 1426–1436.

11. Laemmli, U.K. (1970) Cleavage of structural proteins during the assembly of the head of bacteriophage T4. *Nature*, **227** (259), 680–685. SDS-PAGE.

12. O'Farrell, P.H. (1975) High resolution two-dimensional electrophoresis of proteins. *Journal of Biological Chemistry*, **250** (10), 4007–4021. 2D gel electrophoresis.

13. Görg, A., Postel, W., Domscheit, A. and Gunther, S. (1988) Two-dimensional electrophoresis with immobilized pH gradients of leaf proteins from barley (*Hordeum vulgare*): method, reproducibility and genetic aspects. *Electrophoresis*, **9** (11), 681–692.

14. Görg, A., Obermaier, C., Boguth, G. *et al.* (2000) The current state of two-dimensional electrophoresis with immobilized pH gradients. *Electrophoresis*, **21** (6), 1037–1053.

15. Vogt, J.A., Schroer, K., Holzer, K. *et al.* (2003) Protein abundance quantification in embryonic stem cells using incomplete metabolic labelling with ^{15}N amino acids, matrix-assisted laser desorption/ionisation time-of-flight mass spectrometry, and analysis of relative isotopologue abundances of peptides. *Rapid Communications in Mass Spectrometry*, **17** (12), 1273–1282.

16. Tonge, R., Shaw, J., Middleton, B. *et al.* (2001) Validation and development of fluorescence two-dimensional differential gel electrophoresis proteomics technology. *Proteomics*, **1** (3), 377–396.

17. Von Eggeling, F., Gawriljuk, A., Fiedler, W. *et al.* (2001) Fluorescent dual colour 2D-protein gel electrophoresis for rapid detection of differences in protein pattern with standard image analysis software. *International Journal of Molecular Medicine*, **8** (4), 373–377.

18. Bengtsson, S., Krogh, M., Szigyarto, C.A. *et al.* (2007) Large-scale proteomics analysis of human ovarian cancer for biomarkers. *Journal of Proteome Research*, **6** (4), 1440–1450 [Epub 2007 Feb 22].

19. Friedman, D.B., Stauff, D.L., Pishchany, G. *et al.* (2006) *Staphylococcus aureus* redirects central metabolism to increase iron availability. *PLoS Pathogens*, **2** (8), e87.

20. Friedman, D.B., Wang, S.E., Whitwell, C.W. *et al.* (2007) Multivariable difference gel electrophoresis and mass spectrometry: a case study on transforming growth factor-beta and ERBB2 signaling. *Molecular & Cellular Proteomics*, **6**, 150–169. DIGE experimental design.

21. Seike, M., Kondo, T., Fujii, K. *et al.* (2004) Proteomic signature of human cancer cells. *Proteomics*, **4** (9), 2776–2788. DIGE experimental design.

22. Suehara, Y., Kondo, T., Fujii, K. *et al.* (2006) Proteomic signatures corresponding to histological classification and grading of soft-tissue sarcomas. *Proteomics*, **6** (15), 4402–4409. DIGE experimental design.

23. Karp, N.A. and Lilley, K.S. (2005) Maximising sensitivity for detecting changes in protein expression: experimental design using minimal CyDyes. *Proteomics*, **5** (12), 3105–3115. Variation in proteomics experiments, and statistical power for DIGE experiments.

24. Alban, A., David, S.O., Bjorkesten, L. *et al.* (2003) A novel experimental design for comparative two-dimensional gel analysis: two-dimensional difference gel electrophoresis incorporating a pooled internal standard. *Proteomics*, **3** (1), 36–44. First description of internal standard methodology for DIGE.

25. Friedman, D.B., Hill, S., Keller, J.W. *et al.* (2004) Proteome analysis of human colon cancer by two-dimensional difference gel electrophoresis and mass spectrometry. *Proteomics*, **4** (3), 793–811. DIGE internal standard and limit of sensitivity.

26. Hatakeyama, H., Kondo, T., Fujii, K. *et al.* (2006) Protein clusters associated with carcinogenesis, histological differentiation and nodal metastasis in esophageal cancer. *Proteomics*, **6** (23), 6300–6316.

27. Kolkman, A., Dirksen, E.H., Slijper, M. and Heck, A.J. (2005) Double standards in quantitative proteomics: direct comparative assessment of difference in gel electrophoresis and metabolic stable isotope labeling. *Molecular & Cellular Proteomics*, **4** (3), 255–266 [Epub 2005 Jan 4].

28. Wu, C.C., Yates, J.R. III, Neville, M.C. and Howell, K.E. (2000) Proteomic analysis of two functional states of the Golgi complex in mammary epithelial cells. *Traffic*, **1** (10), 769–782.

29. Ohi, M.D., Link, A.J., Ren, L. *et al.* (2002) Proteomics analysis reveals stable multiprotein complexes in both fission and budding yeasts containing Myb-related Cdc5p/Cef1p, novel pre-mRNA splicing factors, and snRNAs. *Molecular and Cellular Biology*, **22** (7), 2011–2024.

30. Sanders, S.L., Jennings, J., Canutescu, A. *et al.* (2002) Proteomics of the eukaryotic transcription machinery: identification of proteins associated with components of yeast TFIID by multidimensional mass spectrometry. *Molecular and Cellular Biology*, **22** (13), 4723–4738.

31. Washburn, M.P., Wolters, D. and Yates, J.R. III (2001) Large-scale analysis of the yeast proteome by multidimensional protein identification technology. *Nature Biotechnology*, **19** (3), 242–247. MudPIT analysis.

32. Florens, L., Washburn, M.P., Raine, J.D. *et al.* (2002) A proteomic view of the *Plasmodium falciparum* life cycle. *Nature*, **419** (6906), 520–526.

33. Gygi, S.P., Rist, B., Gerber, S.A. *et al.* (1999) Quantitative analysis of complex protein mixtures using isotope-coded affinity tags. *Nature Biotechnology*, **17** (10), 994–999. ICAT labeling for quantitative LC/MS/MS.

34. Ross, P.L., Huang, Y.N., Marchese, J.N. *et al.* (2004) Multiplexed protein quantitation in *Saccharomyces cerevisiae* using amine-reactive isobaric tagging reagents. *Molecular & Cellular Proteomics*, **3** (12), 1154–1169 [Epub 2004 Sep 22]. iTRAQ labeling for quantitative LC/MS/MS.

35. Dayon, L., Hainard, A., Licker, V. *et al.* (2008) Relative quantification of proteins in human cerebrospinal fluids by MS/MS using 6-plex isobaric tags. *Analytical Chemistry*, **80** (8), 2921–2931 [Epub 008 Mar 1].

36. Old, W.M., Meyer-Arendt, K., Aveline-Wolf, L. *et al.* (2005) Comparison of label-free methods for quantifying human proteins by shotgun proteomics. *Molecular & Cellular Proteomics*, **4** (10), 1487–1502 [Epub 2005 Jun 23]. Spectral counting for LC/MS/MS.

37. Zhang, B., VerBerkmoes, N.C., Langston, M.A. *et al.* (2006) Detecting differential and correlated protein expression in label-free shotgun proteomics. *Journal of Proteome Research*, **5** (11), 2909–2918.

38. Anderson, L. and Hunter, C.L. (2006) Quantitative mass spectrometric multiple reaction monitoring assays for major plasma proteins. *Molecular & Cellular Proteomics.*, **5** (4), 573–588 [Epub 2005 Dec 6].

39. Wolf-Yadlin, A., Hautaniemi, S., Lauffenburger, D.A. and White, F.M. (2007) Multiple reaction monitoring for robust quantitative proteomic analysis of cellular signaling networks. *Proceedings of the National Academy of Sciences of the United States of America*, **104** (14), 5860–5865 [Epub 2007 Mar 26].

40. Anderle, M., Roy, S., Lin, H. *et al.* (2004) Quantifying reproducibility for differential proteomics: noise analysis for protein liquid chromatography–mass spectrometry of human serum. *Bioinformatics*, **20** (18), 3575–3582 [Epub 2004 Jul 29]. Variation in proteomics experiments.

41. Cho, H., Smalley, D.M., Theodorescu, D. *et al.* (2007) Statistical identification of differentially labeled peptides from liquid chromatography tandem mass spectrometry. *Proteomics*, **7** (20), 3681–3692. Variation in proteomics experiments.

42. Todd, P.J., Schhaaff, T.G., Chaurand, P. and Caprioli, R.M. (2001) Organic ion imaging of biological tissue with secondary ion mass spectrometry and matrix-assisted laser desorption/ionization. *Journal of Mass Spectrometry*, **36** (4), 355–369.

43. Chaurand, P., Schwartz, S.A. and Caprioli, R.M. (2002) *Current Opinion in Chemical Biology*, **6** (5), 676–681. Review of MALDI-IMS.

44. Schwartz, S.A., Reyzer, M.L. and Caprioli, R.M. (2003) *Journal of Mass Spectrometry*, **38**, 699–708. MALDI-IMS sample preparation.

45. Sanders, M.E., Dias, E.C., Xu, B.J. *et al.* (2008) Differentiating proteomic biomarkers in breast cancer by laser capture microdissection and MALDI MS. *Journal of Proteome Research*, **7** (4), 1500–1507 [Epub 2008 Apr 4]. MALDI-IMS directed by histology.

46. Yildiz, P.B., Shyr, Y., Rahman, J.S. *et al.* (2007) Diagnostic accuracy of MALDI mass spectrometric analysis of unfractionated serum in lung cancer. *Journal of Thoracic Oncology*, **2** (10), 893–901.

47. Chaurand, P. and Caprioli, R.M. (2002) Direct profiling and imaging of peptides and proteins from mammalian cells and tissue sections by mass spectrometry. *Electrophoresis*, **23** (18), 3125–3135.

48. Cornett, D.S., Reyzer, M.L., Chaurand, P. and Caprioli, R.M. (2007) MALDI imaging mass spectrometry: molecular snapshots of biochemical systems. *Nature Methods*, **4** (10), 828–833.

49. Seeley, E.H. and Caprioli, R.M. (2008) Molecular imaging of proteins in tissues by mass spectrometry. *Proceedings of the National Academy of Sciences of the United States of America*, **105** (47), 18126–18131 [Epub 2008 Sep 5].

50. Cornett, D.S., Mobley, J.A., Dias, E.C. *et al.* (2006) A novel histology-directed strategy for MALDI-MS tissue profiling that improves throughput and cellular specificity in human breast cancer. *Molecular & Cellular Proteomics*, **5** (10), 1975–1983 [Epub 2006 Jul 18].

51. Yanagisawa, K., Shyr, Y., Xu, B.J. *et al.* (2003) Proteomic patterns of tumour subsets in non-small-cell lung cancer. *Lancet*, **362** (9382), 433–439.

52. Schwartz, S.A., Weil, R.J., Thompson, R.C. *et al.* (2005) Proteomic-based prognosis of brain tumor patients using direct-tissue matrix-assisted laser desorption ionization mass spectrometry. *Cancer Research*, **65** (17), 7674–7681.

53. Annan, R.S. and Carr, S.A. (1997) The essential role of mass spectrometry in characterizing protein structure: mapping posttranslational modifications. *Journal of Protein Chemistry*, **16** (5), 391–402.

54. Chaurand, P., DaGue, B.B., Pearsall, R.S. *et al.* (2001) Profiling proteins from azoxymethane-induced colon tumors at the molecular level by matrix-assisted laser desorption/ionization mass spectrometry. *Proteomics*, **1** (10), 1320–1326.

55. Rahman, S.M., Shyr, Y., Yildiz, P.B. *et al.* (2005) Proteomic patterns of preinvasive bronchial lesions. *American Journal of Respiratory and Critical Care Medicine*, **172** (12), 1556–1562 [Epub 2005 Sep 22].

56. Schwartz, S.A., Weil, R.J., Johnson, M.D. *et al.* (2004) Protein profiling in brain tumors using mass spectrometry: feasibility of a new technique for the analysis of protein expression. *Clinical Cancer Research*, **10** (3), 981–987.

57. Schwartz, S.A., Weil, R.J., Thompson, R.C. *et al.* (2005) Proteomic-based prognosis of brain tumor patients using direct-tissue matrix-assisted laser desorption ionization mass spectrometry. *Cancer Research*, **65** (17), 7674–7681.

58. Xie, L., Xu, B.J., Gorska, A.E. *et al.* (2005) Genomic and proteomic analysis of mammary tumors arising in transgenic mice. *Journal of Proteome Research*, **4** (6), 2088–2098.

59. Simpkins, F., Czechowicz, J.A., Liotta, L. and Kohn, E.C. (2005) SELDI-TOF mass spectrometry for cancer biomarker discovery and serum proteomic diagnostics. *Pharmacogenomics*, **6** (6), 647–653.

60. Albrethsen, J., Bøgebo, R., Olsen, J. *et al.* (2006) Preanalytical and analytical variation of surface-enhanced laser desorption-ionization time-of-flight mass spectrometry of human serum. *Clinical Chemistry and Laboratory Medicine*, **118**, 1 2 4 3–1252.

61. Roboz, J. (2005) Mass spectrometry in diagnostic oncoproteomics. *Cancer Investigation*, **23** (5), 465–478.

62. Baggerly, K.A., Morris, J.S., Edmonson, S.R. and Coombes, K.R. (2005) Signal in noise: evaluating reported reproducibility of serum proteomic tests for ovarian cancer. *Journal of the National Cancer Institute*, **97** (4), 307–309.

63. Baggerly, K.A., Morris, J.S. and Coombes, K.R. (2004) Reproducibility of SELDI-TOF protein patterns in serum: comparing datasets from different experiments. *Bioinformatics*, **20** (5), 777–785.

64. Diamandis, E.P. (2004) Analysis of serum proteomic patterns for early cancer diagnosis: drawing attention to potential problems. *Journal of the National Cancer Institute*, **96** (5), 353–356.

65. Xu, B.J., Li, J., Beauchamp, R.D. *et al.* (2009) Identification of early intestinal neoplasia protein biomarkers using laser capture microdissection and MALDI MS. *Molecular & Cellular Proteomics*, **8**, 936–945.

66. Groseclose, M.R., Andersson, M., Hardesty, W.M. and Caprioli, R.M. (2007) Identification of proteins directly from tissue: in situ tryptic digestions coupled with imaging mass spectrometry. *Journal of Mass Spectrometry*, **42** (2), 254–262. Protein identification from MALDI-IMS.

67. Anderson, N.L., Polanski, M., Pieper, R. *et al.* (2004) The human plasma proteome: a nonredundant list developed by combination of four separate sources. *Molecular & Cellular Proteomics*, **3** (4), 311–326 [Epub 2004 Jan 12].

68. Karp, N.A. and Lilley, K.S. (2009) Investigating sample pooling strategies for DIGE experiments to address biological variability. *Proteomics*, **9** (2), 388–397.

Index

Keep up with critical fields

Would you like to receive up-to-date information on our books, journals and databases in the areas that interest you, direct to your mailbox?

Join the **Wiley e-mail service** - a convenient way to receive updates and exclusive discount offers on products from us.

Simply visit **www.wiley.com/email** and register online

We won't bombard you with emails and we'll only email you with information that's relevant to you. We will ALWAYS respect your e-mail privacy and NEVER sell, rent, or exchange your e-mail address to any outside company. Full details on our privacy policy can be found online.

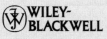

WILEY-
BLACKWELL

www.wiley.com/email